Triumph Spitfire Owners Workshop Manual

Peter G. Strasman
and J H Haynes Member of the Guild of Motoring Writers

Models covered
Triumph Spitfire Mk I, II, III, IV and 1500
1147 cc (70 cu in), 1296 cc (79 cu in), 1493 cc (91 cu in)

ISBN 1 85010 022 5

© Haynes Publishing Group 1983, 1986, 1987, 1988

All rights reserved. No part of this book may be reproduced or transmitted in any form or by any means, electronic or mechanical, including photocopying, recording or by any information storage or retrieval system, without permission in writing from the copyright holder.

Printed in England   *(113-6P5)*

Haynes Publishing Group
Sparkford Nr Yeovil
Somerset BA22 7JJ England

Haynes Publications, Inc
861 Lawrence Drive
Newbury Park
California 91320 USA

Acknowledgements

Thanks are due to BL Cars Ltd for the supply of technical information. The Champion Sparking Plug Company supplied the illustrations showing the various spark plug conditions. Duckhams Oils supplied the lubrication information and Sykes-Pickavant Ltd provided some of the workshop tools. Special thanks are due to all those people at Sparkford who helped in the production of this manual.

About this manual

Its aim

The aim of this manual is to help you get the best value from your vehicle. It can do so in several ways. It can help you decide what work must be done (even should you choose to get it done by a garage), provide information on routine maintenance and servicing, and give a logical course of action and diagnosis when random faults occur. However, it is hoped that you will use the manual by tackling the work yourself. On simpler jobs it may even be quicker than booking the car into a garage and going there twice, to leave and collect it. Perhaps most important, a lot of money can be saved by avoiding the costs a garage must charge to cover its labour and overheads.

The manual has drawings and descriptions to show the function of the various components so that their layout can be understood. Then the tasks are described and photographed in a step-by-step sequence so that even a novice can do the work.

Its arrangement

The manual is divided into twelve Chapters, each covering a logical sub-division of the vehicle. The Chapters are each divided into Sections, numbered with single figures, eg 5; and the Sections into paragraphs (or sub-sections), with decimal numbers following on from the Section they are in, eg 5.1, 5.2, 5.3 etc.

It is freely illustrated, especially in those parts where there is a detailed sequence of operations to be carried out. There are two forms of illustration: figures and photographs. The figures are numbered in sequence with decimal numbers, according to their position in the Chapter – eg Fig. 6.4 is the fourth drawing/illustration in Chapter 6. Photographs carry the same number (either individually or in related groups) as the Section or sub-section to which they relate.

There is an alphabetical index at the back of the manual as well as a contents list at the front. Each Chapter is also preceded by its own individual contents list.

References to the 'left' or 'right' of the vehicle are in the sense of a person in the driver's seat facing forwards.

Unless otherwise stated, nuts and bolts are removed by turning anti-clockwise, and tightened by turning clockwise.

Vehicle manufacturers continually make changes to specifications and recommendations, and these, when notified, are incorporated into our manuals at the earliest opportunity.

Whilst every care is taken to ensure that the information in this manual is correct, no liability can be accepted by the authors or publishers for loss, damage or injury caused by any errors in, or omissions from, the information given.

Introduction to the Triumph Spitfire

The Triumph Spitfire was first introduced in October 1962 in 1147 cc form, being mechanically similar to the Herald except for its twin carburettors and two seat, two door sports car body styling. The Mk 2, introduced in March 1965, was outwardly little different from its predecessor but a water-heated inlet manifold, new camshaft and diaphragm clutch were incorporated.

March 1967 saw the introduction of the Mk 3 model fitted with the 1296 cc engine, followed in October 1970 by the Mk 4 which retained the same engine, but incorporated several other modifications. These included the fitment of a new type of gearbox, improved suspension, and front and rear end body restyling.

1973 saw further modifications to the suspension system and in December 1974 the Spitfire 1500 was introduced in the UK, although this had been available from 1973 in the USA. This model uses a modified version of the Triumph 1500 engine and a new type gearbox.

Contents

	Page
Acknowledgements	2
About this manual	2
Introduction to the Triumph Spitfire	2
General dimensions, weights and capacities	8
Use of English	9
Buying spare parts and vehicle identification numbers	10
Tools and working facilities	11
Jacking and towing	13
Recommended lubricants and fluids	15
Safety first!	16
Routine maintenance	17
Fault diagnosis	20
Chapter 1 Engine	24
Chapter 2 Cooling and heating systems	75
Chapter 3 Fuel, carburation and emission control systems	85
Chapter 4 Ignition system	121
Chapter 5 Clutch	133
Chapter 6 Gearbox and overdrive	139
Chapter 7 Propeller shaft, driveshaft, universal joints	182
Chapter 8 Differential and final drive	188
Chapter 9 Braking system	196
Chapter 10 Electrical system	210
Chapter 11 Suspension and steering	268
Chapter 12 Bodywork and chassis	288
Conversion factors	305
Index	306

Triumph Spitfire Mk III

Triumph Spitfire 1500

Triumph Spitfire Mk IV

Triumph Spitfire Mk II

General dimensions, weights and capacities

Dimensions
Overall length
Mk I and II .... 145 in (368 cm)
Mk III .... 147 in (373 cm)
Mk IV .... 145 in (368 cm)
1500
    (except N. America) .... 156.3 in (397 cm)
    (N. America) .... 157.50 in (400 cm)
Overall width
Mks I, II, III and IV .... 57 in (145 cm)
1500 .... 58.5 in (148.8 cm)
Overall height
With hood erected .... 47.5 in (120.5 cm)
Without hood erected .... 44.5 in (112.5 cm)
Ground clearance .... 4.4 in (111.8 mm)
Wheelbase .... 83 in (210.8 cm)
Track (front) .... 49 in (124.5 cm)
Track (rear)
Mks I, II, III and IV .... 48 in (121.9 cm)
1500 .... 50 in (127 cm)

Weights (kerb)
Mk I and II .... 1624 lb (736 kg)
Mk III .... 1652 lb (748 kg)
Mk IV .... 1624 lb (720 kg)
1500 .... 1856 lb (843 kg)

Capacities
Fuel tank
Mks I, II, III and IV .... 8.25 Imp gal, 9.9 US gal, 37.6 l
Engine sump with filter change .... 8.0 Imp pts, 9.6 US pts, 4.5 l
Engine sump without filter change .... 7.0 Imp pts, 8.4 US pts, 4.0 l
Gearbox .... 1.5 Imp pts, 1.8 US pts, 0.85 l
Gearbox with overdrive .... 2.38 Imp pts, 2.85 US pts, 1.35 l
Rear axle .... 1.0 Imp pts, 1.2 US pts, 0.57 l
Cooling system .... 8.0 Imp pts, 9.6 US pts, 4.5 l

LCS-13-44-SP Amended June 1985

Use of English

As this book has been written in England, it uses the appropriate English component names, phrases, and spelling. Some of these differ from those used in America. Normally, these cause no difficulty, but to make sure, a glossary is printed below. In ordering spare parts remember the parts list may use some of these words:

English	American	English	American
Accelerator	Gas pedal	Locks	Latches
Aerial	Antenna	Methylated spirit	Denatured alcohol
Anti-roll bar	Stabiliser or sway bar	Motorway	Freeway, turnpike etc
Big-end bearing	Rod bearing	Number plate	License plate
Bonnet (engine cover)	Hood	Paraffin	Kerosene
Boot (luggage compartment)	Trunk	Petrol	Gasoline (gas)
Bulkhead	Firewall	Petrol tank	Gas tank
Bush	Bushing	'Pinking'	'Pinging'
Cam follower or tappet	Valve lifter or tappet	Prise (force apart)	Pry
Carburettor	Carburetor	Propeller shaft	Driveshaft
Catch	Latch	Quarterlight	Quarter window
Choke/venturi	Barrel	Retread	Recap
Circlip	Snap-ring	Reverse	Back-up
Clearance	Lash	Rocker cover	Valve cover
Crownwheel	Ring gear (of differential)	Saloon	Sedan
Damper	Shock absorber, shock	Seized	Frozen
Disc (brake)	Rotor/disk	Sidelight	Parking light
Distance piece	Spacer	Silencer	Muffler
Drop arm	Pitman arm	Sill panel (beneath doors)	Rocker panel
Drop head coupe	Convertible	Small end, little end	Piston pin or wrist pin
Dynamo	Generator (DC)	Spanner	Wrench
Earth (electrical)	Ground	Split cotter (for valve spring cap)	Lock (for valve spring retainer)
Engineer's blue	Prussian blue	Split pin	Cotter pin
Estate car	Station wagon	Steering arm	Spindle arm
Exhaust manifold	Header	Sump	Oil pan
Fault finding/diagnosis	Troubleshooting	Swarf	Metal chips or debris
Float chamber	Float bowl	Tab washer	Tang or lock
Free-play	Lash	Tappet	Valve lifter
Freewheel	Coast	Thrust bearing	Throw-out bearing
Gearbox	Transmission	Top gear	High
Gearchange	Shift	Torch	Flashlight
Grub screw	Setscrew, Allen screw	Trackrod (of steering)	Tie-rod (or connecting rod)
Gudgeon pin	Piston pin or wrist pin	Trailing shoe (of brake)	Secondary shoe
Halfshaft	Axleshaft	Transmission	Whole drive line
Handbrake	Parking brake	Tyre	Tire
Hood	Soft top	Van	Panel wagon/van
Hot spot	Heat riser	Vice	Vise
Indicator	Turn signal	Wheel nut	Lug nut
Interior light	Dome lamp	Windscreen	Windshield
Layshaft (of gearbox)	Countershaft	Wing/mudguard	Fender
Leading shoe (of brake)	Primary shoe		

Buying spare parts and vehicle identification numbers

Buying spare parts

Spare parts are available from many sources, for example: British Leyland garages, other garages and accessory shops, and motor factors. Our advice regarding spare part sources is as follows:

Officially appointed BL garages – This is the best source for parts which are otherwise not generally available (eg complete cylinder heads, internal gearbox components, badges, interior trim etc). It is also the only place at which you should buy parts if your car is still under warranty – non-manufacturers components may invalidate the warranty. To be sure of obtaining the correct parts it will always be necessary to give the storeman your car's vehicle identification number, and if possible, to take the 'old' part along for positive identification. Remember that many parts are available on a factory exchange scheme – any parts returned should always be clean! It obviously makes good sense to go straight to the specialists on your car for this type of part for they are best equipped to supply you.

Other garages and accessory shops – These are often very good places to buy materials and components needed for the maintenance of your car (eg oil filters, spark plugs, bulbs, fan belts, oils and greases, touch-up paint, filler paste etc). They also sell general accessories, usually have convenient opening hours, charge lower prices and can often be found not far from home.

Motor factors – Good factors will stock all the more important components which wear out relatively quickly (eg clutch components, pistons, valves, exhaust systems, brake cylinders/pipes/hoses/seals/shoes and pads etc). Motor factors will often provide new or reconditioned components on a part exchange basis – this can save a considerable amount of money.

Vehicle identification numbers

Modifications are a continuing and unpublished process in vehicle manufacture. Spare part manuals and lists are compiled upon a numerical basis, the individual vehicle number being essential to correct identification of the component required.

The vehicle number may be obtained from the identification plate within the engine compartment and the *engine number* is stamped on the cylinder block. Both numbers are also shown in the vehicle registration document.

Typical vehicle identification plate

Location of engine number

Tools and working facilities

Introduction

A selection of good tools is a fundamental requirement for anyone contemplating the maintenance and repair of a motor vehicle. For the owner who does not possess any, their purchase will prove a considerable expense, offsetting some of the savings made by doing-it-yourself. However, provided that the tools purchased meet the relevant national safety standards and are of good quality, they will last for many years and prove an extremely worthwhile investment.

To help the average owner to decide which tools are needed to carry out the various tasks detailed in this manual, we have compiled three lists of tools under the following headings: *Maintenance and minor repair*, *Repair and overhaul*, and *Special*. The newcomer to practical mechanics should start off with the *Maintenance and minor repair* tool kit and confine himself to the simpler jobs around the vehicle. Then, as his confidence and experience grow, he can undertake more difficult tasks, buying extra tools as, and when, they are needed. In this way, a *Maintenance and minor repair* tool kit can be built-up into a *Repair and overhaul* tool kit over a considerable period of time without any major cash outlays. The experienced do-it-yourselfer will have a tool kit good enough for most repair and overhaul procedures and will add tools from the *Special* category when he feels the expense is justified by the amount of use to which these tools will be put.

It is obviously not possible to cover the subject of tools fully here. For those who wish to learn more about tools and their use there is a book entitled *How to Choose and Use Car Tools* available from the publishers of this manual.

Maintenance and minor repair tool kit

The tools given in this list should be considered as a minimum requirement if routine maintenance, servicing and minor repair operations are to be undertaken. We recommend the purchase of combination spanners (ring one end, open-ended the other); although more expensive than open-ended ones, they do give the advantages of both types of spanner.

Combination spanners - $\frac{7}{16}$, $\frac{1}{2}$, $\frac{9}{16}$, $\frac{5}{8}$, $\frac{13}{16}$, $\frac{15}{16}$ in AF
Adjustable spanner - 9 inch
Engine sump/gearbox/rear axle drain plug key
Spark plug spanner (with rubber insert)
Spark plug gap adjustment tool
Set of feeler gauges
Brake adjuster spanner
Brake bleed nipple spanner
Screwdriver - 4 in long x $\frac{1}{4}$ in dia (flat blade)
Screwdriver - 4 in long x $\frac{1}{4}$ in dia (cross blade)
Combination pliers - 6 inch
Hacksaw (junior)
Tyre pump
Tyre pressure gauge
Grease gun
Oil can
Fine emery cloth (1 sheet)
Wire brush (small)
Funnel (medium size)

Repair and overhaul tool kit

These tools are virtually essential for anyone undertaking any major repairs to a motor vehicle, and are additional to those given in the *Maintenance and minor repair* list. Included in this list is a comprehensive set of sockets. Although these are expensive they will be found invaluable as they are so versatile - particularly if various drives are included in the set. We recommend the $\frac{1}{2}$ in square-drive type, as this can be used with most proprietary torque wrenches. If you cannot afford a socket set, even bought piecemeal, then inexpensive tubular box spanners are a useful alternative.

The tools in this list will occasionally need to be supplemented by tools from the *Special* list.

Sockets (or box spanners) to cover range in previous list
Reversible ratchet drive (for use with sockets)
Extension piece, 10 inch (for use with sockets)
Universal joint (for use with sockets)
Torque wrench (for use with sockets)
'Mole' wrench - 8 inch
Ball pein hammer
Soft-faced hammer, plastic or rubber
Screwdriver - 6 in long x $\frac{5}{16}$ in dia (flat blade)
Screwdriver - 2 in long x $\frac{5}{16}$ in square (flat blade)
Screwdriver - 1$\frac{1}{2}$ in long x $\frac{1}{4}$ in dia (cross blade)
Screwdriver - 3 in long x $\frac{1}{8}$ in dia (electricians)
Pliers - electricians side cutters
Pliers - needle nosed
Pliers - circlip (internal and external)
Cold chisel - $\frac{1}{2}$ inch
Scriber
Scraper
Centre punch
Pin punch
Hacksaw
Valve grinding tool
Steel rule/straight-edge
Allen keys
Selection of files
Wire brush (large)
Axle-stands
Jack (strong scissor or hydraulic type)

Special tools

The tools in this list are those which are not used regularly, are expensive to buy, or which need to be used in accordance with their manufacturers' instructions. Unless relatively difficult mechanical jobs are undertaken frequently, it will not be economic to buy many of these tools. Where this is the case, you could consider clubbing together with friends (or joining a motorists' club) to make a joint purchase, or borrowing the tools against a deposit from a local garage or tool hire specialist.

The following list contains only those tools and instruments freely available to the public, and not those special tools produced by the vehicle manufacturer specifically for its dealer network. You will find occasional references to these manufacturers' special tools in the text of this manual. Generally, an alternative method of doing the job without the vehicle manufacturers' special tool is given. However, sometimes, there is no alternative to using them. Where this is the case and the relevant tool cannot be bought or borrowed, you will have to entrust the work to a franchised garage.

Valve spring compressor (where applicable)
Piston ring compressor
Balljoint separator
Universal hub/bearing puller
Impact screwdriver
Micrometer and/or vernier gauge
Dial gauge
Stroboscopic timing light
Dwell angle meter/tachometer
Universal electrical multi-meter

Cylinder compression gauge
Lifting tackle (photo)
Trolley jack
Light with extension lead

Buying tools

For practically all tools, a tool factor is the best source since he will have a very comprehensive range compared with the average garage or accessory shop. Having said that, accessory shops often offer excellent quality tools at discount prices, so it pays to shop around.

There are plenty of good tools around at reasonable prices, but always aim to purchase items which meet the relevant national safety standards. If in doubt, ask the proprietor or manager of the shop for advice before making a purchase.

Care and maintenance of tools

Having purchased a reasonable tool kit, it is necessary to keep the tools in a clean serviceable condition. After use, always wipe off any dirt, grease and metal particles using a clean, dry cloth, before putting the tools away. Never leave them lying around after they have been used. A simple tool rack on the garage or workshop wall, for items such as screwdrivers and pliers is a good idea. Store all normal wrenches and sockets in a metal box. Any measuring instruments, gauges, meters, etc, must be carefully stored where they cannot be damaged or become rusty.

Take a little care when tools are used. Hammer heads inevitably become marked and screwdrivers lose the keen edge on their blades from time to time. A little timely attention with emery cloth or a file will soon restore items like this to a good serviceable finish.

Working facilities

Not to be forgotten when discussing tools, is the workshop itself. If anything more than routine maintenance is to be carried out, some form of suitable working area becomes essential.

It is appreciated that many an owner mechanic is forced by circumstances to remove an engine or similar item, without the benefit of a garage or workshop. Having done this, any repairs should always be done under the cover of a roof.

Wherever possible, any dismantling should be done on a clean, flat workbench or table at a suitable working height.

Any workbench needs a vice: one with a jaw opening of 4 in (100 mm) is suitable for most jobs. As mentioned previously, some clean dry storage space is also required for tools, as well as for lubricants, cleaning fluids, touch-up paints and so on, which become necessary.

Another item which may be required, and which has a much more general usage, is an electric drill with a chuck capacity of at least $\frac{5}{16}$ in (8 mm). This, together with a good range of twist drills, is virtually essential for fitting accessories such as mirrors and reversing lights.

Last, but not least, always keep a supply of old newspapers and clean, lint-free rags available, and try to keep any working area as clean as possible.

Spanner jaw gap comparison table

Jaw gap (in)	Spanner size
0.250	$\frac{1}{4}$ in AF
0.276	7 mm
0.313	$\frac{5}{16}$ in AF
0.315	8 mm
0.344	$\frac{11}{32}$ in AF; $\frac{1}{8}$ in Whitworth
0.354	9 mm
0.375	$\frac{3}{8}$ in AF
0.394	10 mm
0.433	11 mm
0.438	$\frac{7}{16}$ in AF
0.445	$\frac{3}{16}$ in Whitworth; $\frac{1}{4}$ in BSF
0.472	12 mm
0.500	$\frac{1}{2}$ in AF
0.512	13 mm
0.525	$\frac{1}{4}$ in Whitworth; $\frac{5}{16}$ in BSF
0.551	14 mm
0.563	$\frac{9}{16}$ in AF
0.591	15 mm
0.600	$\frac{5}{16}$ in Whitworth; $\frac{3}{8}$ in BSF
0.625	$\frac{5}{8}$ in AF
0.630	16 mm
0.669	17 mm
0.686	$\frac{11}{16}$ in AF
0.709	18 mm
0.710	$\frac{3}{8}$ in Whitworth; $\frac{7}{16}$ in BSF
0.748	19 mm
0.750	$\frac{3}{4}$ in AF
0.813	$\frac{13}{16}$ in AF
0.820	$\frac{7}{16}$ in Whitworth; $\frac{1}{2}$ in BSF
0.866	22 mm
0.875	$\frac{7}{8}$ in AF
0.920	$\frac{1}{2}$ in Whitworth; $\frac{9}{16}$ in BSF
0.938	$\frac{15}{16}$ in AF
0.945	24 mm
1.000	1 in AF
1.010	$\frac{9}{16}$ in Whitworth; $\frac{5}{8}$ in BSF
1.024	26 mm
1.063	$1\frac{1}{16}$ in AF; 27 mm
1.100	$\frac{5}{8}$ in Whitworth; $\frac{11}{16}$ in BSF
1.125	$1\frac{1}{8}$ in AF
1.181	30 mm
1.200	$\frac{11}{16}$ in Whitworth; $\frac{3}{4}$ in BSF
1.250	$1\frac{1}{4}$ in AF
1.260	32 mm
1.300	$\frac{3}{4}$ in Whitworth; $\frac{7}{8}$ in BSF
1.313	$1\frac{5}{16}$ in AF
1.390	$\frac{13}{16}$ in Whitworth; $\frac{15}{16}$ in BSF
1.417	36 mm
1.438	$1\frac{7}{16}$ in AF
1.480	$\frac{7}{8}$ in Whitworth; 1 in BSF
1.500	$1\frac{1}{2}$ in AF
1.575	40 mm; $\frac{15}{16}$ in Whitworth
1.614	41 mm
1.625	$1\frac{5}{8}$ in AF
1.670	1 in Whitworth; $1\frac{1}{8}$ in BSF
1.688	$1\frac{11}{16}$ in AF
1.811	46 mm
1.813	$1\frac{13}{16}$ in AF
1.860	$1\frac{1}{8}$ in Whitworth; $1\frac{1}{4}$ in BSF
1.875	$1\frac{7}{8}$ in AF
1.969	50 mm
2.000	2 in AF
2.050	$1\frac{1}{4}$ in Whitworth; $1\frac{3}{8}$ in BSF
2.165	55 mm
2.362	60 mm

Jacking and towing

Jacking

The jack supplied with the car is designed for roadwise wheel changing only. Whenever the car is being raised for repair or overhaul operations, use a hydraulic or screw type jack or better still a trolley jack. Always supplement the jacks with safety (axle) stands before crawling underneath.

Use the correct jacking points (which take the form of bolt heads) located under the front and rear of the side sills or alternatively place the jack under the chassis box members.

Roadwheel – removal and refitting

The spare wheel is located with the luggage compartment on a bracket.

Pressed steel type

Gently lever off the hub cap. These are spring loaded. Release but do not remove the roadwheel nuts before raising the car. Raise the car and ensure that it is securely supported. Remove the roadwheel nuts and roadwheel. Refit the wheel and nuts, then check their security. Lower the car to the ground.

Wire type

This type of roadwheel is retained on a hub adaptor by a large retaining nut. It is most important to realise that the adaptors are left-hand and right-hand threaded. The wheel nut tightens in the direction opposite to that of the roadwheel forward rotation. Tighten the nut using a copper-faced hammer when the car is on the ground. The adaptor splines and threads should be kept lightly smeared with grease, but do not apply grease to the wheel or adaptor taper.

Towing

Towing another vehicle is not recommended, but if being towed, attach the tow rope by winding it around the chassis front crossmember or the front suspension lower wishbone; do not attach it to the front bumper. When being towed, remember to release the steering lock using the ignition key.

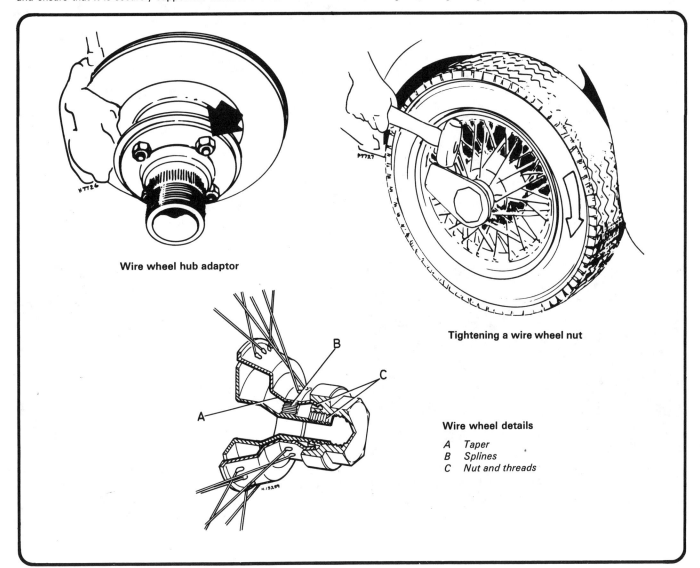

Wire wheel hub adaptor

Tightening a wire wheel nut

Wire wheel details

A Taper
B Splines
C Nut and threads

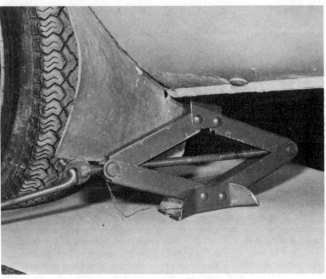
Using tool kit jack at front position

Using tool kit jack at rear position

Removing a wheel nut

Spring-loaded hub cap

Spare wheel mounting bracket

Wheel brace and spare wheel cover

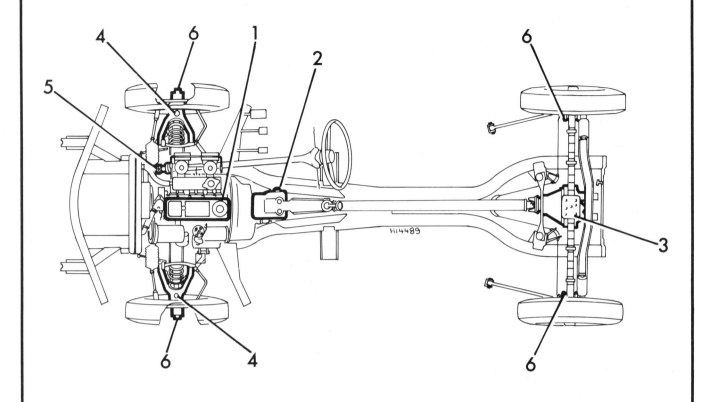

Recommended lubricants and fluids

Component or system	Lubricant type/specification	Duckhams recommendation
1 Engine	Multigrade engine oil, viscosity SAE 20W/50	Duckhams Hypergrade
2 Gearbox	Hypoid gear oil, viscosity SAE 90EP	Duckhams Hypoid 90S
3 Rear axle	Hypoid gear oil, viscosity SAE 90EP	Duckhams Hypoid 90S
4 Steering lower swivels	Hypoid gear oil, viscosity SAE 90EP	Duckhams Hypoid 90S
5 Steering rack	General purpose grease	Duckhams LB 10
6 Wheel hubs	General purpose grease	Duckhams LB 10

Safety first!

Professional motor mechanics are trained in safe working procedures. However enthusiastic you may be about getting on with the job in hand, do take the time to ensure that your safety is not put at risk. A moment's lack of attention can result in an accident, as can failure to observe certain elementary precautions.

There will always be new ways of having accidents, and the following points do not pretend to be a comprehensive list of all dangers; they are intended rather to make you aware of the risks and to encourage a safety-conscious approach to all work you carry out on your vehicle.

Essential DOs and DON'Ts

DON'T rely on a single jack when working underneath the vehicle. Always use reliable additional means of support, such as axle stands, securely placed under a part of the vehicle that you know will not give way.
DON'T attempt to loosen or tighten high-torque nuts (e.g. wheel hub nuts) while the vehicle is on a jack; it may be pulled off.
DON'T start the engine without first ascertaining that the transmission is in neutral (or 'Park' where applicable) and the parking brake applied.
DON'T suddenly remove the filler cap from a hot cooling system – cover it with a cloth and release the pressure gradually first, or you may get scalded by escaping coolant.
DON'T attempt to drain oil until you are sure it has cooled sufficiently to avoid scalding you.
DON'T grasp any part of the engine, exhaust or catalytic converter without first ascertaining that it is sufficiently cool to avoid burning you.
DON'T allow brake fluid or antifreeze to contact vehicle paintwork.
DON'T syphon toxic liquids such as fuel, brake fluid or antifreeze by mouth, or allow them to remain on your skin.
DON'T inhale dust – it may be injurious to health (see *Asbestos* below).
DON'T allow any spilt oil or grease to remain on the floor – wipe it up straight away, before someone slips on it.
DON'T use ill-fitting spanners or other tools which may slip and cause injury.
DON'T attempt to lift a heavy component which may be beyond your capability – get assistance.
DON'T rush to finish a job, or take unverified short cuts.
DON'T allow children or animals in or around an unattended vehicle.
DO wear eye protection when using power tools such as drill, sander, bench grinder etc, and when working under the vehicle.
DO use a barrier cream on your hands prior to undertaking dirty jobs – it will protect your skin from infection as well as making the dirt easier to remove afterwards; but make sure your hands aren't left slippery. Note that long-term contact with used engine oil can be a health hazard.
DO keep loose clothing (cuffs, tie etc) and long hair well out of the way of moving mechanical parts.
DO remove rings, wristwatch etc, before working on the vehicle – especially the electrical system.
DO ensure that any lifting tackle used has a safe working load rating adequate for the job.
DO keep your work area tidy – it is only too easy to fall over articles left lying around.
DO get someone to check periodically that all is well, when working alone on the vehicle.
DO carry out work in a logical sequence and check that everything is correctly assembled and tightened afterwards.
DO remember that your vehicle's safety affects that of yourself and others. If in doubt on any point, get specialist advice.
IF, in spite of following these precautions, you are unfortunate enough to injure yourself, seek medical attention as soon as possible.

Asbestos

Certain friction, insulating, sealing, and other products – such as brake linings, brake bands, clutch linings, torque converters, gaskets, etc – contain asbestos. *Extreme care must be taken to avoid inhalation of dust from such products since it is hazardous to health.* If in doubt, assume that they *do* contain asbestos.

Fire

Remember at all times that petrol (gasoline) is highly flammable. Never smoke, or have any kind of naked flame around, when working on the vehicle. But the risk does not end there – a spark caused by an electrical short-circuit, by two metal surfaces contacting each other, by careless use of tools, or even by static electricity built up in your body under certain conditions, can ignite petrol vapour, which in a confined space is highly explosive.

Always disconnect the battery earth (ground) terminal before working on any part of the fuel or electrical system, and never risk spilling fuel on to a hot engine or exhaust.

It is recommended that a fire extinguisher of a type suitable for fuel and electrical fires is kept handy in the garage or workplace at all times. Never try to extinguish a fuel or electrical fire with water.

Note: *Any reference to a 'torch' appearing in this manual should always be taken to mean a hand-held battery-operated electric lamp or flashlight. It does NOT mean a welding/gas torch or blowlamp.*

Fumes

Certain fumes are highly toxic and can quickly cause unconsciousness and even death if inhaled to any extent. Petrol (gasoline) vapour comes into this category, as do the vapours from certain solvents such as trichloroethylene. Any draining or pouring of such volatile fluids should be done in a well ventilated area.

When using cleaning fluids and solvents, read the instructions carefully. Never use materials from unmarked containers – they may give off poisonous vapours.

Never run the engine of a motor vehicle in an enclosed space such as a garage. Exhaust fumes contain carbon monoxide which is extremely poisonous; if you need to run the engine, always do so in the open air or at least have the rear of the vehicle outside the workplace.

If you are fortunate enough to have the use of an inspection pit, never drain or pour petrol, and never run the engine, while the vehicle is standing over it; the fumes, being heavier than air, will concentrate in the pit with possibly lethal results.

The battery

Never cause a spark, or allow a naked light, near the vehicle's battery. It will normally be giving off a certain amount of hydrogen gas, which is highly explosive.

Always disconnect the battery earth (ground) terminal before working on the fuel or electrical systems.

If possible, loosen the filler plugs or cover when charging the battery from an external source. Do not charge at an excessive rate or the battery may burst.

Take care when topping up and when carrying the battery. The acid electrolyte, even when diluted, is very corrosive and should not be allowed to contact the eyes or skin.

If you ever need to prepare electrolyte yourself, always add the acid slowly to the water, and never the other way round. Protect against splashes by wearing rubber gloves and goggles.

When jump starting a car using a booster battery, for negative earth (ground) vehicles, connect the jump leads in the following sequence: First connect one jump lead between the positive (+) terminals of the two batteries. Then connect the other jump lead first to the negative (–) terminal of the booster battery, and then to a good earthing (ground) point on the vehicle to be started, at least 18 in (45 cm) from the battery if possible. Ensure that hands and jump leads are clear of any moving parts, and that the two vehicles do not touch. Disconnect the leads in the reverse order.

Mains electricity and electrical equipment

When using an electric power tool, inspection light etc, always ensure that the appliance is correctly connected to its plug and that, where necessary, it is properly earthed (grounded). Do not use such appliances in damp conditions and, again, beware of creating a spark or applying excessive heat in the vicinity of fuel or fuel vapour. Also ensure that the appliances meet the relevant national safety standards.

Ignition HT voltage

A severe electric shock can result from touching certain parts of the ignition system, such as the HT leads, when the engine is running or being cranked, particularly if components are damp or the insulation is defective. Where an electronic ignition system is fitted, the HT voltage is much higher and could prove fatal.

Routine maintenance

Maintenance is essential for ensuring safety and desirable for the purpose of getting the best in terms of performance and economy from your car. Over the years the need for periodic lubrication — oiling, greasing, and so on — has been drastically reduced if not totally eliminated. This has unfortunately tended to lead some owners to think that because no such action is required, components either no longer exist, or will last for ever. This is a serious delusion. It follows therefore that the largest initial element of maintenance is visual examination. This may lead to repairs or renewals.

Every 250 miles (400 km) or weekly — whichever comes first

Check tyre pressures and inflate if necessary (photo)
Check and top-up the engine oil (photos)
Check and top-up battery electrolyte level (photo)
Check and top-up windscreen washer fluid level (photo)
Check and top-up coolant level (photo)
Check operation of all lights, wipers and horn
Check brake and clutch fluid levels (photo)

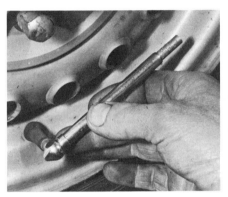
Checking a tyre pressure

Withdrawing engine oil dipstick

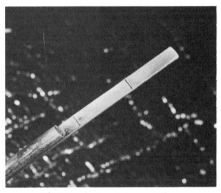
Engine oil dipstick markings

Topping up engine oil

Battery vent cap removed

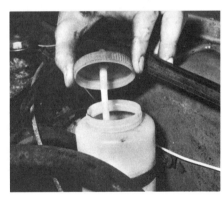
Windscreen washer fluid reservoir

Topping up clutch master cylinder fluid reservoir

Topping up coolant

Routine maintenance

Every 6000 miles (9600 km) or at six monthly intervals

Engine
Check drivebelts for tension and condition
Renew engine oil* (photo)

Steering and suspension
Check for wear in balljoints and suspension bushes and for split gaiters or dust excluders
Check tyres for wear or damage (photo)

Braking system
Check disc brake pads for wear
Check brake hydraulic pipes and hoses for condition
Check exhaust system for corrosion or leaks

*On later models, the oil changing interval has been increased to 12 000 miles (19 000 km) by the manufacturers.

Every 12 500 miles (20 000 km) or annually

Engine
Renew engine oil filter
Lubricate engine controls (throttle, choke etc)
Check carburettor adjustment
Top up oil in carburettor dampers
Clean and adjust sparking plugs
Clean or renew and adjust distributor contact points
Check ignition timing
Clean air cleaner elements

Transmission
Check gearbox oil level (photo)
Check rear axle oil level (photo)

Steering and suspension
Lubricate rack-and-pinion gear (early models)
Lubricate steering swivels (oil)
Check wheel alignment

Braking system
Check rear drum shoe lining wear
Check handbrake and foot brake adjustment
Lubricate handbrake linkage

Electrical system
Clean battery terminals and apply petroleum jelly
Renew wiper blades
Check headlamp beam alignment

General
Lubricate door locks, hinges and controls

Every 25 000 miles (40 000 km) or every two years

Engine
Renew air cleaner element
Clean crankcase breather system
Renew spark plugs
Clean fuel pump filter
Adjust valve clearances

Transmission and final drive
Renew lubricant (see Chapter 6, Section 2 and Chapter 8, Section 2) (photo)
Check torque of propeller and drive shaft bolts

Engine sump drain plug

Checking tyre tread depth

Gearbox filler/level plug

Differential filler/level plug

Typical gearbox drain plug

Routine Maintenance

Steering and suspension
    Adjust front hub bearings
    Check for wear in suspension bushes
    Lubricate rear hubs

At two yearly intervals, the brake hydraulic fluid should be renewed and the coolant (antifreeze) renewed

Additional maintenance operations at this service interval for vehicles operating in North America
    Check operation of deceleration bypass valve
    Check security of EGR system hoses and clean valve
    Check security of evaporation system hoses
    Renew carbon canister filter pad (early models)
    Check air injection system hoses for security and condition
    Check operation of air cleaner temperature control system
    Renew in-line fuel filter
    Check condition of fuel filler cap seal
    Check operation of seat belt warning system
    Renew catalytic converter (early models only)

Every 50 000 miles (80 000 km)

    Renew carbon canister (all models)
    Check operation of EGR system

Fault diagnosis

Introduction

The vehicle owner who does his or her own maintenance according to the recommended schedules should not have to use this section of the manual very often. Modern component reliability is such that, provided those items subject to wear or deterioration are inspected or renewed at the specified intervals, sudden failure is comparatively rare. Faults do not usually just happen as a result of sudden failure, but develop over a period of time. Major mechanical failures in particular are usually preceded by characteristic symptoms over hundreds or even thousands of miles. Those components which do occasionally fail without warning are often small and easily carried in the vehicle.

With any fault finding, the first step is to decide where to begin investigations. Sometimes this is obvious, but on other occasions a little detective work will be necessary. The owner who makes half a dozen haphazard adjustments or replacements may be successful in curing a fault (or its symptoms), but he will be none the wiser if the fault recurs and he may well have spent more time and money than was necessary. A calm and logical approach will be found to be more satisfactory in the long run. Always take into account any warning signs or abnormalities that may have been noticed in the period preceding the fault – power loss, high or low gauge readings, unusual noises or smells, etc – and remember that failure of components such as fuses or spark plugs may only be pointers to some underlying fault.

The pages which follow here are intended to help in cases of failure to start or breakdown on the road. There is also a Fault Diagnosis Section at the end of each Chapter which should be consulted if the preliminary checks prove unfruitful. Whatever the fault, certain basic principles apply. These are as follows:

Verify the fault. This is simply a matter of being sure that you know what the symptoms are before starting work. This is particularly important if you are investigating a fault for someone else who may not have described it very accurately.

Don't overlook the obvious. For example, if the vehicle won't start, is there petrol in the tank? (Don't take anyone else's word on this particular point, and don't trust the fuel gauge either!) If an electrical fault is indicated, look for loose or broken wires before digging out the test gear.

Cure the disease, not the symptom. Substituting a flat battery with a fully charged one will get you off the hard shoulder, but if the underlying cause is not attended to, the new battery will go the same way. Similarly, changing oil-fouled spark plugs for a new set will get you moving again, but remember that the reason for the fouling (if it wasn't simply an incorrect grade of plug) will have to be established and corrected.

Don't take anything for granted. Particularly, don't forget that a 'new' component may itself be defective (especially if it's been rattling round in the boot for months), and don't leave components out of a fault diagnosis sequence just because they are new or recently fitted. When you do finally diagnose a difficult fault, you'll probably realise that all the evidence was there from the start.

Electrical faults

Electrical faults can be more puzzling than straightforward mechanical failures, but they are no less susceptible to logical analysis if the basic principles of operation are understood. Vehicle electrical wiring exists in extremely unfavourable conditions – heat, vibration and chemical attack – and the first things to look for are loose or corroded connections and broken or chafed wires, especially where the wires

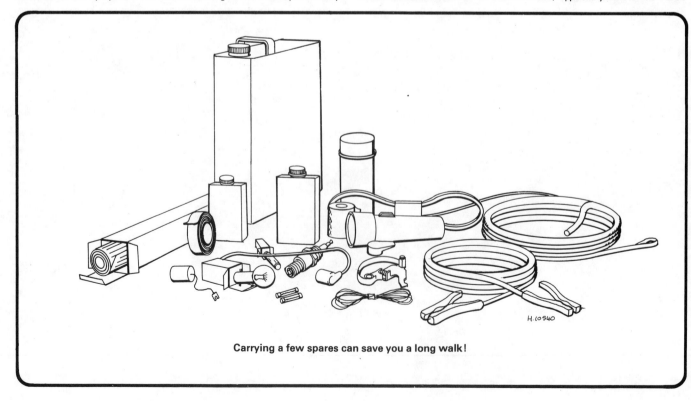

Carrying a few spares can save you a long walk!

Fault diagnosis

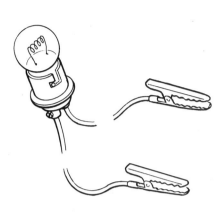

A simple test lamp is useful for tracing electrical faults

pass through holes in the bodywork or are subject to vibration.

All metal-bodied vehicles in current production have one pole of the battery 'earthed', ie connected to the vehicle bodywork, and in nearly all modern vehicles it is the negative (−) terminal. The various electrical components — motors, bulb holders etc — are also connected to earth, either by means of a lead or directly by their mountings. Electric current flows through the component and then back to the battery via the bodywork. If the component mounting is loose or corroded, or if a good path back to the battery is not available, the circuit will be incomplete and malfunction will result. The engine and/or gearbox are also earthed by means of flexible metal straps to the body or subframe; if these straps are loose or missing, starter motor, generator and ignition trouble may result.

Assuming the earth return to be satisfactory, electrical faults will be due either to component malfunction or to defects in the current supply. Individual components are dealt with in Chapter 10. If supply wires are broken or cracked internally this results in an open-circuit, and the easiest way to check for this is to bypass the suspect wire temporarily with a length of wire having a crocodile clip or suitable connector at each end. Alternatively, a 12V test lamp can be used to verify the presence of supply voltage at various points along the wire and the break can be thus isolated.

If a bare portion of a live wire touches the bodywork or other earthed metal part, the electricity will take the low-resistance path thus formed back to the battery: this is known as a short-circuit. Hopefully a short-circuit will blow a fuse, but otherwise it may cause burning of the insulation (and possibly further short-circuits) or even a fire. This is why it is inadvisable to bypass persistently blowing fuses with silver foil or wire.

Spares and tool kit

Most vehicles are supplied only with sufficient tools for wheel changing; the *Maintenance and minor repair* tool kit detailed in *Tools and working facilities*, with the addition of a hammer, is probably sufficient for those repairs that most motorists would consider attempting at the roadside. In addition a few items which can be fitted without too much trouble in the event of a breakdown should be carried. Experience and available space will modify the list below, but the following may save having to call on professional assistance:

Spark plugs, clean and correctly gapped
HT lead and plug cap — long enough to reach the plug furthest from the distributor
Distributor rotor, condenser and contact breaker points
Drivebelt(s) — emergency type may suffice
Spare fuses
Set of principal light bulbs
Tin of radiator sealer and hose bandage
Exhaust bandage
Roll of insulating tape
Length of soft iron wire
Length of electrical flex
Torch or inspection lamp (can double as test lamp)
Battery jump leads
Tow-rope
Ignition waterproofing aerosol
Litre of engine oil
Sealed can of hydraulic fluid
Emergency windscreen
'Jubilee' clips
Tube of filler paste

If spare fuel is carried, a can designed for the purpose should be used to minimise risks of leakage and collision damage. A first aid kit and a warning triangle, whilst not at present compulsory in the UK, are obviously sensible items to carry in addition to the above.

When touring abroad it may be advisable to carry any additional spares which, even if you cannot fit them yourself, could save having to wait while parts are obtained. The items below may be worth considering:

Throttle cable
Cylinder head gasket
Dynamo or alternator brushes
Fuel pump repair kit
Tyre valve core

One of the motoring organisations will be able to advise on availability of fuel etc in foreign countries.

Engine will not start

Engine fails to turn when starter operated
  Flat battery (recharge, use jump leads, or push start)
  Battery terminals loose or corroded
  Battery earth to body defective
  Engine earth strap loose or broken
  Starter motor (or solenoid) wiring loose or broken
  Ignition/starter switch faulty
  Major mechanical failure (seizure)
  Starter or solenoid internal fault (see Chapter 10)

Starter motor turns engine slowly
  Partially discharged battery (recharge, use jump leads, or push start)
  Battery terminals loose or corroded
  Battery earth to body defective
  Engine earth strap loose
  Starter motor (or solenoid) wiring loose
  Starter motor internal fault (see Chapter 10)

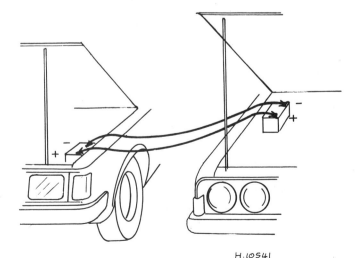

Correct way to connect jump leads. Do not allow car bodies to touch

Fault diagnosis

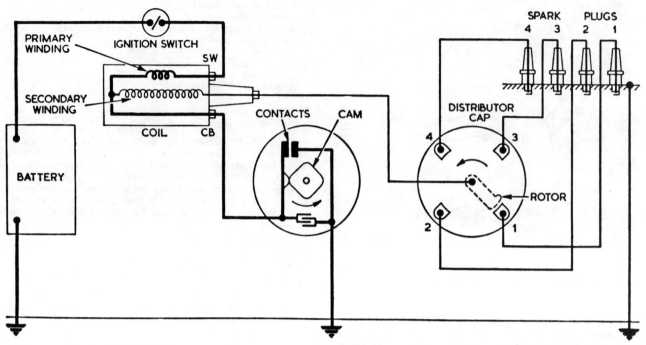

Ignition system schematic diagram. Some later models have a 6V coil and ballast resistor

Starter motor spins without turning engine
Flat battery
Starter motor pinion sticking on sleeve
Flywheel gear teeth damaged or worn
Starter motor mounting bolts loose

Engine turns normally but fails to start
Damp or dirty HT leads and distributor cap (crank engine and check for spark)
Dirty or incorrectly gapped distributor points (if applicable)
No fuel in tank (check for delivery at carburettor)
Excessive choke (hot engine) or insufficient choke (cold engine)
Fouled or incorrectly gapped spark plugs (remove, clean and regap)
Other ignition system fault (see Chapter 4)
Other fuel system fault (see Chapter 3)
Poor compression (see Chapter 1)
Major mechanical failure (eg camshaft drive)

Engine fires but will not run
Insufficient choke (cold engine)
Air leaks at carburettor or inlet manifold
Fuel starvation (see Chapter 3)
Ballast resistor defective, or other ignition fault (see Chapter 4)

Engine cuts out and will not restart

Engine cuts out suddenly – ignition fault
Loose or disconnected LT wires
Wet HT leads or distributor cap (after traversing water splash)
Coil or condenser failure (check for spark)
Other ignition fault (see Chapter 4)

Engine misfires before cutting out – fuel fault
Fuel tank empty
Fuel pump defective or filter blocked (check for delivery)
Fuel tank filler vent blocked (suction will be evident on releasing cap)
Carburettor needle valve sticking
Carburettor jets blocked (fuel contaminated)
Other fuel system fault (see Chapter 3)

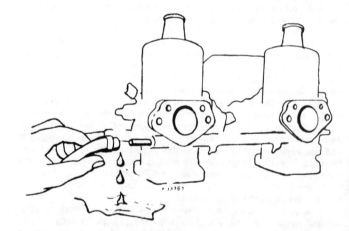

Remove fuel pipe from carburettor and check that fuel is being delivered

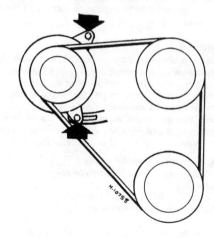

A slack drivebelt may cause overheating and battery charging problems. Slacken bolts (arrowed) to adjust

Fault diagnosis

Engine cuts out – other causes
    Serious overheating
    Major mechanical failure (eg camshaft drive)

Engine overheats

Ignition (no-charge) warning light illuminated
    Slack or broken drivebelt – retension or renew (Chapter 2)

Ignition warning light not illuminated
    Coolant loss due to internal or external leakage (see Chapter 2)
    Thermostat defective
    Low oil level
    Brakes binding
    Radiator clogged externally or internally
    Engine waterways clogged
    Ignition timing incorrect or automatic advance malfunctioning
    Mixture too weak

Note: *Do not add cold water to an overheated engine or damage may result*

Low engine oil pressure

Gauge reads low or warning light illuminated with engine running
    Oil level low or incorrect grade
    Defective gauge or sender unit
    Wire to sender unit earthed
    Engine overheating
    Oil filter clogged or bypass valve defective
    Oil pressure relief valve defective
    Oil pick-up strainer clogged
    Oil pump worn or mountings loose
    Worn main or big-end bearings

Note: *Low oil pressure in a high-mileage engine at tickover is not necessarily a cause for concern. Sudden pressure loss at speed is far more significant. In any event, check the gauge or warning light sender before condemning the engine.*

Engine noises

Pre-ignition (pinking) on acceleration
    Incorrect grade of fuel
    Ignition timing incorrect
    Distributor faulty or worn

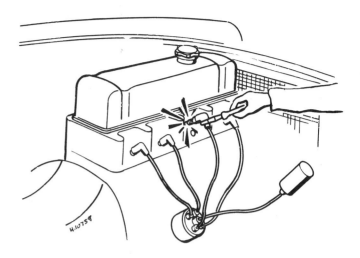

Crank engine and check for a spark. Note use of insulated pliers – dry cloth or a rubber glove will suffice

    Worn or maladjusted carburettor
    Excessive carbon build-up in engine

Whistling or wheezing noises
    Leaking vacuum hose
    Leaking carburettor or manifold gasket
    Blowing head gasket

Tapping or rattling
    Incorrect valve clearances
    Worn valve gear
    Worn timing chain or belt
    Broken piston ring (ticking noise)

Knocking or thumping
    Unintentional mechanical contact (eg fan blades)
    Worn fanbelt
    Peripheral component fault (generator, water pump etc)
    Worn big-end bearings (regular heavy knocking, perhaps less under load)
    Worn main bearings (rumbling and knocking, perhaps worsening under load)
    Piston slap (most noticeable when cold)

Chapter 1 Engine

Contents

Part A – General all models
Ancillary engine components – removal	9
Dismantling the engine – general	8
Engine – removal and refitting with gearbox	7
Engine – removal and refitting without gearbox	6
Engine oil – topping up, oil and filter renewal	2
General description	1
Major operations with engine in place	3
Major operations with engine removed	4
Methods of engine removal	5

Part B – 1147 cc and 1296 cc engines
Camshaft – removal and refitting (engine in car)	14
Crankcase ventilation system	21
Cylinder head – decarbonising	19
Cylinder head – removal and refitting (engine in car)	10
Dismantling sub-assemblies	17
Engine – complete dismantling (engine removed)	16
Engine – complete reassembly	23
Engine lubrication system – description	20
Engine reassembly – general	22
Engine sump – removal and refitting (engine in car)	11
Examination and renovation	18
Fault diagnosis – engine	26
Oil pump – removal and refitting (engine in car)	15
Pistons, connecting rods and big-end bearings – removal and refitting (engine in car)	12
Starting-up after major overhaul	25
Timing chain, gears and cover oil seal – removal and refitting (engine in car)	13
Valve clearances – adjustment	24

Part C – 1493 cc engine
Camshaft – removal and refitting	32
Cylinder head – decarbonising	37
Cylinder head – removal and refitting	27
Dismantling sub assemblies	35
Engine – complete dismantling (engine removed)	34
Engine – complete reassembly	40
Engine reassembly – general	39
Engine sump – removal and refitting	29
Examination and renovation	36
Fault diagnosis – engine	43
Lubrication and crankcase ventilation systems – description	38
Oil pump – removal and refitting (engine in car)	33
Pistons, connecting rods and big-end bearings – removal and refitting (engine in car)	30
Rocker gear – dismantling and reassembly	28
Starting up after major overhaul	42
Timing cover, sprockets and chain – removal and refitting	31
Valve clearances – adjustment	41

Specifications

1147 cc engine
Type	4 cylinder in line OHV
Bore	2.728 in (69.3 mm)
Stroke	2.992 in (76.0 mm)
Cubic capacity	1147 cc (70.0 cu in)
Compression ratio	
High compression	9.0 to 1
Low compression	7.0 to 1
Compression pressure	
High compression	144 lbf/in² (10.10 kgf/cm²)
Low compression	130 lbf/in² (9.12 kgf/cm²)
Torque	
High compression	67 lbf ft (8.47 kgf m) at 3760 rpm
Low compression	60 lbf ft (8.3 kgf m) at 3760 rpm
Max BHP	
High compression	67 at 6000 rpm
Low compression	57 at 5800 rpm
Firing order	1,3,4,2
Location of No. 1 cylinder	Next to timing cover

Camshaft
Camshaft drive	From crankshaft by single roller chain
Camshaft journal diameter	1.8402 to 1.8407 in (46.74 to 46.75 mm)
Diametrical clearance	0.0026 to 0.0046 in (0.07 to 0.12 mm)
Endfloat	0.004 to 0.008 in (0.10 to 0.20 mm)

Chapter 1 Engine

Connecting rods and big and little end bearings
Type .... Angular split big-end. Fully floating small-end
Big-end bearing – Type .... Shell
Big-end bearing – Internal diameter .... 1.626 to 1.627 in (41.30 to 41.32 mm)
Endfloat on crankpin .... 0.0105 to 0.0126 in (0.266 to 0.320 mm)
Undersizes available .... 0.010, 0.020, 0.030 in (0.254, 0.508, 0.762 mm)
Internal diameter of small-end bush .... 0.8122 to 0.8126 in (20.63 to 20.64 mm)

Crankshaft and main bearings
Main journal diameter .... 2.0005 to 2.001 in (50.81 to 50.83 mm)
Crankpin diameter .... 1.6250 to 1.6255 in (41.27 to 44.28 mm)
Crankshaft endthrust .... Taken by thrust washers on rear main bearing
Endfloat .... 0.004 to 0.008 in (0.10 to 0.20 mm)
Main bearing internal diameter .... 2.0015 to 2.0037 in (50.84 to 50.89 mm)
Main bearing housing internal diameter .... 2.146 to 2.1465 in (54.51 to 54.52 mm)
Rear journal width .... 1.2976 to 1.2995 in (32.95 to 33.01 mm)
Thickness of thrust washers .... 0.091 to 0.093 in (2.31 to 2.36 mm)
Oversize thrust washers .... 0.096 to 0.098 in (2.44 to 2.49 mm)
Undersize bearings available .... -0.010, -0.020, -0.030, -0.040 in (-0.254, -0.508, -0.762, -1.016 mm)

Cylinder block
Type .... Cylinder cast integral with top half of crankcase
Oversize bores:
    First .... 0.010 in (0.254 mm)
    Max .... 0.030 in (0.762 mm)

Cylinder head
Type .... Cast iron with vertical valves
Port arrangements .... Inlet and exhaust ports on same side
Number of ports – Exhaust .... 4 separate
Number of ports – Inlet .... 2 siamesed

Gudgeon pins
Type .... Fully floating
Fit in piston .... Light push fit at 68°F (20°C)
Outer diameter .... 0.8123 to 0.8125 in (20.63 to 20.64 mm)

Lubrication system
Type .... Pressure and splash, wet sump
Oil filter .... Full flow
Oil type/specification .... Multigrade engine oil, viscosity SAE 20W/50 (Duckhams Hypergrade)
Sump capacity .... 7 pints (8.4 US pints, 4.0 litres)
Oil pump type .... Eccentric rotor
Oil pressure relief valve spring
    Free length .... 1.54 in (39.11 mm)
    Fitted length .... 1.25 in (31.75 mm)
    Load at fitted length .... 14.5 lb (6.58 kg)
Normal oil pressure at 2000 rpm .... 40 to 60 lbf/in^2 (2.8 – 4.2 kgf/cm^2)

Pistons
Type .... Aluminium alloy – split skirt up to engine No. FC 24449
    Aluminium alloy – solid skirt from engine No. FC 24500
Number of rings .... 2 compression, 1 oil control
Clearance in cylinder:
Pistons made by Automotive Engineering Co Ltd
    Top .... 0.0029 to 0.0033 in (0.07 to 0.08 mm)
    Bottom .... 0.0011 to 0.0015 in (0.03 to 0.04 mm)
Pistons made by British Piston Ring Co Ltd
    Top .... 0.0163 to 0.0193 in (0.41 to 0.49 mm)
    Bottom .... 0.0012 to 0.0015 in (0.03 to 0.04 mm)
Pistons made by Wellworthy
    Top .... 0.0038 to 0.0041 in (0.09 to 0.10 mm)
    Bottom .... 0.0012 to 0.0015 in (0.03 to 0.04 mm)
Width of ring grooves
    Compression rings .... 0.0802 to 0.0812 in (2.03 to 2.06 mm)
Split skirt pistons
    Oil control ring .... 0.157 to 0.158 in (3.99 to 4.01 mm)
Solid skirt pistons
    Compression rings .... 0.0797 to 0.0807 in (2.02 to 2.05 mm)
    Oil control ring .... 0.157 to 0.158 in (3.99 to 4.01 mm)
Piston oversizes available .... +0.010, +0.020, +0.030 in (+0.254, +0.508, +0.762 mm)

Piston rings
Top compression ring .... Parallel – chromium plated

2nd compression ring	Early models parallel, later models tapered
Compression ring width	0.077 to 0.078 in (1.97 to 1.99 mm)
Fitted cap	0.008 to 0.013 in (0.20 to 0.33 mm)
Oil control ring	Slotted scraper
Oil control ring width	0.154 to 0.156 in (3.90 to 3.96 mm)
Oil control ring fitted gap	0.008 to 0.013 in (0.20 to 0.33 mm)

Tappets
Type	Barrel with flat base
Outside diameter	0.6867 to 0.6871 in (17.45 to 17.46 mm)
From engine No. FC 61023	0.7996 to 0.8000 in (20.29 to 20.32 mm)

Rocker gear
Diameter of rocker shaft	0.5607 to 0.5612 in (14.24 to 14.26 mm)
Bore of rockers	0.562 to 0.563 in (14.27 to 14.30 mm)

Valves
Head diameter – Inlet	1.241 to 1.245 in (31.52 to 31.62 mm)
Head diameter – Exhaust	1.148 to 1.182 in (29.16 to 29.26 mm)
Stem diameter – Inlet	0.310 to 0.311 in (7.87 to 7.89 mm)
Stem diameter – Exhaust	0.308 to 0.309 in (7.82 to 7.85 mm)
Stem to guide clearance	
Inlet	0.001 to 0.003 in (0.025 to 0.075 mm)
Exhaust	0.003 to 0.005 in (0.076 to 0.13 mm)
Valve stem to rocker arm clearance	0.010 in (0.25 mm) cold

Valve guides
Length	2.25 in (57.15 mm)
Outside diameter – Inlet and exhaust	0.501 to 0.502 in (12.72 to 12.75 mm)
Inside diameter – Inlet and exhaust	0.312 to 0.313 in (7.92 to 7.95 mm)
Fitted height above head – Inlet and exhaust	0.749 to 0.751 in (19.025 to 19.075 mm)

Valve timing
Spitfire Mk I	
Inlet opens	18° BTDC
Inlet closes	58° ABDC
Exhaust opens	58° BBDC
Exhaust closes	18° ATDC
Spitfire Mk II	
Inlet opens	25° BTDC
Inlet closes	65° ABDC
Exhaust opens	65° BBDC
Exhaust closes	25° ATDC
Timing marks	Scribed lines on camshaft and crankshaft sprockets

Valve springs
Type	Single valve springs
Fitted length	1.38 in (35.03 mm)
Fitted load	32 to 42 lbs (14.51 to 19.05 kgs)
Number of coils	6

1296 cc engine. The engine specification and data are identical to the 1147 cc engine except for the following differences.

Type	4 cylinder in line OHV
Bore	2.900 in (73.7 mm)
Cubic capacity	1296 cc (79.2 cu in)
Compression ratio	
High compression – Emission controlled	8.5 to 1
Low compression	7.5 to 2
Compression pressure	
High compression – Emission controlled	140 lbf/in^2 (9.82 kgf/cm^2)
Torque	
High compression	75 lbf ft (10.37 kgf/m) at 4000 rpm
Low compression	68 lbf ft (9.4 kgf/m) at 4000 rpm
High compression – Emission controlled	73 lbf ft (10.13 kgf/m) at 3000 rpm
Max BHP	
High compression	75 at 6000 rpm
Low compression	65 at 5800 rpm
High compression – Emission controlled	68 at 5500 rpm

Camshaft
Camshaft journal diameter	1.9649 to 1.9654 in (49.91 to 49.92 mm)
Endfloat	0.0042 to 0.0085 in (0.11 to 0.216 mm)

Chapter 1 Engine

Connecting rods and big and little end bearing
Type .... Angular split big-end. Fully floating or interference fit small-end
Endfloat on crankpin .... 0.0025 to 0.0086 in (0.063 to 0.218 mm)
Internal diameter of small-end bush .... 0.811 to 0.8115 in (20.60 to 20.612 mm)

Crankshaft and main bearings
Main bearings internal diameter .... 2.002 to 2.0025 in (50.85 to 50.86 mm)

Cylinder head
Number of ports – Exhaust .... 4 separate
Number of ports – Inlet .... 4 separate

Gudgeon pin
Type .... Full floating or interference fit
Fit in piston (fully floating) .... Light push fit at 68°F (20°C)

Pistons
Type .... Aluminium alloy, solid skirt
Clearance in cylinder
Pistons made by Brico Co. Ltd.
    Top .... 0.020 to 0.025 in (0.51 to 0.64 mm)
    Bottom .... 0.0019 to 0.0024 in (0.05 to 0.06 mm)
Pistons made by Hepworth Co. Ltd.
    Top .... 0.0201 to 0.0248 in (0.51 to 0.063 mm)
    Bottom .... 0.0019 to 0.0024 in (0.05 to 0.06 mm)
Width of ring grooves:
    Compression rings .... 0.0797 to 0.0807 in (2.02 to 2.05 mm)
    Oil control ring .... 0.157 to 0.158 in (3.99 to 4.01 mm)

Piston rings
Oil control ring .... Three-piece slotted scraper
Fitted gap: Compression & oil control .... 0.012 to 0.022 in (0.30 to 0.56 mm)

Tappets
Outside diameter .... 0.7996 to 0.8000 in (20.294 to 20.320 mm)

Valves
Head diameter
    Inlet .... 1.304 to 1.308 in (33.12 to 33.22 mm)
    Exhaust .... 1.168 to 1.172 in (29.66 to 29.76 mm)
Stem diameter
    Inlet .... 0.310 to 0.3112 in (7.87 to 7.90 mm)
    Exhaust .... 0.310 to 0.3105 in (7.878 to 7.887 mm)
Stem to guide clearance
    Inlet .... 0.0008 to 0.0023 in (0.02 to 0.06 mm)
    Exhaust .... 0.0015 to 0.003 in (0.038 to 0.076 mm)

Valve guides
Length .... 2.0625 in (52.387 mm)

Valve timing
Low and high compression
    Inlet opens .... 25° BTDC
    Inlet closes .... 65° ABDC
    Exhaust opens .... 65° BBDC
    Exhaust closes .... 25° ATDC
Emission controlled
    Inlet opens .... 10° BTDC
    Inlet closes .... 50° ABDC
    Exhaust opens .... 50° BBDC
    Exhaust closes .... 10° ATDC

1493 cc engine
Type .... 4 cylinder in-line OHV
Bore .... 2.9 in (73.7 mm)
Stroke .... 3.44 in (87.5 mm)
Capacity .... 1493 cc (91 cu in)
Compression ratio
    UK .... 9 to 1
    USA and Canada .... 7.5 to 1
Maximum BHP
    Except N. America .... 71 bhp DIN at 5500 rpm
    N. America .... 61 bhp DIN at 5000 rpm

Chapter 1 Engine

Maximum torque
    Except N. America .... 82 lbf ft at 3000 rpm
    N. America .... 81 lbf ft 2700 rpm
Firing order 1,3,4,2
Oversize bores .... +0.020 in (+0.51 mm)
Valve operation .... Overhead valves, pushrod and rocker operated

Crankshaft
Main journal diameter .... 2.3115 to 2.3120 in (58.713 to 58.725 mm)
Minimum main journal regrind diameter .... 2.2815 to 2.280 in (57.935 to 57.948 mm)
Crankpin journal diameter .... 1.8750 to 1.8755 in (47.625 to 47.638 mm)
Minimum crankpin journal regrind diameter .... 1.8450 to 1.8455 in (46.865 to 46.878 mm)
Crankshaft and thrust .... Via rear main bearing thrust washer
Crankshaft endfloat .... 0.006 to 0.014 in (0.152 to 0.355 mm)

Main bearings
Number 3
Type .... Thin wall
Length (front, centre and rear) .... 0.840 to 0.855 in (21.34 to 21.72 mm)
End thrust .... Via rear main bearing thrust washers
Thrust washer oversize .... 0.005 in (0.13 mm)
Diametrical clearance .... 0.0005 to 0.002 in (0.013 to 0.050 mm)
Undersizes .... 0.010 to 0.020, 0.030 in (0.25, 0.51, 0.76 mm)

Connecting rods
Length between centres .... 5.748 to 5.752 in (145.90 to 146.10 mm)
Small-end bush diameter (reamed in position) .... 0.8126 to 0.8129 in (20.64 to 20.65 mm)

Big-end bearings
Length .... 0.672 to 0.692 in (17.07 to 17.58 mm)
Diametrical clearance .... 0.001 to 0.003 in (0.03 to 0.08 mm)
Undersizes .... 0.010, 0.020, 0.030 in (0.25, 0.51, 0.76 mm)

Gudgeon pin
Type .... Fully floating
Connecting rod fit .... Hand push
Outside diameter .... 0.8123 to 0.8125 in (20.63 to 20.64 mm)

Pistons
Type .... Solid skirt, aluminium alloy
Bore size:
    F grade .... 2.8995 to 2.900 in (73.647 to 73.66 mm)
    G grade .... 2.9001 to 2.9006 in (73.663 to 73.673 mm)
Bottom diameter of piston
    F grade .... 2.8984 to 2.8989 in (73.619 to 73.632 mm)
    G grade .... 2.8990 to 2.8995 in (73.635 to 73.647 mm)
Clearance of skirt in bore
    Top .... 0.002 to 0.003 in (0.051 to 0.076 mm)
    Bottom .... 0.0002 to 0.0016 in (0.005 to 0.041 mm)
Number of rings .... 2 compression and 1 oil control
Width of ring grooves
    Top/second .... 0.064 to 0.065 in (1.625 to 1.650 mm)
    Oil control .... 0.1578 to 0.1588 in (3.99 to 4.01 mm)
Gudgeon pin bore .... 0.8124 to 0.8126 in (20.63 to 20.64 mm)
Piston oversize .... 0.020 in (0.508 mm)

Piston rings
Compression
    Top .... Plain type – chrome plated
    Second .... Tapered periphery
    Width (top/second) .... 0.0615 to 0.0625 in (1.575 to 1.578 mm)
    Fitted gap (top/second) .... 0.012 to 0.022 in (0.305 to 0.559 mm)
Ring/groove clearance
    Top and second .... 0.0015 to 0.0035 in (0.038 to 0.089 mm)
Oil control
    Type .... Expander with two chrome-faced rings
    Fitted gap (chrome-faced rings) .... 0.015 to 0.055 in (0.38 to 1.40 mm)

Camshaft
Front and rear journal diameters .... 1.9659 to 1.9664 in (49.93 to 49.95 mm)
Centre journal diameter .... 1.9649 to 1.9654 in (49.90 to 49.92 mm)
Bore in block .... 1.9680 to 1.9695 in (49.980 to 50.025 mm)
Diametrical clearances
    Front and rear .... 0.0016 to 0.0036 in (0.04 to 0.09 mm)

Chapter 1 Engine

Centre	0.0026 to 0.0046 in (0.07 to 0.12 mm)
End thrust	At front end location plate
Endfloat	0.0045 to 0.0085 in (0.014 to 0.216 mm)
Drive	Chain and gear from crankshaft
Camshaft sprocket adjusting shims	0.004 to 0.006 in (0.102 to 0.152 mm)

Rocker gear
Rocker shaft outside diameter	0.5607 to 0.5612 in (14.27 to 14.35 mm)
Rocker arm bore diameter	0.563 to 0.564 in (14.30 to 14.33 mm)

Valves
Rocker arm-to-valve clearance (inlet and exhaust)	0.010 in (0.25 mm) cold
Seat angle (inlet and exhaust)	45.5°
Valve face angle (inlet and exhaust)	45°
Head diameter	
Inlet	1.377 to 1.383 in (34.97 to 35.01 mm)
Exhaust	1.168 to 1.172 in (29.66 to 29.76 mm)
Stem diameter	
Inlet	0.3107 to 0.3113 in (7.89 to 7.91 mm)
Exhaust	0.3100 to 0.3105 in (7.874 to 7.887 mm)
Stem guide clearance	
Inlet	0.0007 to 0.0023 in (0.02 to 0.06 mm)
Exhaust	0.0015 to 0.0030 in (0.04 to 0.07 mm)

Valve guides
Length (inlet and exhaust)	2.06 in (52.223 mm)
Fitted height	0.75 in (19.050 mm)
Diameter (inlet and exhaust)	
Outside	0.5015 to 0.502 in (12.73 to 12.75 mm)
Inside	0.312 to 0.313 in (7.92 to 7.95 mm)

Valve springs – UK
Free length	
Inner	1.14 in (30 mm)
Outer	1.52 in (38.6 mm)

Valve springs – USA and Canada
Free length (inner and outer)	1.52 in (38.6 mm)

Valve timing
Timing mark	Notch on crankshaft pulley – pointers on timing chest
Rocker arm/valve clearances – numbers 7 and 8 valves (for valve timing only)	0.050 in (1.27 mm)
Inlet valve opens	18° BTDC
Inlet valve closes	58° ABDC
Exhaust valve opens	58° BBDC
Exhaust valve closes	18° ATDC

Lubrication system
Type	Wet sump pressure feed
Oil type/specification	Multigrade engine oil, viscosity SAE 20W/50 (Duckhams Hypergrade)
Sump capacity	7 Imp pts (8.4 US pts, 4.0 l)
System pressure (running)	40 to 60 lbf/in² (2.81 to 4.2 kgf/cm²)
Oil pump type	Hobourn-Eaton eccentric lobe
Oil pump clearances	
Inner rotor endfloat	0.0004 in (0.1 mm)
Outer rotor endfloat	0.0004 in (0.1 mm)
Rotor lobe clearance	0.010 in (0.25 mm)
Outer rotor-to-body clearance	0.008 in (0.2 mm)
Oil filter type	Full-flow, disposable cartridge
Bypass valve opens	8 to 12 lbf/in² (0.56 to 0.84 kgf/cm²)
Oil pressure relief valve	53 lbf/in² (3.71 kgf/cm²)
Relief valve spring free length	1.53 in (38.86 mm)

Flywheel
Run-out (maximum)	0.002 in (0.051 mm) measured at 3.0 in (76.2 mm) radius from axis
Concentricity	0.004 in (0.100 mm) maximum error

Distributor driveshaft
Endfloat	0.003 to 0.007 in (0.08 to 0.18 mm)

Torque wrench settings
1147 cc and 1296 cc engines

	lbf ft	Nm
Big-end bolts	40	55
Clutch attachment	20	27

Cylinder head nuts	45	61
Engine mounting bolts	20	27
Flywheel securing bolts	45	61
Fuel pump bolts	12	16
Main bearing bolts	50	68
Manifold nuts	25	34
Oil gallery set screws	20	27
Oil pump to block	8	11
Rocker pedestals	25	34
Sump to crankcase	16	22
Timing cover (5/16 in setscrew)	16	22
Timing cover (5/16 in slotted setscrew)	10	14
Water outlet elbow nuts	16	22
Water pump to cylinder head	20	27

1493 cc engine

Alternator to mounting bracket and front engine plate	22	30
Alternator to adjusting link	20	27
Clutch attachment to flywheel	22	30
Connecting rod bolt:		
Colour dyed	50	68
Phosphated	46	63
Chainwheel to camshaft	24	33
Crankshaft pulley nut	150	204
Cylinder block drain plug	35	48
Cylinder head to block	46	63
Distributor to pedestal	20	27
Fan attachment	9	12
Flywheel to crankshaft		
Cadmium plated	40	54
Parkerised	45	61
Fuel pump to cylinder block	14	19
Gearbox and rear engine plate to block	14	19
Manifold inlet-to-exhaust	14	19
Manifold to head	25	34
Main bearing cap bolts	65	88
Oil sump drain plug	25	34
Oil sump to block	20	27
Oil pressure switch plug to cylinder head	14	19
Oil seal retainer	14	19
Rocker cover to head	2	3
Rocker pedestal to cylinder head	32	44
Crankshaft rear seal housing	20	27
Rear engine mounting platform on frame	20	27
Sealing block to engine plate	20	27
Spark plug to head	20	27
Starter motor attachment	34	46
Timing cover to front engine plate		
Small	10	14
Large	20	27
Water elbow to water pump	20	27
Water pump bearing housing to pump	14	19
Water pump to cylinder head	20	27

PART A – GENERAL (ALL MODELS)

1 General description

1 The engine is a four-cylinder, overhead valve type. It is supported by rubber mountings in the interests of silence and lack of vibration.
2 Two valves per cylinder are mounted vertically in the cast iron cylinder head and run in pressed-in valve guides. They are operated by rocker arms, pushrods and tappets from the camshaft which is located at the base of the cylinder bores in the left-hand side of the engine. The correct valve stem to rocker arm pad clearance can be obtained by the adjusting screws in the ends of the rocker arms.
3 On 1147 cc models the cylinder head has two siamesed inlet, and four separate ports on the right-hand side. 1296 cc models have four inlet and four exhaust valves.
4 The cylinder block and the upper half of the crankcase are cast together. The bottom half of the crankcase consists of a pressed steel sump.
5 The pistons are made from anodised aluminium alloy with split or solid skirts. Two compression rings and a slotted oil control ring are fitted. The gudgeon pin is retained in the little end of the connecting rod by circlips or by interference fit.
6 Renewable shell type big-end bearings are fitted.
7 At the front of the engine a single chain drives the camshaft via the camshaft and crankshaft chain wheels which are enclosed in a pressed steel cover.
8 The chain is tensioned automatically by a spring blade which presses against the non-driving side of the chain so avoiding any lash or rattle.
9 The camshaft is supported by four bearings bored directly into the cylinder block except on certain later engines which are fitted with special replaceable bearings. Endfloat is controlled by a forked locating plate positioned on the front end plate.
10 The statically and dynamically balanced forged steel crankshaft is supported by three renewable thinwall shell main bearings which are in turn supported by substantial webs which form part of the crankcase. Crankshaft endfloat is controlled by semi-circular thrust washers located on each side of the rear main bearing.
11 The centrifugal water pump and radiator cooling fan are driven, together with the dynamo, from the crankshaft pulley wheel by a rubber/fabric belt. The distributor is mounted in the middle of the left-hand side of the cylinder block and advances and retards the ignition timing by mechanical and vacuum means. The distributor is driven at

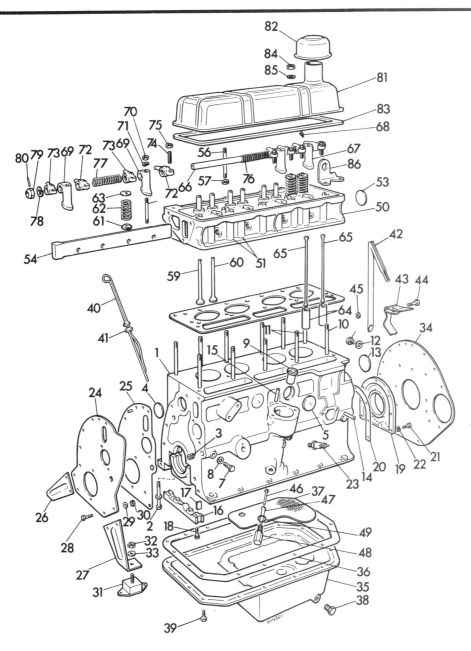

Fig. 1.1 Exploded view of typical smaller capacity engine (Sec 1)

1. Cylinder block
2. Bolt and lockwasher
3. Oil gallery plug
4. Welch plug
5. Welch plug
6. Welch plug
7. Oil gallery bolt
8. Washer
9. Oil pump shaft bush
10. Cylinder head stud
11. Cylinder head stud
12. Coolant drain plug
13. Fibre washer
14. Fuel pump stud
15. Distributor mounting stud
16. Front sealing block
17. Filler piece
18. Screw
19. Oil seal retainer
20. Gasket
21. Bolt
22. Spring washer
23. Oil pressure switch
24. Engine front plate
25. Gasket
26. Typical engine mounting bracket
27. Typical engine mounting bracket
28. Bolt
29. Spring washer
30. Nut
31. Engine mounting pad
32. Nut
33. Spring washer
34. Engine rear plate
35. Sump
36. Gasket
37. Oil strainer gauze
38. Oil drain plug
39. Bolt
40. Dipstick
41. Felt washer
42. Breather pipe
43. Deflector
44. Bolt
45. Nut
46. Oil pressure relief valve plunger
47. Plunger spring
48. Retaining plug
49. Washer
50. Cylinder head
51. Pushrod tubes
52. Valve guide
53. Core plug
54. Coolant distributor tube
55. Rocker pedestal stud
56. Rocker cover stud
57. Nut
58. Gasket
59. Inlet valve
60. Exhaust valve
61. Spring lower seat
62. Valve spring
63. Valve spring upper seat
64. Tappet (cam follower)
65. Pushrods
66. Rocker shaft
67. Drilled rocker pedestal
68. Bolt and lockwasher
69. Undrilled rocker pedestal
70. Nut
71. Lockwasher
72. Rocker arm
73. Rocker arm
74. Ball pin
75. Locknut
76. Rocker shaft spring
77. Rocker shaft springs
78. Spring washer
79. Collar
80. Roll pin
81. Rocker cover
82. Oil filler cap with breather
83. Rocker cover gasket
84. Nut
85. Plain and fibre washers

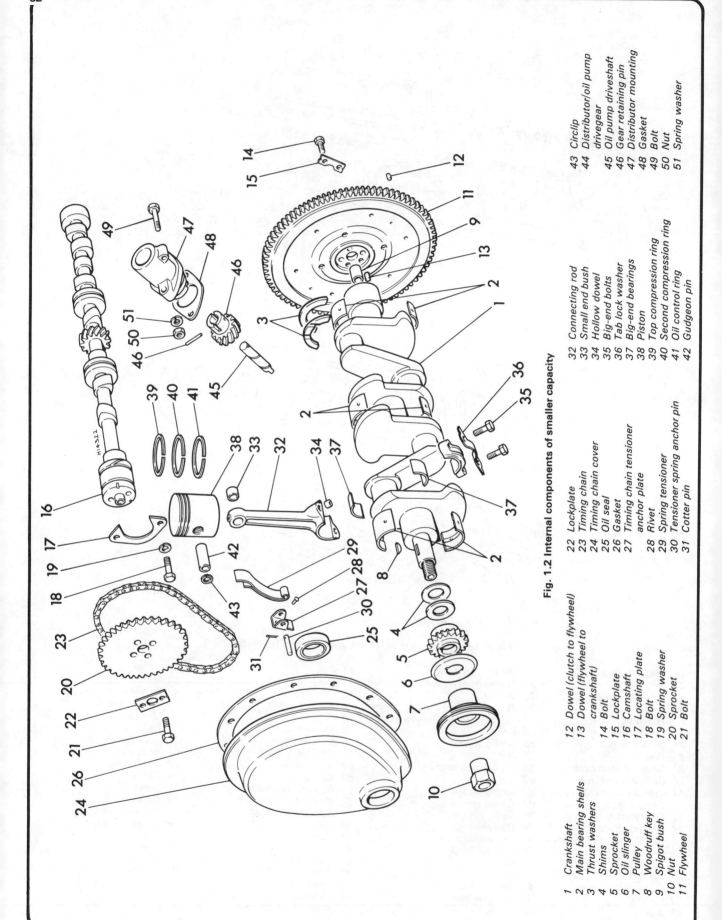

Fig. 1.2 Internal components of smaller capacity

1 Crankshaft
2 Main bearing shells
3 Thrust washers
4 Shims
5 Sprocket
6 Oil slinger
7 Pulley
8 Woodruff key
9 Spigot bush
10 Nut
11 Flywheel
12 Dowel (clutch to flywheel)
13 Dowel (flywheel to crankshaft)
14 Bolt
15 Lockplate
16 Camshaft
17 Locating plate
18 Bolt
19 Spring washer
20 Sprocket
21 Bolt
22 Lockplate
23 Timing chain
24 Timing chain cover
25 Oil seal
26 Gasket
27 Timing chain tensioner anchor plate
28 Rivet
29 Spring tensioner
30 Tensioner spring anchor pin
31 Cotter pin
32 Connecting rod
33 Small end bush
34 Hollow dowel
35 Big-end bolts
36 Tab lock washer
37 Big-end bearings
38 Piston
39 Top compression ring
40 Second compression ring
41 Oil control ring
42 Gudgeon pin
43 Circlip
44 Distributor/oil pump drivegear
45 Oil pump driveshaft
46 Gear retaining pin
47 Distributor mounting
48 Gasket
49 Bolt
50 Nut
51 Spring washer

Chapter 1 Engine

half crankshaft speed by a short shaft and skew gear from a skew gear on the camshaft located between the second and third journals.
12 The oil pump is located in the crankcase and is driven by a short shaft from the skew gear on the camshaft.
13 Attached to the end of the crankshaft by four bolts and two dowels is the flywheel to which is bolted the Borg & Beck clutch. Attached to the engine end plate is the gearbox bellhousing.

2 Engine oil – topping-up, oil and filter renewal

1 The engine oil level should be maintained at the HIGH mark on the dipstick. Every week, with the engine switched off, withdraw the dipstick, wipe it clean and re-insert it. Withdraw it for the second time and read off the oil level.
2 Add oil of correct grade as necessary to bring the level up to the HIGH mark. The oil filler cap is located at the rear end of the rocker cover.
3 At the intervals specified in *Routine Maintenance*, with the engine hot, remove the sump drain plug and catch the oil in a suitable container.
4 While the engine oil is draining, if this is the service interval for also changing the oil filter (refer to *Routine Maintenance*, unscrew and discard the old filter. If it is stuck tight, use an oil filter wrench or small chain wrench. If you do not have one of these tools, drive a screwdriver through the filter and use this as a lever to unscrew it.
5 Smear the sealing ring of the new filter with oil and screw it into position using hand pressure only (photo).
6 Refit the drain plug and pour in the specified quantity and grade of engine oil.
7 Start the engine. It will take a few seconds for the oil warning light to go out, this is normal and is due to the time taken to fill the new filter cartridge.
8 Check for leaks from the filter after a few minutes and then switch off and check the oil level.

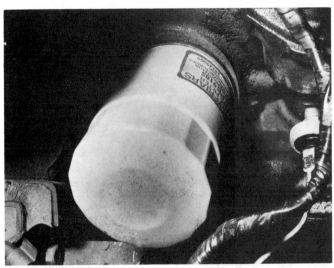

2.5 Oil filter

3 Major operations with engine in place

The following major operations can be carried out to the engine with it in place in the car:-

(a) Removal and replacement of the cylinder head assembly
(b) Removal and replacement of the sump
(c) Removal and replacement of the big-end bearings
(d) Removal and replacement of the pistons and connecting rods
(e) Removal and replacement of the timing chain and gears and the timing cover oil seal
(f) Removal and replacement of the camshaft
(g) Removal and replacement of the oil pump
(h) Before commencing work, disconnect the battery earth lead

4 Major operations with engine removed

The following major operations can be carried out with the engine out of the car and on the bench or floor

(a) Removal and replacement of the main bearings
(b) Removal and replacement of the crankshaft
(c) Removal and replacement of the flywheel

5 Methods of engine removal

There are two methods of engine removal. The engine can either be removed complete with gearbox, or the engine can be removed without the gearbox by separation at the gearbox bellhousing. Both methods are described. Irrespective of whether the Triumph engine is removed with or without the gearbox, it will be found to be one of the easiest units to take out and replace. No pit or ramps are necessary as the jobs usually done underneath such as propshaft/gearbox separation are all done from inside the car. As a further bonus engine accessibility is excellent.

6 Engine – removal and refitting without gearbox

1147 and 1296 cc engines

1 Depending upon the type of lifting gear available, it is possible to remove the engine without having to remove the bonnet first, but it is much easier to work without the bonnet in the way, so details of removal are given.
2 Remove the front overriders and radiator grille assembly. Mark the position of the bonnet hinge pivot bolt heads. Release the wiring harness from the clips at the inside front edge of the bonnet and unbolt the bonnet stay.
3 With the help of two assistants to take the weight of the bonnet, remove the pivot bolts from the hinge brackets and lift the bonnet from the car.
4 Unbolt and remove the shields from both sides of the engine compartment. Pull the flexible air intake ducts from the air cleaner (photos).
5 Drain the cooling system either by means of the radiator drain tap, or if one is not fitted, disconnect the radiator bottom hose. Remove the cylinder block drain plug. Drain the engine oil.
6 Disconnect the battery earth lead.
7 Disconnect the LT wires from the distributor and the HT wire from the centre socket of the ignition coil.

6.3a Engine side shield bulkhead bolts

6.3b Engine side shield and strut

6.3c Removing side shield lower bolt

8 Pull the wires by their tags off the Lucar connector on the back of the generator.
9 Pull the wire off the connector on the oil pressure warning sender unit.
10 Undo the union nut from the petrol input pipe on the fuel pump. On some models the pipe is connected by a rubber connector which is simply pulled off. Block the open end of the pipe to prevent petrol from the tank running out.
11 Different types of air cleaner have been used at various times. If the air cleaner(s) gets in the way it must be removed. The small pancake type can always be left in place.
12 Undo the screw from the clamp which holds the accelerator cable to the carburettor or in some instances undo the nut and washer holding the ball joint rod to the accelerator linkage arm. Release the return spring(s).
13 The next stage is to release the choke cable by undoing the clamp nut which holds it to the linkage.
14 Pull off the wire by its tag from the Lucar terminal on the thermostat housing.
15 Undo the clips on the two radiator hoses and pull the hoses off the thermostat outlet pipe, and water pump inlet pipe.
16 Undo the clips on the heater inlet hoses and pull them off the inlet and outlet pipes on the rear of the engine.
17 The next step is to undo the nuts and bolts from either side of the radiator. Make a special note if the horns or any other items are hung on the radiator bolts.

18 With the bolts removed the radiator can be lifted out.
19 Undo the nuts from the three studs which hold the exhaust pipe to the exhaust manifold.
20 Pull the exhaust pipe away from the exhaust manifold and undo the petrol drain pipe (if fitted) from the bottom of the inlet manifold (photo).
21 Undo the nut and bolt holding the exhaust downpipe to the clamp attached to the bellhousing bottom bolt.
22 Take off the nut retaining the starter lead to the starter terminal post and pull off the starter motor lead.
23 Undo the two bolts holding the starter motor in place and lift out the starter motor together with the distance piece.
24 Where fitted undo the knurled nut which holds the tachometer drive cable to the rear of the distributor.
25 Pull off the advance and retard pipe from the distributor or on certain models undo the union nut which holds the pipe in place.
26 Undo the nuts from the bolts and studs which hold the engine endplate to the bellhousing. Note the clip which secures the clutch hydraulic pipe to the top centre stud.
27 The engine earth lead must be taken off next. It will either be attached to a bolt on the front right engine mounting bracket or to one of the bellhousing bolts.
28 Attach a lifting chain or strong rope to the two lifting hook holes provided on the engine. Take the weight of the engine on suitable lifting tackle. Place a jack under the gearbox to take the weight of this unit.
29 Undo the nuts which hold the front engine mountings in place. Several different types of mounting with different attachment arrangements have been used at different times. On some models two nuts have to be undone and the bolt removed, while on others it is necessary to remove only one nut.
30 Slightly raise the engine, and the jack, and pull the engine forwards and up until the clutch is free from the first motion shaft in the gearbox. It is important that no excess load is placed on the clutch so take great care at this stage. Once clear of the bellhousing pull the engine forwards tilting the front further upwards to clear the front crossmember and the engine out of the car.

6.19 Manifold fuel drain pipe

1493 cc engine

31 Remove the bonnet as described in paragraphs 1 and 2 of this Section.
32 Disconnect the battery leads from the terminals.
33 Unscrew the heater air intake hose retaining clip and remove the hose.
34 Refer to Chapter 2 and drain the cooling system, then remove the radiator.
35 Undo the four cooling fan bolts and remove the fan. On 1979 and later USA models an electric fan is used.
36 Drain the engine oil.

Chapter 1 Engine

37 Refer to Chapter 3 and remove the carburettor(s) and air cleaner assembly. After disconnecting the accelerator and choke cables, move them to one side out of the way.
38 Now disconnect the exhaust manifold-to-downpipe clamp bolts.
39 On USA models, the running-on control valve vacuum pipe is next disconnected from the inlet manifold.
40 The heater hose can be disconnected from the water return pipe and the heater control valve hose from the inlet manifold. On 1979 USA models, disconnect the coolant hose from the carburettor auto choke (photo).
41 From the thermostat housing, remove the water temperature gauge capillary tube, or disconnect the electrical lead from the coolant temperature transmitter, as appropriate.
42 On cars with emission control, disconnect the air supply hose from the check valve and undo the clips supporting the air injection pipe. Unscrew the check valve from the air injection pipe, taking care not to twist the pipe. Disconnect the four unions connecting the pipe to the cylinder head (later systems) or the single union connecting the pipe to the exhaust manifold (earlier systems) and remove the air injection pipe.
43 Disconnect the alternator leads plug from the alternator.
44 Remove the distributor cap and disconnect the HT leads from the spark plugs and coil. Remove the assembly, and then the rotor arm. Disconnect the LT lead from the distributor.
45 On cars with emission control remove the air pump, referring to Chapter 3 if necessary.
46 Disconnect the oil pressure gauge pipe from its engine union or, alternatively, disconnect the lead from the oil pressure transmitter switch.
47 Undo the bolt securing the earth strap to the engine and move the strap out of the way.
48 Remove the supply lead from the starter motor and remove the starter by undoing the two retaining nuts and bolts to release the starter, adaptor plate and shim.
49 Disconnect the fuel feed pipe at the connection adjacent to the bellhousing/engine flange joint.
50 Disconnect the heater hose connection from the heater unit and the inlet manifold T-pipe connection at the rear of the manifold.
51 Remove the nuts, bolts and washers securing the clutch bellhousing to the engine and, on USA models, move the restraint cable out of the way.
52 Position a jack under the gearbox to take its weight and fit slings or suitable lifting tackle to the engine.
53 Take the weight of the engine and undo the engine mountings. Make a thorough, progressive check around the engine that all connections with the rest of the car have been undone and carefully lift the engine. It will be necessary to move the engine forward to withdraw the gearbox input shaft from the clutch, and care must be taken to avoid placing load on the shaft as this is done.
54 Remove the engine from the car.

6.39 Heater hose connection

Refitting 1147 cc and 1296 cc engines

55 Although the engine can be replaced with one man and a suitable winch, it is easier if two are present. One to lower the engine into the engine compartment and the other to guide the engine into position and to ensure that it does not foul anything. Generally speaking, engine replacement is a reversal of the procedures used when removing the engine, but one or two added tips may come in useful.
56 Ensure all the loose leads, cables etc., are tucked out of the way. If not, it is easy to trap one and so cause much additional work after the engine is replaced.
57 Fit the starter motor and oil filter before lowering the engine and gearbox into place.
58 After the generator has been replaced it is advisable to fit a new fan belt.
59 Carefully lower the engine into position and then refit the following:

(a) Gearbox mounting nuts and washers
(b) Front mountings
(c) Propeller shaft to gearbox
(d) Reconnect the clutch pipe to the master cylinder
(e) Speedometer cable
(f) Gearchange remote control and lever (solenoid wires if fitted)
(g) Gearbox cover and carpets etc (refill gearbox first)
(h) Oil pressure switch
(i) Rev counter drive (if fitted)
(j) Wires to coil, distributor and generator
(k) Carburettor controls
(l) Fuel pipe to pump and carburettors
(m) Air cleaner(s)
(n) Exhaust manifold to pipe and bracket
(o) Earth and starter motor cables
(p) Radiator and hoses and any items hung on radiator attachment bolts (shield, horns etc)
(q) Heater hoses
(r) Water temperature cable
(s) Vacuum advance pipe
(t) Battery
(u) Bonnet and engine shields

60 Finally check that the drain plugs are tight and taps closed, then fill with the correct quantities of engine oil and coolant.
61 The engine is now ready for starting (see Section 25).

Refitting 1493 cc engine

62 Although the engine can be refitted by one man with a suitable winch, it is advisable to have two assistants at hand for certain operations. These are principally to help guide the engine into position and ensure that it does not foul anything and for refitting the bonnet.
63 Generally speaking, engine refitting is a direct reversal of the removal sequence, but several items should be mentioned and these are as follows.
64 Ensure that all loose leads, cables etc are tucked out of the way.
65 Ensure that the engine lifting sling is correctly located around the engine and that the lifting equipment is in good condition.
66 Carefully lower the engine into the car and locate to the gearbox and front engine mountings with the respective bolts.
67 Then reconnect the following components:

(a) Oil pressure switch
(b) Wiring to coil, distributor and alternator
(c) Carburettor fittings
(d) Fuel pipe to pump
(e) Air cleaner(s)
(f) Exhaust manifold to downpipe
(g) Earth strap and starter motor cables
(h) Radiator, hoses and associate fittings
(i) Heater hoses
(j) Water temperature capillary tube and sender unit or the electrical leads as appropriate
(k) Vacuum advance retard
(l) Battery

68 On USA models the emission control components will also have to be reconnected. If necessary refer to Chapter 3 for details.
69 Finally check that the cooling system hoses are correctly fitted and refill the cooling system (see Chapter 2).

70 Refill the engine with the correct grade and quantity of oil. The engine is now ready to start (see Section 25).
71 Refit the engine side shields and the bonnet.

7 Engine – removal and refitting with gearbox

1147cc and 1296 cc engines

1 It is possible to remove the engine and gearbox as a combined unit without the necessity of first removing the transmission cover and fascia support bracket, the latter being a laborious and intricate task (see Chapter 6, Section 3).
2 Remove the gear lever knob and then remove the front compartment carpet by sliding the gear lever gaiter off the gear lever.
3 Check that the gear lever is in neutral and release the bayonet fitting lever retaining cap within the rear end of the transmission tunnel cover. Withdraw the gear lever taking care not to lose the spring and plunger.
4 Working within the engine compartment, remove the bonnet and side shields as described in Section 6. Carry out the operation described in paragraphs 4 to 17 in Section 6.
5 Disconnect the exhaust downpipe from the manifold and then working under the car disconnect the exhaust pipe brackets with the exception of the rearmost one. The exhaust pipe should now be pulled downward at its front end and tied with wire. This will give access to the propeller shaft and gearbox mounting bolts and to the clutch slave cylinder.
6 Release the clutch slave cylinder clamp bolt and pull the cylinder from its housing and move it to one side without straining the pipeline.
7 Working under the car, unscrew the retainer and disconnect the speedometer cable from the gearbox.
8 Carefully scribe a mating mark across the universal joint and gearbox drive flanges.
9 Undo the four nuts and bolts which hold the universal joint to the gearbox mainshaft flange.
10 Then undo the single nut and washer from each of the two gearbox extension mounting rubbers, so the rubbers can be freed from the mounting bracket.
11 At the front of the engine undo the front mounting nuts. On early models one nut has to be undone on each side. See also Section 6, paragraph 28.
12 Now note the following points:- as soon as the engine is lifted an inch or two, free the rear mounting bolts from the mounting bracket.
13 The engine and gearbox should be lifted out at an angle of about 45° once the gearbox extension has cleared the chassis crossmember.
14 Unless a movable hoist is being used, it will be necessary to lift the engine about 5 feet to enable the car to be wheeled out from under the engine.
15 Separate engine and gearbox by removing the connecting bolts which hold the bellhousing to the engine plate.
16 Also undo the two bolts holding the starter motor in place and lift off the motor. Note and retain the distance piece and any shims which may be fitted.
17 Carefully pull the gearbox complete with bellhousing off the engine.

1493 cc engine

18 The operations are very similar to those described for the smaller capacity engines in earlier paragraphs of this Section, but observe the following differences and additional operations.
19 Disconnect the vent pipes which run from the carburettor float chamber lids and are retained to the engine side shield by spring clips. Disconnect the leads from the reverse lamp switch (photo).
20 The speedometer drive cable is retained to the gearbox by a forked clamp plate (photo).
21 On some North American models, a restraining cable is attached to the gearbox. This should be disconnected.
22 Next remove the two securing bolts from the rear mounting bracket to the gearbox tunnel.
23 Place the engine lifting slings securely in position to take the weight of the engine unit. The engine will have to be tilted to approximately 70° when lifted, and the slings should therefore be located accordingly with a shorter sling to the front.
24 With the slings in position, lift to take up any slack and then remove the right- and left-hand engine mounting nuts or bolts (photo).
25 Check that all of the engine and gearbox attachments to the body and surrounding components are disconnected and carefully lift the engine. An assistant will be needed to lift the gearbox rear coupling over the gearbox tunnel crossmember (photos).
26 With the engine and gearbox lifted clear push the car rearwards and lower the engine/gearbox unit.
27 With the engine and gearbox removed from the car, disconnect the starter motor unit (if this has not already been done) by unscrewing the two retaining nuts and washers (photo).
28 Withdraw the starter motor from the clutch housing complete with shim and adaptor plate.
29 Undo the respective gearbox-to-engine retaining bolts around the clutch housing and withdraw the engine from the gearbox. Do not let the gearbox hang on the first motion shaft unsupported, or damage may occur.

Reconnection and refitting

30 Reconnection of engine to gearbox and refitting of the complete assembly are reversals of removal and separation, but make sure that the clutch driven plate has been centralised (see Chapter 5).
31 When the engine and gearbox have been lowered into position, reconnect the gearbox mounting brackets.
32 Reconnect the propeller shaft to gearbox driveshaft flange ensuring that the mating marks align.
33 Relocate the clutch slave cylinder. Note the pipeline clip at the top of the bellhousing (photo).
34 Reconnect the speedometer cable.
35 Reconnect the reversing light switch cable, and on top of the gearbox, refit the fuel line pipe clip to the cylinder head (photo).
36 Reinsert the gearchange remote control and lever, then refill the gearbox with the correct grade and quantity of oil to the level plug. This will have to be filled from underneath the car on the right-hand side of the gearbox.
37 Refit the carpets and gearchange gaiter.

8 Dismantling the engine – general

1 It is best to mount the engine on a dismantling stand, but if one is not available, then stand the engine on a strong bench so as to be at a comfortable working height. Failing this, the engine can be stripped down on the floor.
2 During the dismantling process the greatest care should be taken to keep the exposed parts free from dirt. As an aid to achieving this, it is a sound scheme to thoroughly clean down the outside of the engine, removing all traces of oil and congealed dirt.
3 Use paraffin or a good grease solvent. The latter compound will make the job much easier, as, after the solvent has been applied and allowed to stand for a time, a vigorous jet of water will wash off the solvent and all the grease and filth. If the dirt is thick and deeply embedded, work the solvent into it with a wire brush.
4 Finally wipe down the exterior of the engine with a rag and only then, when it is quite clean should the dismantling process begin. As the engine is stripped, clean each part in a bath of paraffin or petrol.
5 Never immerse parts with oilways in paraffin, but to clean wipe down carefully with a petrol dampened rag. Oilways can be cleaned out with wire. If an air line is present all parts can be blown dry and the oilways blown through as an added precaution.
6 Re-use of old engine gaskets is a false economy and can give rise to oil and water leaks, if nothing worse. To avoid the possibility of trouble after the engine has been reassembled always use new gaskets throughout.
7 Do not throw the old gaskets away as it sometimes happens that an immediate replacement cannot be found and the old gasket is then very useful as a template. Hang up the old gaskets as they are removed on a suitable hook or nail.
8 To strip the engine it is best to work from the top down. The sump provides a firm base on which the engine can be supported in an upright position. When the stage where the sump must be removed is reached, the engine can be turned on its side and all other work carried out with it in this position.
9 Wherever possible, replace nuts, bolts and washers finger-tight from wherever they were removed. If they cannot be replaced then lay them out in such a fashion that it is clear from where they came.

Chapter 1 Engine

7.19 Carburettor vent pipes

7.20 Speedo cable retainer

7.24 1493 cc engine front mounting

7.25a Removing 1493 cc engine with gearbox (RHS)

7.25b Removing 1493 cc engine with gearbox (LHS)

7.25c 1493 cc engine removal, viewed from front

7.27 1493 cc engine/gearbox removed from car

7.33 Clutch pipeline on top of bell housing

7.35 Fuel line clip at cylinder head

9 Ancillary engine components – removal

1147 cc and 1296 cc engines

1 Before basic engine dismantling begins the engine should be stripped of all its ancillary components. These items should also be removed if a factory exchange reconditioned unit is being purchased. The items comprise:-

Generator and brackets
Water pump and thermostat housing
Starter motor
Distributor and sparking plugs
Inlet and exhaust manifold and carburettors
Fuel pump and fuel pipes
Oil filter and dipstick
Oil filler cap
Clutch assembly
Breather pipe and gauge (where fitted)
Auxiliary header tank (where fitted)

2 Without exception all these items can be removed with the engine in the car if it is merely an individual item which requires attention. (It is necessary to remove the gearbox if the clutch is to be renewed with the engine in situ).
3 Starting work on the left-hand side of the engine slacken off the generator retaining bolts and remove the unit and then the support brackets.
4 Take off the distributor and housing after undoing the two nuts and washers (photo) which hold the bottom flange of the distributor housing to the cylinder block. Retain and note the shims between the housing and the block. Do not loosen the square nut on the clamp at the base of the distributor body or the timing will be lost. Undo the sparking plugs.

5 Note that the fuel pump is held in place by two studs. A nut fits on the stud on the left and a special nut over the stud on the right (photo).
6 Undo the nuts and lift off the fuel pump (photo).
7 Undo and remove the low oil pressure warning sender unit.
8 Undo and remove the oil filter. The complete body just screws off anti-clockwise (photo).
9 Moving to the front of the engine undo the left-hand thermostat housing cover bolt to free the clip which carried the fuel and vacuum advance/retard lines.
10 Undo the nuts and washers which hold the inlet and exhaust manifolds to the cylinder head. The inner nuts are very difficult to get at, and are best loosened with a thin ring spanner.
11 Lift off the inlet and exhaust manifolds together with the carburettor(s). If stiff tap the manifolds gently with a piece of wood.
12 Undo the bolts which hold the water pump in place on the front face of the block.
13 Undo the rear right cylinder head nut from the clip which holds the main heater pipe in place.
14 The coolant pump is removed with the main heater pipe as one unit (photo).
15 Where a breather pipe is fitted note that it is a press fit in the block and should be carefully twisted and pulled out.
16 Undo a quarter of a turn at a time the six bolts which hold the clutch pressure plate assembly to the flywheel (photo).
17 Lift off the pressure plate together with the loose friction plate. Check that all the items listed in paragraph 1 of this section have been removed. The engine is now stripped and ready for major dismantling to begin.

1493 cc engine

18 Before basic engine dismantling, the following ancillary items must be removed:

(a) Alternator
(b) Distributor
(c) Thermostat housing and water pump
(d) Oil filter if it is not already removed with the air pump (USA models)
(e) Inlet and exhaust manifolds, and carburettor(s)
(f) Fuel pump
(g) Starter motor, if not already removed

19 Remove the alternator from its bracket complete with spacer.
20 Rotate the crankshaft to align the timing notches on the crankshaft pulley to the 10° BTDC mark on the timing indicator or 2° ATDC for California models only. Check that the distributor rotor arm is in the firing position for number 1 spark plug, and lightly mark the distributor body opposite the rotor arm pointer. Undo the two flange nuts and spring washers and remove the distributor. Note and retain the shims between the housing and the block. Do not loosen the square nut on the clamp on the base of the distributor or the timing will be lost. Remove the spark plugs.
21 The distributor/oil pump driveshaft can also be withdrawn from the block at this stage.
22 Unscrew and remove the oil filter.
23 Undo the water pump retaining nuts and lightly tap it away from the block at the front, complete with thermostat.
24 If they have not already been removed, disconnect the inlet and exhaust manifolds together with the carburettor(s) from the cylinder head. Note the bridging 'plate washers' between the two manifolds on the top row of the studs. The inner nuts are best removed with a thin ring spanner.
25 Remove the two manifolds together/ If stiff, lightly tap the manifolds with a piece of wood.
26 Remove the fuel pump from the block by unscrewing the two retaining nuts and washers.
27 Remove the starter motor, if applicable, by unscrewing the two retaining nuts and washers, and withdrawing the motor complete with shims and adaptor plate.

9.4 Removing distributor mounting nuts

9.5 Fuel pump special mounting nut

9.6 Removing fuel pump

9.8 Unscrewing oil filter

9.14 Removing coolant pump with heater pipe

9.16 Releasing a clutch pressure plate cover bolt

Chapter 1 Engine

PART B – 1147 CC AND 1296 CC ENGINES

10 Cylinder head – removal and refitting (engine in car)

To remove the cylinder head with the engine still in the car the following additional procedure must first be carried out.

1 Disconnect the battery by removing the earth terminal.
2 Drain the water by turning the taps at the base of the radiator, and at the bottom left-hand corner of the cylinder block.
3 Loosen the clip at the thermostat housing end on the top water hose, and pull the hose from the thermostat housing pipe.
4 Slacken the securing bolts and move the generator away from the cylinder head, at the same time remove the fan belt.
5 Disconnect the fuel line at the carburettor end, also the vacuum advance/retard pipe at the distributor and undo the three bolts holding the water pump and thermostat housing in place.
6 Remove the main heater pipe from the rear cylinder head stud and pull the water pump clear of the cylinder head.
7 Disconnect the controls from the carburettor(s), and undo the three nuts holding the exhaust manifold to the exhaust down pipe. Leave the inlet and exhaust manifolds in place as they provide useful leverage when removing the cylinder head.
8 Unscrew the two rocker cover nuts and lift the rocker cover with its gasket away (photo).
9 Unscrew the rocker pedestal nuts and lift off the rocker assembly (photo).
10 Unscrew the cylinder head nuts, progressively, half a turn at a time in the reverse order to that shown.
11 Withdraw the pushrods, keeping them in their originally fitted order. A simple way to do this is to push each rod through a sheet of card on which the numbers 1 to 8 have been marked.
12 The cylinder head can now be removed by lifting upwards. If the head is jammed, try to rock it to break the seal. Under no circumstances try to prise it apart from the block with a screwdriver or cold chisel as damage may be done to the faces of the head or block. If the head will not readily free, turn the engine over by the flywheel as the compression in the cylinders will often break the cylinder head joint. If this fails to work, strike the head sharply with a plastic headed hammer, or with a wooden hammer, or with a metal hammer with an interposed piece of wood to cushion the blows. Under no circumstances hit the head directly with a metal hammer as this may cause the iron casting to fracture. Several sharp taps wth the hammer at the same time pulling upwards should free the head. Lift the head off and place on one side (photo).
13 Clean the cylinder block and head mating faces.
14 Refer to Section 17 and 19 for cylinder head dismantling and decarbonising.

10.8 Removing rocker cover nut

10.9 Removing rocker assembly

10.12a Using a block of wood and hammer to release cylinder head from block

10.12b Removing cylinder head

Chapter 1 Engine

10.15 Locating cylinder head gasket

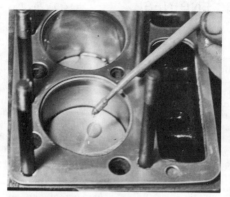
10.16 Oiling cylinder bores

10.17 Lowering cylinder head into position

10.19 Tightening cylinder head nuts

10.20 Inserting push-rods

10.21 Seating rocker arms with push-rod cups

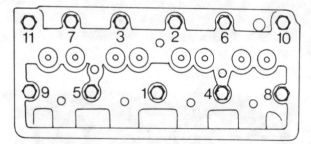

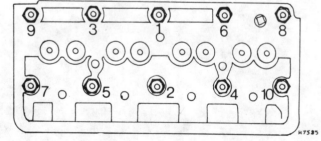

Fig. 1.3 Cylinder head nut tightening sequence (alternative arrangements shown) (Sec 10)

15 Fit a new gasket in place (photo). If one side of the gasket is marked TOP it must naturally be fitted with this side facing upwards.
16 Generously lubricate each cylinder with engine oil (photo).
17 Ensure that the cylinder head face is perfectly clean and then lower the cylinder head into place, keeping it parallel to the block to avoid binding on any of the studs (photo).
18 With the head in place fit the lifting eye over the two rear right-hand studs and the accelerator cable attachment bracket to the next stud along.
19 Fit the cylinder head nuts and washers and tighten down the nuts half a turn at a time in the order shown in Fig. 1.3 to the specified torque (photo).
20 Insert the pushrods into the block so the ball end rests in the tappet. Ensure the pushrods are replaced in the same order in which they were removed (photo).
21 Then refit the rocker shaft ensuring that the rocker arm balljoints seat in the pushrod cups (photo).
22 Replace the four rocker pedestal nuts and washers and tighten them down evenly.

11 Engine sump – removal and refitting (engine in car)

1 Drain the engine oil.
2 Unscrew and remove the sump retaining bolts and lift the sump away.
3 Remove the flange gasket.
4 Clean sludge and debris from the sump. Scrape and wipe the flange mating surfaces clean.
5 Refitting is a reversal of removal. Use a new gasket and tighten fixing bolts to the specified torque and remember that the special bracket is located under the third bolt on the right-hand side (photos).

Chapter 1 Engine

11.5a Fitting sump and gasket

11.5b Special bracket under sump bolt

12 Pistons, connecting rods and big-end bearings – removal and refitting (engine in car)

1 Remove the cylinder head as described in Section 10.
2 Remove the sump as described in the preceding Section.
3 Knock back with a cold chisel the locking tabs on the big-end retaining bolts, and remove the bolts and locking tabs.
4 Remove the big-end caps one at a time, taking care to keep them in the right order and the correct way round. Also ensure that the shell bearings are kept with their correct connecting rods and caps unless they are to be renewed. Normally, the numbers 1 to 4 are stamped on adjacent sides of the big-end caps and connecting rods, indicating which cap fits on which rod and which way round the cap fits. If no numbers or lines can be found then with a sharp screwdriver or file scratch mating marks across the joint from the rod to the cap. One line for connecting rod No 1, two for connecting rod No 2 and so on. This will ensure there is no confusion later as it is most important that the caps go back in the correct position on the connecting rods from which they were removed.
5 If the big-end caps are difficult to remove they may be gently tapped with a soft hammer.
6 To remove the shell bearings, press the bearing opposite the groove in both the connecting rod, and the connecting rod caps and the bearings will slide out easily.
7 Withdraw the pistons and connecting rods upwards and ensure they are kept in the correct order for replacement in the same bore. Refit the connecting rod caps and bearings to the rods if the bearings do not require renewal to minimise the risk of getting the caps and rods muddled.
8 Refer to Examination and Renovation – Section 18 or to Section 17 if the pistons are to be disconnected from the connecting rods.
9 The pistons, complete with connecting rods, can be fitted to the cylinder bores in the following sequence.
10 With a wad of clean rag wipe the cylinder bores clean.
11 The pistons, complete with connecting rods, are fitted to their bores from the top of the block (photo).
12 As each piston is inserted into its bore ensure that it is the correct piston/connecting rod assembly for that particular bore and that the connecting rod is the right way round, and that the front of the piston is towards the front of the bore, ie towards the front of the engine.
13 The piston will only slide into the bore as far as the oil control ring. It is then necessary to compress the piston rings into the clamp (photo) and to gently tap the piston into the cylinder bore with a wooden or plastic hammer.
14 Wipe the connecting rod half of the big-end bearing cap and the underside of the shell bearing clean, and fit the shell bearing in position with its locating tongue engaged with the corresponding rod (photo).
15 If the old bearings are nearly new and are being refitted then

12.11 Installing a piston/rod assembly

12.13 Compressing piston rings with clamp

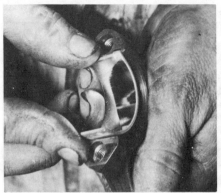
12.14 Fitting connecting rod shell bearing

12.16 Lubricating crank pins

12.18 Fitting big-end bearing cap

12.19 Tightening a big-end bearing cap bolt

12.20 Locking a big-end bearing cap bolt

13.4 Bending up crankshaft pulley start dog lock tab

13.8a Tightening timing cover retaining screw

13.8b Timing cover special short screw

13.9 Removing timing cover

13.11 Flattening camshaft sprocket bolt lockplate tabs

13.13 Worn timing chain tensioner slipper

13.14 Removing timing chain spring tensioner

ensure they are replaced in their correct locations on the correct rods.
16 Generously lubricate the crankpins with engine oil (photo), and turn the crankshaft so that the crankpin is in the most advantageous position for the connecting rod to be drawn onto it.
17 Wipe the connecting rod bearing cap and back of the shell bearing clean and fit the shell bearing in position ensuring that the locating tongue at the back of the bearing engages with the locating groove in connecting rod cap.
18 Generously lubricate the shell bearing and offer up the connecting rod bearing cap to the connecting rod (photo).
19 Fit the connecting rod bolts with the one-piece locking tab under them and tighten the bolts with a torque spanner to the specified torque.
20 With a hammer or pair of pliers knock up the locking tabs against the bolt head (photo).
21 When all the connecting rods have been fitted, rotate the crankshaft to check that everything is free, and that there are no high spots causing binding.
22 Refit the sump and the cylinder head as previously described.

13 Timing chain, gears and cover oil seal – removal and refitting (engine in car)

1 Drain the cooling system as described in Chapter 2.
2 Remove the radiator also as described in Chapter 2.
3 Slacken the dynamo or alternator mounting and adjuster link bolts, swivel the unit in towards the engine and slip the drivebelt from the pulleys.
4 Bend back the locking tab of the crankshaft pulley locking washer under the crankshaft pulley retaining bolt, or starter dog bolt (photo).
5 This bolt is very large and it is unlikely that the average owner will possess a spanner large enough to undo it. To free the bolt place a metal drift in the position shown and hit the drift with a hammer until the bolt starts to turn.
6 The crankshaft pulley wheel may pull off quite easily. If not place two large screwdrivers behind the camshaft pulley wheel at 180° to each other, and carefully lever off the wheel. It is preferable to use a proper pulley extractor if this is available, but large screwdrivers or tyre levers are quite suitable, providing care is taken not to damage the pulley flange.
7 Remove the Woodruff key from the crankshaft nose with a pair of pliers and note how the channel in the pulley is designed to fit over it. Place the Woodruff key in a glass jar as it is a very small part and can easily become lost.
8 Unscrew the screws holding the timing cover to the block. Note the special short screw adjacent to the crankshaft nose (photos).
9 Pull off the timing cover and gasket. The chain in the photo is very badly worn. On a less worn chain check for wear by measuring how much the chain can be depressed. More than $\frac{1}{4}$ in means a new chain can be depressed. More than $\frac{1}{2}$ in means a new chain must be fitted on reassembly.
10 With the timing cover off, take off the oil thrower. Note that the concave side faces the gearwheel.
11 With a drift or screwdriver tap back the tabs on the lockwasher under the two camshaft gearwheel retaining bolts (photo) and undo the bolts.
12 To remove the camshaft and crankshaft timing wheels complete with chain, ease each wheel forward a little at a time levering behind each gearwheel in turn with two large screwdrivers at 180° to each other. If the gearwheels are locked solid then it will be necessary to use a proper gearwheel and pulley extractor, and if one is available this should be used anyway in preference to screwdrivers. With both gearwheels safely off, remove the Woodruff key from the crankshaft with a pair of pliers and place them in the jar for safe keeping. Note the number of very thin packing washers behind the crankshaft gearwheel and remove them very carefully.
13 With time the spring blade timing chain tensioner will become worn and it should be renewed at the same time as the timing chain. Wear can be clearly seen as two grooves on the face of the tensioner slipper where it presses against the chain (photo).
14 To remove the tensioner bend it back and then pull out from its securing pins (photo).
15 On replacement fit the open end of the tensioner over the pin and press the blade into place with the aid of a screwdriver until it snaps into place (photo).

13.15 Snapping tensioner into place

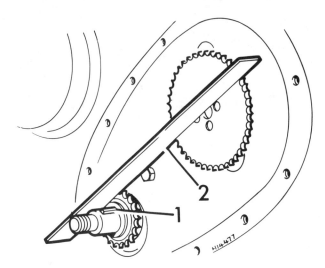

Fig. 1.4 Aligning chain sprockets (Sec 13)

1 Woodruff key   2 Straight edge

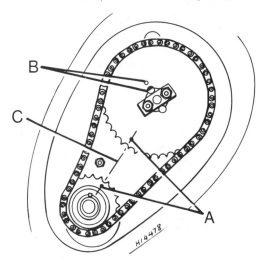

Fig. 1.5 Timing sprocket marks

A   Sprocket marks   C   Scribed line
B   Tooth offset adjustment marks

16 Refer to Examination and Renovation, Section 18.

17 Place the gearwheels in position without the timing chain and place the straight edge of a steel ruler from the side of the camshaft gearteeth to the crankshaft gearwheel, and measure the gap (if any) between the steel rule and the crankshaft gearwheel. If a gap exists a suitable number of packing washers must be placed on the crankshaft nose to bring the crankshaft gearwheel onto the same plane as the camshaft gearwheel.

18 Fit the Woodruff key to the slot in the crankshaft nose.

19 It is all too easy to fit the sprocket wheel 180° out on the camshaft. The best way of ensuring that the wheel is fitted the right way round is to ensure the two different slot marks on the back of the wheel correspond with the slots on the front of the camshaft flange (photo).

20 Lay the camshaft gearwheels on a clean surface so that the two timing marks are adjacent to each other. Slip the timing chain over them and pull the gearwheels back into mesh with the chain so that the timing marks, although further apart, are still adjacent to each other. A special point to note is that should the chain have a removable link, always position it so the spring clip closed end is in the running direction.

21 With the timing marks adjacent to each other hold the gearwheels above the crankshaft and camshaft. Turn the camshaft and crankshaft so that the Woodruff key will enter the slot in the crankshaft gearwheel, and the camshaft gearwheel is in the correct position relative to the camshaft.

22 Fit the timing chain and gearwheel assembly onto the camshaft and crankshaft, keeping the timing marks in alignment and the chain taut on the drive side (photo). On 1296 cc engined models, the camshaft sprocket is drilled with four holes which are equally spaced, but offset from a tooth centre. To carry out fractional adjustment of the camshaft sprocket, provision is made by this offset to move the sprocket position by part turn corrections. Half tooth adjustment is carried out by rotating the camshaft sprocket through 90° from its original position. Quarter turn adjustment is obtained by turning the sprocket back to front from its original position. Three quarter turn

13.19 Checking camshaft sprocket slot markings

13.22 Timing sprocket alignment marks

13.23 Bending up camshaft sprocket bolt lockplate tab

13.24 Removing timing cover oil seal

13.25 Using a vice to press in a new timing cover oil seal

13.26 Fitting crankshaft oil thrower

13.27a Depressing spring tensioner prior to fitting timing cover

13.27b Fitting timing cover

13.29 Tightening timing cover bolt. Arrow indicates position of short screw-headed bolt

adjustment is obtained by rotating the sprocket (in its back to front position) through 90°.

23 Fit a new lockplate to the camshaft sprocket, fit and tighten the bolts and bend up the tabs of the lockplate (photo).
24 The oil seal in the front of the timing cover should be renewed. To remove it, carefully drive it out with a screwdriver taking care not to damage the timing cover in the process (photo).
25 Evenly press a new seal into the cover using a vice (photo) ensuring that the seal lip is towards the crankshaft sprocket wheel.
26 Fit the oil thrower in place on the nose of the crankshaft (photo) making sure that the dished periphery is towards the cover (if dished type fitted).
27 Lubricate the front cover oil seal, fit a new gasket in place on the end plate, and fit the cover at an angle (photo), so as to catch the spring tensioner against the side of the chain. Swing the cover into its correct position and insert one or two bolts finger tight.
28 Note that the short screw headed bolt must be fitted to the hole indicated by the arrow.
29 Now tighten down all the bolts and screws evenly (photo).
30 Then fit the crankshaft nose pulley wheel.
31 The next step is to replace the pulley wheel nut or starter dog using a new lockwasher.
32 Tighten the nut or starter dog and prevent the crankshaft from moving by temporarily refitting two bolts and holding a screwdriver between them.
33 Finally knock down one of the lockwasher tabs.
34 Refit the radiator, drivebelt and fill the cooling system.

14 Camshaft – removal and refitting (engine in car)

1 Drain the cooling system and remove the radiator as described in Chapter 2.
2 Remove the drivebelt for the generator.
3 Unbolt and remove the cylinder head as described in Section 10.
4 Remove the fuel pump by unscrewing the nuts which hold it to the crankcase.
5 Withdraw the tappets (cam followers) and keep them in their originally fitted order.
6 Remove the timing cover, gears and chain as described in Section 13.
7 Remove the distributor drive gear. To do this, first unscrew the two nuts which hold the distributor housing in place (photo).
8 Lift off the distributor and distributor housing and then with a pair of long nosed pliers lift out the drive shaft (photo). As the shaft is removed turn it slightly to allow the shaft skew gears to disengage with the camshaft skew gear.
9 First measure the camshaft endfloat with a feeler gauge placed between the keeper plate and the flange. If endfloat exceeds 0.008 in (0.20 mm) it will be necessary to fit a new plate. Then remove the two bolts and spring washers which hold the camshaft locating plate to the block. The bolts are normally covered by the camshaft gearwheel (photo).
10 Remove the plate. The camshaft can now be withdrawn. Take great care to remove the camshaft gently, and in particular ensure that the cam peaks do not damage the camshaft bearings as the shaft is pulled forward (photo).
11 Refer to Examination and Renovation, Section 18.
12 Wipe the camshaft bearing journals clean and lubricate them generously with engine oil.
13 Insert the camshaft into the crankcase gently taking care not to damage the camshaft bearings with the cams.
14 Replace the camshaft locating plate and tighten down the two retaining bolts and washers.
15 Refit the distributor drive gear in the following way: it is important to set the distributor drive correctly as otherwise the ignition timing will be totally incorrect. It is easy to set the distributor drive in apparently the right position, but, exactly 180° out by omitting to select the correct cylinder which must not only be at TDC but must almost be on its firing stroke with both valves closed. The distributor drive should therefore not be fitted until the cylinder head is in position and the valves can be observed. Alternatively, if the timing cover has not been replaced, the distributor drive can be replaced when the marks on the timing wheels are adjacent to each other.
16 Rotate the crankshaft so that No 1 piston is at TDC and on its firing stroke (the marks in the timing gears will be adjacent to each

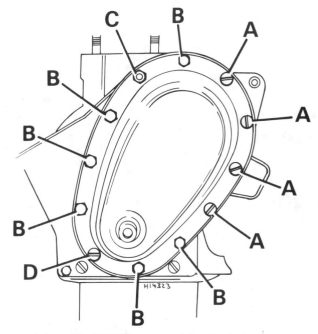

Fig. 1.6 Timing cover fixing (Sec 13)

A Screws
B Bolts
C Nut
D Short screw

other). When No 1 piston is at TDC the inlet valve on No 4 cylinder is just opening and the exhaust valve closing.
17 When the dimple on the crankshaft pulley wheel is in line with the pointer on the timing gear cover, then Nos. 1 and 4 pistons are at TDC (photo).
18 Insert the distributor drive into its housing so that when fully home the slot in the top of the drive shaft is positioned with the larger segment facing downwards in the exact position as shown. The end of the shaft engages with a slot in the top of the pump rotor shaft. It may be necessary to turn the pump rotor shaft to allow the distributor drive to engage fully.
19 It is essential that between 0.002 in and 0.007 in (0.051 and 0.178 mm) endfloat exists between the top side of the gear driven by the skew gear on the camshaft and the underside of the pedestal boss. If the original components are being used it will be safe to assume that the endfloat is correct but ensure the same number of packing washers are used (if any), and always fit a new gasket.

14.7 Unscrewing distributor mounting nut

Chapter 1 Engine

14.8 Removing distributor driveshaft

14.9 Camshaft locating plate

14.10 Removing camshaft

14.17 No. 1 piston at TDC (1147 cc and 1296 cc engines)

20 If the drive gears are assembled without endfloat wear on the crankshaft gearwheels, chain and distributor drive gear will be very heavy. If new components are being fitted then cut a small notch in the outer edge of the distributor housing flange gasket and bolt the housing down firmly. Measure the thickness of the gasket with a feeler gauge placed in the notch. Then remove the distributor housing and gasket and replace the housing without the gasket. Measure the gap between the underside of the housing flange and the block (photo), and subtract this latter figure from the former to determine the endfloat with the standard gasket.

21 Turn the distributor so the rotor arm is pointing to the terminal in the cap which carries the lead to No 1 cylinder, and fit the distributor to the distributor housing. The lip on the distributor should mate perfectly with the slot in the distributor drive shaft. Fit the bolt which holds the distributor clamp plate to the housing.

22 Tighten down the two nuts and washers which hold the distributor housing in place.

23 If the clamp bolt on the clamping plate was not previously loosened and the distributor body was not turned in the clamping plate, then the ignition timing will be as previously. If the clamping bolt has been loosened, then it will be necessary to retime the ignition as described in Chapter 4.

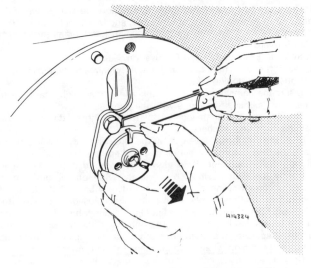
Fig. 1.7 Measuring camshaft endfloat (Sec 14)

Fig. 1.8 Distributor drive gear (Sec 14)

1　Mounting nuts
2　Pedestal
3　Driveshaft and gear

Fig. 1.9 Distributor driveshaft set to receive distributor (Sec 14)

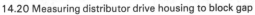

14.20 Measuring distributor drive housing to block gap

15.2 Removing oil pump mounting bolt

15.5 Fitting pump body and driveshaft

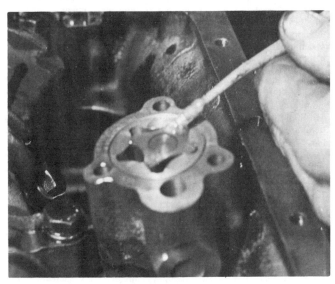

15.6 Priming oil pump with engine oil

15 Oil pump – removal and refitting (engine in car)

1 Remove the sump as described in Section 11.
2 Undo the three bolts and spring washers which hold the pump to the block (photo).
3 Removal of the bolts also releases the end cover so the pump can be taken from the engine and the outer and inner rotors pulled off together with the pump shaft.
4 Refer to Examination and Renovation, Section 18.
5 Fit the pump and drive shaft to the crankcase, (photo).
6 Prime the pump to preclude any possibility of oil starvation when the engine starts (photo).
7 Refit the cover to the pump and tighten down the three securing bolts and washers.
8 Refit the sump and gasket.

16 Engine – complete dismantling (engine removed)

1 Remove the rocker gear and cylinder head as described in Section 10, Paragraphs 8 to 22.
2 Remove the timing gear as described in Section 13.
3 Remove the camshaft as described in Section 14.
4 Remove the distributor drivegear as described in Section 14.
5 Remove the sump, piston/rod assemblies as described in Sections 11 and 12 respectively.
6 Remove the oil pump as described in Section 15.
7 Remove the flywheel and engine rear plate in the following way.
8 Bend back the locking tabs from the four bolts which hold the flywheel to the flywheel flange on the rear of the crankshaft.
9 Unscrew the bolts and remove them, complete with the locking plates.
10 Lift the flywheel away from the crankshaft flange. Some difficulty may be experienced in removing the bolts by the rotation of the crankshaft every time pressure is put on the spanner. To lock the crankshaft in position while the bolts are removed, use a screwdriver as a wedge between a backplate stud and the ring gear as shown in the photo. Alternatively a wooden wedge can be inserted between the crankshaft and the side of the block inside the crankcase.
11 The engine endplate is held in position by a number of bolts and spring washers of varying size. Release the bolts noting where different sizes fit and place them together to ensure none of them become lost. Lift away the endplate from the block complete with the paper gasket (photo).
12 The front engine endplate is removed in identical fashion.
13 The crankshaft and main bearings should now be removed.
14 Undo by one turn the nuts which hold the three main bearing caps in place.
15 Unscrew the nuts and remove them together with the washers.
16 At the rear of the engine undo the seven bolts which hold the special oil retaining cover in place and remove the cover.
17 Remove the main bearing caps and the bottom half of each bearing shell, taking care to keep the bearing shells in the right caps and the caps identified as to position and which way round they are fitted.
18 When removing the rear bearing cap, note the bottom semi-circular halves of the thrust washers, one half lying on either side of the main bearing. Lay them with the centre bearing along the correct side.
19 Slightly rotate the crankshaft to free the upper halves of the bearing shells and thrust washers which should now be extracted and placed over the correct bearing cap.
20 Remove the crankshaft by lifting it away from the crankcase (photo).

17 Dismantling sub-assemblies

Rocker gear

1 To dismantle the rocker assembly, release the rocker shaft locating screw, remove the pins and caps, and spring washers from each end of the shaft and slide from the shaft the pedestals, rocker arms, and rocker spacing springs.
2 From the end of the shaft undo the plug which gives access to the inside of the rocker which can now be cleaned of sludge etc. Ensure the rocker arm lubricating holes are clear.

16.11a Removing engine end plate bolts

16.11b Removing engine end plate

16.20 Removing crankshaft

Chapter 1 Engine

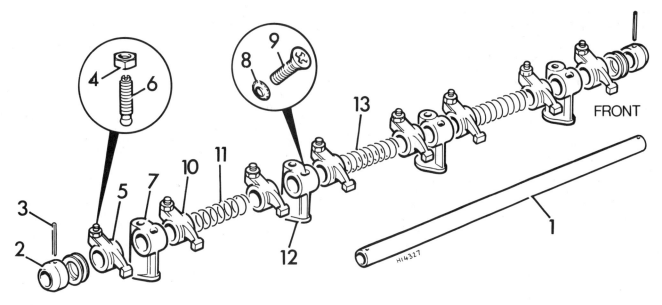

Fig. 1.10 Exploded view of rocker gear (Sec 17)

1	Shaft	6	Ball pin	10	Rocker arm
2	End stop	7	Pedestal	11	Spring
3	Roll pin	8	Shakeproof washer	12	Pedestal
4	Locknut	9	Crosshead screw	13	Spring
5	Rocker arm				

Cylinder head

3 Two types of valve spring retainers are used on Triumph engines. The commonest type fitted is a special cap which has two interconnecting holes drilled in it, the larger of which slides over the valve stem. The double hole type is simply removed by pressing against the valve head with one hand, the other hand pressing the valve cap down at the same time pressing it across so the larger hole is directly around the valve stem so allowing the cap to come off.

4 The other type fitted is of normal split collet pattern. These are removed as follows.

5 Compress each spring in turn with a valve spring compressor until the two halves of the collets can be removed. Release the compressor and remove the spring, shroud, and valve.

6 If, when the valve spring compressor is screwed down, the valve spring retaining cap refuses to free and expose the split collet, do not continue to screw down on the compressor as there is a likelihood of damaging it.

7 Gently tap the top of the tool directly over the cap with a light hammer. This will free the cap. To avoid the compressor jumping off the valve spring retaining cap when it is tapped, hold the compressor firmly in position with one hand. Drop each valve out through the combustion chamber.

8 It is essential that the valves are kept in their correct sequence unless they are so badly worn that they are to be renewed. If they are going to be kept and used again, place them in a sheet of card having eight holes number 1 to 8 corresponding with the relative positions the valves were in when fitted. Also keep the valve springs, washers, etc, in the correct order.

9 If it is wished to remove the valve guides they can be removed from the cylinder head in the following manner. Place the cylinder head with the gasket face on the bench and with a suitable hard steel punch drift the guides out of the cylinder head.

Pistons/connecting rods

10 To remove the piston rings, slide them carefully over the top of the piston, taking care not to scratch the aluminium alloy. Never slide them off the bottom of the piston skirt. It is very easy to break the iron piston rings if they are pulled off roughly so this operation should be done with extreme caution. It is helpful to make use of feeler gauges as slides by inserting them at equidistant points behind the rings.

11 Lift one end of the piston ring to be removed out of its groove and insert the end of the feeler gauge under it.

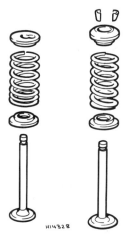

Fig. 1.11 Alternative types of valve spring retainers (Sec 17)

12 Turn the feeler gauge slowly round the piston and as the ring comes out of its groove apply slight upward pressure so that it rests on the land above. It can then be eased off the piston with the feeler gauge stopping it from slipping into any empty groove if it is any but the top piston ring that is being removed.

13 To remove the gudgeon pin to free the piston from the connecting rod remove one of the circlips at either end of the pin with a pair of circlip pliers.

14 Press out the pin from the rod and piston with your finger.

15 If the pin shows reluctance to move, then on no account force it out, as this could damage the piston. Immerse the piston in a pan of boiling water for three minutes. On removal the expansion of the aluminium should allow the gudgeon pin to slide out easily.

16 Make sure the pins are kept with the same piston for ease of refitting.

17 Certain models use gudgeon pins which are an interference fit in the little end of the connecting rod. The tightness of fit is their sole means of retention. These pins must be pressed out and replaced using a special tool so as not to damage the connecting rods or the pistons. This is really a job best left to your local Triumph dealer or engineering works.

18 Examination and renovation

1 With the engine stripped down and all parts thoroughly cleaned, it is now time to examine everything for wear. The following items should be checked and where necessary renewed or renovated as described in the following sections.

Crankshaft

2 Examine the crankpin and main journal surfaces for signs of scoring or scratches. Check the ovality of the crankpins at different positions with a micrometer. If more than 0.001 in (0.025 mm) out of round, the crankpins will have to be reground. It will also have to be reground if there are any scores or scratches present. Also check the journals in the same fashion. On highly tuned engines the centre main bearing has been known to break up. This is not always immediately apparent, but slight vibration in an otherwise normally smooth engine and a very slight drop in oil pressure under normal conditions are clues. If the centre main bearing is suspected of failure it should be immediately investigated by dropping the sump and removing the centre main bearing cap. Failure to do this will result in a badly scored centre main journal. If it is necessary to regrind the crankshaft and fit new bearings your local Triumph garage or engineering works will be able to decide how much metal to grind off and the correct undersize shells to fit. Don't forget to check the small spigot bush in the rear end of the crankshaft flange for wear. If the bush must be renewed, extract it by tapping a thread into it and screwing in a large bolt. Alternatively fill it with grease and drive a close fitting rod into it. Hydraulic pressure will force the bush out.

Big-end and main bearings

3 Big-end bearing failure is accompanied by a noisy knocking from the crankcase, and a slight drop in oil pressure. Main bearing failure is accompanied by vibration which can be quite severe as the engine speed rises and falls and a drop in oil pressure.

4 Bearings which have not broken up, but are badly worn will give rise to low oil pressure and some vibration. Inspect the big-ends, main bearings, and thrust washers for signs of general wear, scoring pitting and scratches. The bearings should be matt grey in colour. With lead-indium bearings should a trace of copper colour be noticed the bearings are badly worn as the lead bearing material has worn away to expose the indium underlay. Renew the bearings if they are in this condition or if there is any sign of scoring or pitting.

5 The undersizes available are designed to correspond with the regrind sizes, ie – 0.010 in (– 0.254 mm) bearings are correct for a crankshaft reground – 0.010 in (– 0.254 mm) undersize. The bearings are in fact, slightly more than the stated undersize as running clearances have been allowed for during their manufacture.

Cylinder bores

6 The cylinder bores must be examined for taper, ovality, scoring and scratches. Start by carefully examining the top of the cylinder bores. If they are at all worn a very slight ridge will be found on the thrust side. This marks the top of the piston ring travel. The owner will have a good indication of the bore wear prior to dismantling the engine, or removing the cylinder head. Excessive oil consumption accompanied by blue smoke from the exhaust is a sure sign of worn cylinder bores and piston rings.

7 Measure the bore diameter just under the ridge with a micrometer and compare it with the diameter at the bottom of the bore, which is not subject to wear. If the difference between the two measurements is more than 0.006 in (0.152 mm) then it will be necessary to fit special pistons and rings or to have the cylinders rebored and fit oversize pistons. If no micrometer is available remove the rings from a piston and place the piston in each bore in turn about $\frac{3}{4}$ in (19.1 mm) below the top of the bore. If an 0.010 in (0.254 mm) feeler gauge can be slid between the piston and the cylinder wall on the thrust side of the bore then remedial action must be taken. Oversize pistons are available in the following sizes:-
+ 0.010 in (0.254 mm), 0.020 in (0.508 mm)
+ 0.030 in (0.762 mm)

8 These are accurately machined to just below these measurements so as to provide correct running clearances in bores bored out to the exact oversize dimensions.

18.10 One method of removing cylinder head stud

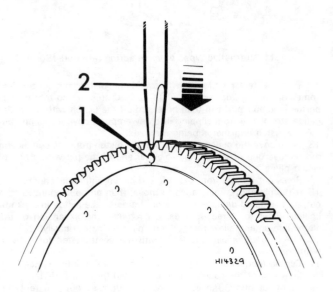

Fig. 1.12 Removing flywheel starter ring gear (Sec 18)

1 Hole   2 Cold chisel

18.11 Removing cylinder bore glaze

Chapter 1 Engine 51

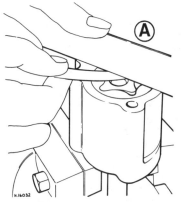 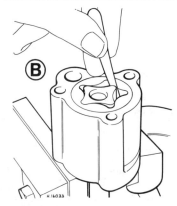 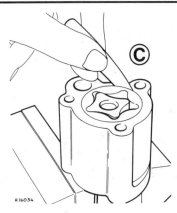

Fig. 1.13 Checking oil pump clearances (Sec 18)

9 If the bores are slightly worn but not so badly worn as to justify reboring them, then special oil control rings and pistons can be fitted which will restore compression and stop the engine burning oil. Several different types are available and the manufacturer's instructions concerning their fitting must be followed closely.

10 If the block is to be sent away for reboring it is essential to remove the cylinder head studs. Lock two nuts together on a stud (photo) and then wind the stud out by turning the bottom nut anti-clockwise.

11 If new pistons are being fitted and the bores have not been reground, it is essential to slightly roughen the hard glaze on the sides of the bores with fine glass paper (photo) so the new piston rings will have a chance to bed in properly. Never use emery cloth.

Pistons and piston rings

12 If the old pistons are to be refitted, carefully remove the piston rings and then thoroughly clean them. Take particular care to clean out the piston ring grooves. At the same time do not scratch the aluminium in any way. If new rings are to be fitted to the old pistons then the top ring should be stepped so as to clear the ridge left above the previous top ring. If a normal but oversize new ring is fitted, it will hit the ridge and break, because the new ring will not have worn in the same way as the old, which will have worn in unison with the ridge.

13 Before fitting the rings on the pistons each should be inserted approximately 3 in (76.0 mm) down the cylinder bore and the gap measured with a feeler gauge. This should be between 0.015 in (0.38 mm) and 0.038 in (0.97 mm). It is essential that the gap should be measured at the bottom of the ring travel, as if it is measured at the top of a worn bore and gives a perfect fit, it could easily seize at the bottom. If the ring gap is too small rub down the ends of the ring with a very fine file until the gap, when fitted, is correct. To keep the rings square in the bore for measurement line each up in turn by inserting an old piston in the bore upside down, and use the piston to push the ring down about 3 in (76.0 mm). Remove the piston and measure the piston ring gap.

14 Whe fitting new pistons and rings to a rebored engine the piston ring gap can be measured at the top of the bore as the bore will not now taper. It is not necessary to measure the side clearance in the piston ring grooves with the rings fitted as the groove dimensions are accurately machined during manufacture. When fitting new oil control rings to old pistons it may be necessary to have the grooves widened by machining to accept the new wider rings. In this instance the manufacturers representative will make this quite clear and will supply the address to which the pistons must be sent for machining.

Camshaft and camshaft bearings

15 On the majority of engines the camshaft runs direct in the block and wear of the journals and bearings is negligible. On certain of the latest models pre-formed camshaft bearings are fitted and these can be replaced. This is an operation for your Triumph dealer or the local engineering works as it demands the use of specialised equipment. The bearings are removed with a special drift after which new bearings are pressed in, care being taken to ensure the oil holes in the bearings line up with those in the block. On no account can the bearings be reamed in position.

16 The camshaft itself should show no sign of wear, but, if very slight scoring on the cams is noticed, the score marks can be removed by very gentle rubbing down with a very fine emery cloth. The greatest care should be taken to keep the cam profiles smooth.

Valves and valve seat

17 Examine the heads of the valves for pitting and burning, especially the heads of the exhaust valves.

18 The valve seatings should be examined at the same time. If the pitting on valve and seat is very slight the marks can be removed by grinding the seats and valves together with coarse, and then fine, valve grinding paste. Where bad pitting has occurred to the valve seats it will be necessary to recut them and fit new valves. If the valve seats are so worn that they cannot be recut, then it will be necessary to fit new valve seat inserts. These latter two jobs should be entrusted to the local Triumph agent or engineering works. In practice it is very seldom that the seats are so badly worn that they require renewal. Normally, it is the exhaust valve that is too badly worn for replacement, and the owner can easily purchase a new set of valves and match them to the seats by valve grinding.

Timing gears and chain

19 Examine the teeth on both the crankshaft gearwheel and the camshaft gearwheel for wear. Each tooth forms an inverted V with the gearwheel periphery, and if worn the side of each tooth under tension will be slightly concave in shape when compared with the other side of the tooth, ie one side of the inverted V will be concave when compared with the other. If any sign of wear is present the gearwheels must be renewed.

20 Examine the links of the chain for side slackness and renew the chain if any slackness is noticeable when compared with a new chain. The actual rollers on a very badly worn chain may be slightly grooved.

Timing chain tensioner

21 If the timing chain is badly worn it is more than likely that the tensioner will be too.

22 Examine the side of the tensioner which bears against the chain and renew it if it is grooved or ridged. See Section 13 for details.

Rockers and rocker shaft

23 Remove the threaded plug with a screwdriver from the end of the rocker shaft and thoroughly clean out the shaft. As it acts as the oil passage for the valve gear also ensure the oil holes in it are quite clear after having cleaned them out. Check the shaft for straightness by rolling it on the bench. If this is not successful purchase a new shaft. The surface of the shaft should be free from any worn ridges caused by the rocker arms. If any wear is present, renew the shaft. Wear is only likely to have occurred if the rocker shaft oil holes have become blocked.

24 Check the rocker arms for wear of the rocker bushes, for wear at the rocker arm face which bears on the valve stem, and for wear of the adjusting ball ended screws. Wear in the rocker arm bush can be checked by gripping the rocker arm tip and holding the rocker arm in place on the shaft, noting if there is any lateral rocker arm shake. If shake is present, and the arm is very loose on the shaft, a new bush or rocker arm must be fitted.

25 Check the tip of the rocker arm where it bears on the valve head for cracking or serious wear on the case hardening. If none is present reuse the rocker arm. Check the lower half of the ball on the end of the rocker arm adjusting screw. On high performance engines wear on the ball and top of the pushrod is easily noted by the unworn 'pip' which fits in the small central oil hole on the ball. The larger this 'pip' the more wear has taken place to both the ball and the pushrod. Check the

18.35 Bending up sump filter screen retaining tabs

18.36 Removing sump filter screen

pushrods for straightness by rolling them on the bench. Renew any that are bent.

Tappets (cam followers)

26 Examine the bearing surface of the tappets which lie on the camshaft. Any indentation in this surface or any cracks indicate serious wear and the tappets should be renewed. Thoroughly clean them out, removing all traces of sludge. It is most unlikely that the sides of the tappets will prove worn, but, if they are a very loose fit in their bores and can readily be rocked, they should be exchanged for new units. It is very unusual to find any wear in the tappets, and any wear present is likely to occur only at very high mileages.

Flywheel starter ring

27 If the teeth on the flywheel starter ring are badly worn, or if some are missing, then it will be necessary to remove the ring. This is achieved by drilling a hole across the ring splitting the ring with a cold chisel. The greatest care should be taken not to damage the flywheel during this process.

28 To fit a new ring heat it gently and evenly with an oxyacetylene flame until a temperature of approximately 662°F (350°C) is reached. This is indicated by a light metallic blue surface colour. With the ring at this temperature, fit it to the flywheel with the front of the teeth facing the flywheel register. The ring should be tapped gently down onto its register and left to cool naturally when the shrinkage of the metal on cooling will ensure that it is a secure and permanent fit. Great care must be taken not to overheat the ring, as if this happens, the temper of the ring will be lost.

Oil pump

29 Thoroughly clean all the component parts in petrol and then check the rotor endfloat and lobe clearances in the following manner.

30 Position the rotors in the pump and place the straight edge of a steel ruler across the joint face of the pump. Measure the gap between the bottom of the straight edge and the top of the rotors with a feeler gauge as at A in Fig. 1.13. If the measurement exceeds 0.005 in (0.127 mm) then check the lobe clearances as described in the following paragraphs. If the lobe clearances are correct then lap the joint face on a sheet of plate glass.

31 Measure with a feeler gauge the gap (B) between the inner and outer rotors. It should not be more than 0.010 in (0.254 mm).

32 Then measure the gap (C) between the outer rotor and the side of the pump body which should not exceed 0.008 in (0.203 mm). It is essential to renew the pump if the measurements are outside these figures. It can be safely assumed that at any major reconditioning the pump will need renewal.

Valve guides

33 Examine the valve guides internally for wear. If the valves are a very loose fit in the guides and there is the slightest suspicion of lateral rocking using a new valve, then new guides will have to be fitted. If the valve guides have been removed compare them internally by visual inspection with a new guide as well as testing them for rocking with a new valve.

Sump

34 It is essential to thoroughly wash out the sump with petrol and this can only be done properly with the gauze removed.

35 With a screwdriver and a pair of pliers carefully pull back the tags which hold the gauze in place (photo).

36 The gauze can then be lifted out (photo) and the inside cleaned out properly. Scrape all traces of the old sump gasket from the flange.

19 Cylinder head – decarbonising

1 This can be carried out with the engine either in or out of the car. With the cylinder head off carefully remove with a wire brush and blunt scraper all traces of carbon deposits from the combustion spaces and the ports. The valve head stems and valve guides should also be freed from any carbon deposits. Wash the combustion spaces and ports down with petrol and scrape the cylinder head surface free of any foreign matter with the side of a steel rule, or a similar article.

19.6 Grinding in a valve

2 Clean the pistons and top of the cylinder bores. If the pistons are still in the block then it is essential that great care is taken to ensure that no carbon gets into the cylinder bores as this could scratch the cylinder walls or cause damage to the piston and rings. To ensure this does not happen, first turn the crankshaft so that two of the pistons are at the top of their bores. Stuff rag into the other two bores or seal them off with paper and masking tape. The waterways should also be covered with small pieces of marking tape to prevent particles or carbon entering the cooling system and damaging the water pump.

3 There are two schools of thought as to how much carbon should be removed from the piston crown. One school recommends that a ring of carbon should be left round the edge of the piston and on the cylinder bore wall as an aid to low oil consumption. Although this is probably true for early engines with worn bores, on later engines the thought of the second school can be applied; which is that for effective decarbonisation all traces of carbon should be removed.

4 If all traces of carbon are to be removed, press a little grease into the gap between the cylinder walls and the two pistons which are to be worked on. With a blunt scraper carefully scrape away the carbon from the piston crown, taking great care not to scratch the aluminium. Also scrape away the carbon from the surrounding lip of the cylinder wall. When all carbon has been removed, scrape away the grease which will now be contaminated with carbon particles, taking care not to press any into the bores. To assist prevention of carbon build-up the piston crown can be polished with a metal polish. Remove the rags or masking tape from the other two cylinders and turn the crankshaft so that the two pistons which were at the bottom are now at the top. Place rag or masking tape in the cylinders which have been decarbonised and proceed as just described.

5 If a ring of carbon is going to be left round the piston then this can be helped by inserting an old piston ring into the top of the bore to rest on the piston and ensure that carbon is not accidentally removed. Check that there are no particles of carbon in the cylinder bores. Decarbonising is now complete.

6 Valve grinding is carried out as follows: smear a trace of coarse carborundum paste on the seat face and apply a suction grinder tool to the valve head. With a semi-rotary motion, grind the valve head to its seat, lifting the valve occasionally to redistribute the grinding paste (photo). When a dull matt even surface finish is produced on both the valve seat and the valve, then wipe off the paste and repeat the process with fine carborundum paste, lifting and turning the valve to redistribute the paste as before. A light spring placed under the valve head will greatly ease this operation. When a smooth unbroken ring of light grey matt finish is produced, on both valve and valve seat faces, the grinding operation is completed.

7 Scrape away all carbon from the valve head and the valve stem. Carefully clean away every trace of grinding compound, taking great care to leave none in the ports or in the valve guides. Clean the valves and valve seats with a paraffin soaked rag then with a clean rag, and finally, if an air line is available, blow the valves, valve guides and valve ports clean.

20 Engine lubrication system – description

1 A forced feed system of lubrication is fitted with oil circulated round the engine from the sump below the block. The level of engine oil in the sump is indicated on the dipstick which is fitted on the right-hand side of the engine. It is marked to indicate the optimum level which is the maximum mark.

2 The level of the oil in the sump, ideally, should not be above or below this line. Oil is replenished via the filler cap on the rocker cover.

3 The eccentric rotor-type oil pump is bolted inside the left-hand side of the crankcase and is driven by a short shaft from the skew gear on the camshaft which also drives the distributor shaft.

4 The pump is the non draining variety to allow rapid pressure build-up when starting from cold.

5 Oil is drawn into the pump from the sump via the pick-up pipe. From the oil pump the lubricant passes through a non-adjustable relief valve to the by-pass (early models only) or full flow filter. Filtered oil enters the main gallery which runs the length of the engine on the left-hand side. Drillings from the main gallery carry the oil to the crankshaft and camshaft journals.

6 The crankshaft is drilled so that oil under pressure reaches the crankpins from the crankshaft journals. The cylinder bores pistons and gudgeon pins are all lubricated by splash and oil mist.

7 Oil is fed to the valve gear via the hollow rocker shaft at a reduced

Fig. 1.14 Location of oil pressure relief valve (Sec 20)

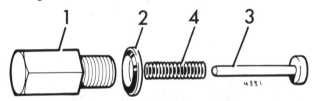

Fig. 1.15 Components of the oil pressure relief valve (Sec 20)

1 Cap
2 Seal
3 Plunger
4 Spring

20.10 Oil pressure relief valve

pressure by means of a scroll and two flats on the camshaft rear journal.

8 Drillings and grooves in the camshaft front journal lubricate the camshaft thrust plate, and the timing chain and gearwheels. Oil returns to the sump by gravity, the pushrods and cam followers being lubricated by oil returning via the pushrod drillings in the block.

9 To prevent excessive oil pressure – for example when the engine is cold – an oil pressure relief valve is built into the left-hand side of the engine immediately above the crankcase flange and in line vertically with the distributor.

10 The relief valve assembly is dismantled by undoing the large hexagon headed bolt which holds the relief valve piston and spring in place (photo).

11 Always renew the spring at a major overhaul. To replace the assembly fit the valve piston into its orifice in the block, then the spring and then the bolt, ensuring that the sealing washer is in place on the latter.

21 Crankcase ventilation system

1 Any one of three types of crankcase ventilation system may be fitted depending on the model and its year of manufacture. The three systems are known as 'Open Ventilation', 'Semi-sealed' and PCV (Positive Crankcase Ventilation).
2 'Open Ventilation' is very straightforward and is only fitted to early models. It comprises an open angled tube fitted on the right-hand side of the engine which relieves crankcase pressure directly into the air.
3 The 'Semi-sealed' is slightly more sophisticated system with crankcase pressure being relieved by means of a rubber pipe from the rocker cover to the air cleaner. The hole for the open road tube is blocked over and the possibility of crankcase fumes entering the car is considerably reduced.
4 The PCV system is similar to the semi-sealed but more efficient and complicated. An emission control valve is positioned on top of the inlet manifold to which it is connected. It is also connected to a tube from the rocker cover. The control valve works by manifold depression so that when the depression is greatest (ie on the over-run) crankcase gas flow is restricted. A special oil filler cap is also used and this contains a non-return valve which ensures that crankcase and atmospheric pressures are kept in balance.
5 At the specified intervals, the crankcase breather valve of the PCV

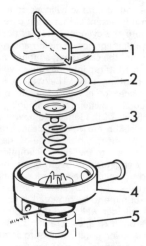

Fig. 1.16 Exploded view of crankcase breather valve (Sec 21)

1 Clip
2 Diaphragm
3 Spring
4 Body
5 Manifold

22.1 Removing a stud with an extractor

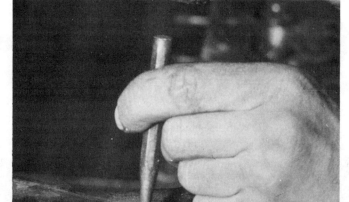

22.3 Piercing an engine core plug

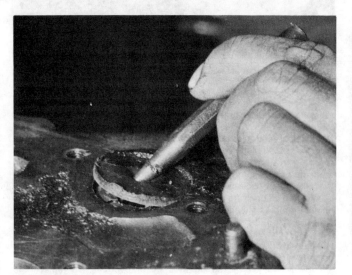

22.4 Prising out an engine core plug

22.5 Fitting an engine core plug

system, mounted on the inlet manifold must be serviced. The valve enables crankcase fumes to be fed into the inlet manifold where they are burnt with the petrol air mixture in the cylinders.

6 Referring to Fig. 1.16 free the clip and lift off the cover, diaphragm, valve pin and spring. Clean all these parts in methylated spirit and renew the diaphragm if torn. Reassembly is a straightforward reversal of the removal sequence.

22 Engine reassembly – general

1 To ensure maximum life with minimum trouble from a rebuilt engine, not only must everything be correctly assembled, but all the parts must be spotlessly clean, all the oilways must be clear, locking washers and spring washers must always be fitted where indicated and all bearing and other working surfaces must be thoroughly lubricated during assembly. Before assembly begins, renew any bolts or studs the threads of which are in any way damaged, and whenever possible use new spring washers (photo).

2 Check the core plugs for signs of weeping and always renew the plug at the front of the engine as it is normally covered by the engine endplate.

3 Drive a punch through the centre of the core plug (photo).

4 Using the punch as a lever lift out the old core plug (photo).

5 Thoroughly clean the core plug orifice and using a light weight hammer as an expander firmly tap a new core plug in place, convex side facing out (photo). Ensure that the crankcase is thoroughly clean and that all oilways are clear. A thin-twist drill or a pipe cleaner is useful for cleaning them out. If possible, blow them out with compressed air. Treat the crankshaft in the same fashion, and then inject engine oil into the crankshaft oilways.

6 Apart from your normal tools, a supply of clean rag, an oil can filled with engine oil (an empty plastic detergent bottle thoroughly cleaned and washed out, will invariably do just as well), a new supply of assorted spring washers, a set of new gaskets, and preferably a torque spanner, should be collected together.

23 Engine – complete reassembly

1 Commence work on rebuilding the engine by replacing the crankshaft and main bearings. Fit the three upper halves of the main bearing shells to their location in the crankcase, after wiping the locations clean.

2 Note that at the back of each bearing is a tab which engages in locating grooves in either the crankcase or the main bearing cap housings.

3 If new bearings are being fitted, carefully clean away all traces of the protective grease with which they are coated.

4 With the three upper bearing shells securely in place, wipe the lower bearing cap housings and fit the three lower shell bearings to their caps ensuring that the right shell goes into the right cap if the old bearings are being refitted.

5 Wipe the recesses either side of the rear main bearing which locate the thrust washers.

6 Generously lubricate the crankshaft journals and the upper and lower main bearing shells and carefully lower the crankshaft into place (photo).

7 Fit the upper halves of the thrust washers into their grooves either side of the rear main bearing (photo), rotating the crankshaft in the direction towards the main bearing tabs (so that the main bearing shells do not slide out). At the same time feed the thrust washers into their locations with their oil grooves outwards away from the bearing.

8 Fit the main bearing caps in position ensuring they locate properly. The mating surfaces must be spotlessly clean or the caps will not seat correctly (photo). As the bearing caps were assembled to the cylinder block and then line bored during manufacture, it is essential that they are returned to the same positions from which they were removed.

9 Refit the main bearing cap bolts and locking tabs (if fitted) and tighten the bolts to the specified torque (photo). Later models have self locking nuts.

10 Test the crankshaft for freedom of rotation, should it be very stiff to turn or possess high spots a most careful inspection must be made, preferably by a qualified mechanic with a micrometer to get to the root of the trouble. It is very seldom that any trouble of this nature will be experienced when fitting the crankshaft.

11 Check the crankshaft endfloat with a feeler gauge measuring the longitudinal movement between the crankshaft and a bearing cap. Endfloat should be between the specified limits. If endfloat is excessive oversize thrust washers can be fitted.

12 Next fit the sealing block over the front main bearing cap. Smear the ends of the block with jointing compound (photo), and fit the block in place. Fit the securing screws but do not tighten fully. Fit new wedge seals at each end and line up the front face of the block with the front of the cylinder block. Tighten the screws fully and cut the wedge seals flush with the crankcase flange (photo).

13 If the same pistons are being used, then they must be mated to the same connecting rod with the same gudgeon pin. If new pistons are being fitted it does not matter which connecting rod they are used with, but, the gudgeon pins should be fitted on the basis of selective assembly.

14 If interference fit gudgeon pins are used on your engine (ie the pins are held firmly in place by tightness of fit in the little end), then the pistons and connecting rods must be assembled by your local Triumph agent who will have the special tool necessary to draw the pin in, in conjunction with a torque wrench.

15 Because aluminium alloy, when hot, expands more than steel, the gudgeon pin may be a very tight fit in the piston when they are cold. To avoid any damage to the piston it is best to heat it in boiling water when the pin will slide in easily.

16 Lay the correct piston adjacent to each connecting rod and remember that the same rod and piston must go back into the same bore. If new pistons are being used it is only necessary to ensure that the right connecting rod is placed in each bore.

17 Fit a gudgeon pin circlip in position at one end of the gudgeon pin hole in the piston.

18 Locate the connecting rod in the piston with the marking 'FRONT' or arrow on the piston crown towards the front of the engine, ie the timing cover end, and the connecting rod cap towards the camshaft side of the engine (photo).

19 Slide the gudgeon pin in through the hole in the piston and through the connecting rod little end until it rests against the previously fitted circlip (photo). Note that the pin should be a push fit.

20 Fit the second circlip in position (photo). Repeat this procedure for all four pistons and connecting rods.

21 Where new piston rings are being fitted to the original pistons, make sure that a step has been machined on the top of the top compression ring so no fouling occurs between the unworn portion at the top of the bore and the piston ring when the latter is at the top of its stroke (photo).

22 Check that the piston ring grooves and oilways are thoroughly clean and unblocked. Piston rings must always be fitted over the head of the piston and never from the bottom.

23 The easiest method to use when fitting rings is to wrap a 0.020 in (0.51 mm) feeler gauge round the top of the piston and place the rings one at a time, starting with the bottom oil control ring, over the feeler gauge.

24 The feeler gauge, complete with ring, can then be slid down the piston over the other piston ring grooves until the correct groove is reached. The piston ring is then slid gently off the feeler gauge into the groove.

25 An alternative method is to fit the rings by holding them slightly open with the thumbs and both of your index fingers. This method requires a steady hand and great care as it is easy to open the ring too much and break it.

26 Fit the piston/rod assemblies as described in Section 12.

27 Fit a new gasket in place over the front of the cylinder block (photo).

28 Lower the front end plate into place noting the hole for the dowel and then fit the securing bolt immediately above the crankshaft nose.

29 Fit the camshaft as described in Section 14.

30 Fit the timing gears, chain, and cover as described in Section 13.

31 Fit the oil pump as described in Section 15.

32 A scroll type crankshaft rear oil seal was used on early models. Later models are fitted with a lip type seal.

33 To fit the the scroll type seal coat a new gasket with jointing compound, position it on the seal housing (photo) and fit the housing to the crankcase, doing up the retaining bolts and spring washers finger tight.

34 Check with a feeler gauge with a gap of 0.003 in (0.076 mm), (aluminium housings only) exists all round the crankshaft journal, tapping the housing with a soft headed hammer until the seal is centralised (photo). Some later models make use of a cast iron

23.6 Oiling crankshaft main bearing shell

23.7 Fitting crankshaft thrust washer

23.8 Fitting a main bearing cap

23.9 Tightening a main bearing cap bolt

23.12a Applying sealant to front main bearing block

23.12b Tightening front main bearing block screws

23.18 Piston front directional mark

23.19 Inserting a gudgeon pin

23.20 Fitting a gudgeon pin circlip

23.21 Step on top compression ring

23.27 Cylinder block front gasket

23.33 Scroll type crankshaft rear oil seal gasket

Chapter 1 Engine

23.34 Centralising scroll type rear oil seal

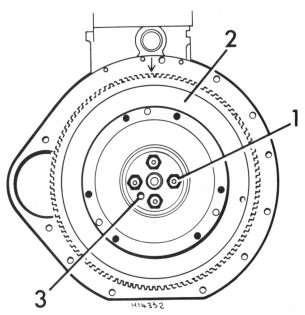

Fig. 1.17 Flywheel (Sec 23)

1 Bolt
2 Clutch contact surface
3 Positioning dowel

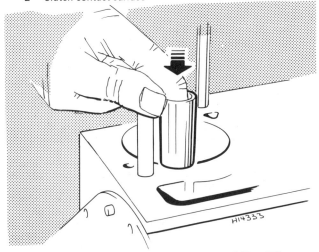

Fig. 1.18 Fitting a tappet (cam follower) (Sec 23)

housing. This is fitted in just the same way but the clearance should be 0.002 in (0.508 mm).

35 To fit the lip type seal first coat both sides of a new gasket with jointing compound and position the gasket on the crankcase joint face.
36 Press a new seal into the crankshaft housing with the lip of the seal facing the crankshaft. Oil the seal and carefully fit the housing making sure the lip of the seal is not turned over. Replace the housing bolts finger tight, turn the crankshaft over several times to centralise the seal, and tighten the bolts down firmly. Irrespective of what type of seal is used now fit the input shaft bush to the hole in the centre of the crankshaft rear journal (if removed).
37 Fit the sump as described in Section 11.
38 No gasket is fitted between the rear end plate and the block. Fit the end plate in place and tighten down the bolts and washers.
39 Make certain that the flange on the crankshaft and the face of the flywheel are perfectly clean and offer up the flywheel to the end of the crankshaft. Ensure that the dowel enters into the special hole in the flywheel. Fit new tab washers, tighten down the four retaining bolts and turn up the lock tags.
40 Smear the crankshaft spigot bush with a small quantity of grease.
41 Fit the sump as described in Section 7 using a new gasket.
42 If the cylinder head was fully dismantled, it should now be reassembled ready for installation.
43 Rest the cylinder head on its side.
44 Fit each valve and valve spring in turn, wiping down and lubricating each valve stem as it is inserted into the same valve guide from which it was removed (photo).
45 Build up each valve assembly by fitting the lower collar (photo).
46 Then fit the valve spring so that the closely coiled portion of the spring is adjacent to the cylinder head (photo).
47 On engines which use the double hole spring retaining collar press the valve in firmly with one hand, and with the other fit the collar over the valve stem by means of the offset hole (photo).
48 Press down hard on the collar to compress the spring and as soon as the collar is in line with the groove in the valve stem push the collar across into the smaller hole so the spring is securely retained.
49 On engines which use split collets to retain the upper retaining collar, move the cylinder head towards the edge of the work bench if it is facing downwards and slide it partially over the edge of the bench so as to fit the bottom half of a valve spring compressor to the valve head. Slide the spring and upper collar over the valve stem.
50 With the base of the valve compressor on the valve head, compress the valve spring until the cotters can be slipped into place in the cotter grooves. Gently release the compressor.
51 Repeat this procedure until all eight valves and valve springs are fitted.
52 Fit an end cap and pin to one end of the shaft and then slide on the springs, rockers, distance springs, and rocker pedestals in their correct order.
53 Make sure that the Phillips screw on the rear rocker pedestal engages properly with the rocker shaft.

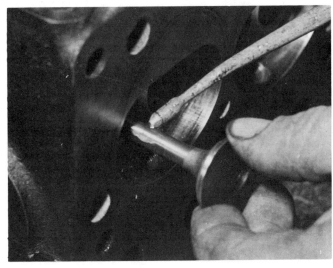

23.44 Oiling a valve stem

23.45 Fitting valve spring lower collar

23.46 Fitting valve spring

23.47 Fitting valve spring retaining collar

23.59 Fitting rocker cover and gasket

23.60 Fitting rocker cover sealing and flat washers

54 When all is correctly assembled fit the remaining end cap and oil the components thoroughly.
55 Generously lubricate the tappets internally and externally and insert them in the bores from which they were removed.
56 Fit the cylinder head as described in Section 10, paragraphs 15 to 22.
57 Fit the distributor drive gear as described in Section 14, paragraphs 15 to 23.
58 Adjust the valve clearances as described in Section 24.
59 Fit a new gasket to the rocker cover and carefully fit the cover in place (photo).
60 Replace the washers over the rocker cover holding down studs ensuring the sealing washer lies under the flat steel washer (photo). Replace the rocker cover nuts.
61 Fit your new sparking plugs (photo). Reconnect the ancillary components to the engine in the reverse order to which they were removed.
62 It should be noted that in all cases it is best to reassemble the engine as far as possible before refitting it to the car. This means that the inlet and exhaust manifolds, starter motor, water thermostat, oil filter, distributor, carburettors and dynamo, should all be in position. If the engine was removed with the gearbox, the clutch assembly and gearbox must be reconnected with the slave cylinder.

Chapter 1 Engine

24 Valve clearances – adjustment

1 The valve adjustments should be made with the engine cold. The importance of correct rocker arm/valve stem clearances cannot be overstressed as they vitally affect the performance of the engine.
2 If the clearances are set too wide, the efficiency of the engine is reduced as the valves open late and close earlier than was intended. If, on the other hand the clearances are set too close there is a danger that the stems will expand upon heating and not allow the valves to close properly which will cause burning of the valve head and seat and possible warping.
3 If the valve clearances are being adjusted at time of routine service, to gain access to the rocker arm adjuster screws and locknuts it is merely necessary to remove the two holding down studs from the rocker cover, and then to lift the rocker cover and gasket away.
4 It is important that the clearance is set when the tappet of the valve being adjusted is on the heel of the cam, (ie opposite the peak). This can be done by carrying out the adjustments in the following order, which also avoids turning the crankshaft more than necessary. The valves are numbered from the timing cover end of the engine.

Valve fully open	Check and adjust
Valve No 8 EX	Valve No 1
Valve No 6 IN	Valve No 3
Valve No 4 EX	Valve No 5
Valve No 7 IN	Valve No 2
Valve No 1 EX	Valve No 8
Valve No 3 IN	Valve No 6
Valve No 5 EX	Valve No 4
Valve No 2 IN	Valve No 7

5 The correct valve clearance of 0.010 in (0.25 mm) is obtained by slackening the hexagon locknut with a spanner while holding the ball pin against rotation with the screwdriver (photo). Then, still pressing down with the screwdriver, insert a feeler gauge in the gap between the valve stem head and the rocker arm and adjust the ball pin until the feeler gauge will just move in and out without nipping, and, still holding the ball pin in the correct position, tighten the locknut.

25 Starting-up after major overhaul

1 Carry out a final inspection to see that all controls and pipes and electrical leads have been correctly connected.
2 Set the engine idling speed slightly higher than normal to offset the stiffness of new internal components.
3 Start the engine. This may require several revolutions of the starter

24.5 Adjusting a valve clearance

motor until the engine fires as the carburettor bowl and fuel pump will have to be filled.
4 Check for oil or coolant leaks. There may be a drip from a hose where the clip requires further tightening, but if new gaskets have been used, nothing more serious should be observed.
5 Start the engine and immediately check for any oil or water leaks.
6 Run the car to normal operating temperature and then check the carburettor, ignition, and emission control settings as described in Chapters 3 and 4.
7 After 500 miles (800 km) check the torque wrench setting of the cylinder head nuts. This must be done cold in the order shown in an earlier illustration. Do not simply place the torque wrench and socket on the nuts and attempt to tighten, but unscrew one nut at a time through a quarter turn and then tighten it to the specified figure. Repeat the operation on the next nut in sequence until they have all been checked.
8 Check the valve clearances after checking the cylinder head nuts.
9 If a number of new internal components have been installed, the speed of the car should be restricted for the first few hundred miles and it is recommended that the engine oil is changed after the first 1000 miles (1600 km).

26 Fault diagnosis – engine

Symptom	Reason(s)
Engine fails to turn over when starter operated	
No current at starter motor	Flat or defective battery
	Loose battery leads
	Defective starter solenoid or switch or broken wiring
	Engine earth strap disconnected
Current at starter motor	Defective starter motor
Engine turns over but will not start	
No spark at spark plug	Ignition damp or wet
	Ignition leads to spark plugs loose
	Shorted or disconnected low tension leads
	*Dirty, incorrectly set, or pitted contact breaker points
	*Faulty condenser
	Defective ignition switch
	Ignition leads connected wrong way round
	Faulty coil
	*Contact breaker point spring earthed or broken

Chapter 1 Engine

Symptom	Reason(s)
Excess of petrol in cylinder or carburettor flooding	Too much choke allowing too rich a mixture to wet plugs Float damaged or leaking or needle not seating Float lever incorrectly adjusted

Engine stalls and will not start

Symptom	Reason(s)
No spark at spark plug	Ignition failure (refer to Chapter 4)
No fuel at jets	No petrol in petrol tank Petrol tank breather choked (early models) Sudden obstruction in carburettor(s) Water in fuel system

Engine misfires or idles unevenly or stops

Symptom	Reason(s)
Intermittent spark at spark plug	Ignition leads loose Battery leads loose on terminals Battery earth strap loose on body attachment point Engine earth lead loose *Low tension leads to + and - terminals on coil loose *Low tension lead from - terminal side to distributor loose Dirty, or incorrectly gapped plugs *Dirty, incorrect set, or pitted contact breaker points Tracking across inside of distributor cover Ignition too retarded Faulty coil
No fuel at carburettor float chamber or at jets	No petrol in petrol tank Vapour lock in fuel line (in hot conditions or at high altitude) Blocked float chamber needle valve Fuel pump filter blocked Choked or blocked carburettor jets Faulty fuel pump
Fuel shortage at engine	Mixture too weak Air leak in carburettor Air leak at inlet manifold to cylinder head, or inlet manifold to carburettor
Mechanical wear	Incorrect valve clearances Burnt out exhaust valves Sticking or leaking valves Weak or broken valve springs Worn valve guides or stems Worn pistons and piston rings

Lack of power and poor compression

Symptom	Reason(s)
Fuel/air mixture leaking from cylinder	Burnt out exhaust valves Sticking or leaking valves Worn valve guides and stems Weak or broken valve springs Blown cylinder head gasket (accompanied by increase in noise) Worn pistons and piston rings Worn or scored cylinder bores
Incorrect adjustments	Ignition timing wrongly set. Too advanced or retarded *Contact breaker points incorrectly gapped Incorrect valve clearances Incorrectly set spark plugs Carburation too rich or too weak
Carburation and ignition faults	*Dirty contact breaker points Distributor automatic balance weights or vacuum advance and retard mechanisms not functioning correctly Faulty fuel pump giving top end fuel starvation

Excessive oil consumption

Symptom	Reason(s)
Oil being burnt by engine	Badly worn, perished or missing valve stem oil seals Excessively worn valve stems and valve guides Worn piston rings Worn pistons and cylinder bores Excessive piston ring gap allowing blow-by Piston oil return holes choked
Oil being lost due to leaks	Leaking oil filter gasket Leaking timing case gasket Leaking sump gasket Loose sump plug

Symptom	Reason(s)
Unusual noises from engine Excessive clearances due to mechanical wear	Worn valve gear (noisy tapping from rocker box) Worn big-end bearing (regular heavy knocking) Worn timing chain or gears (rattling from front of engine) Worn main bearings (rumblings and vibration) Worn crankshaft (knocking, rumbling and vibration)

Not applicable to later N. American models with electronic ignition

PART C – 1493 cc ENGINE

27 Cylinder head – removal and refitting (engine in car)

1 Owing to the various different arrangements of emission control systems that have been introduced, and the changes that have occurred over the years, it is not possible to provide a straightforward removal procedure to suit all installations. However, by using the broad outline of the basic work involved which follows as a guide, it should be possible, by modifying it to suit your own particular car's arrangement, to remove the cylinder head with the engine in the car with few problems.
2 Disconnect the battery connections.
3 Drain the cooling system as described in Chapter 2.
4 Disconnect and remove the carburettor(s).
5 Remove the temperature transmitter fron the thermostat housing.
6 Remove the fanguard and slacken the alternator adjusting link bolt.
7 Remove the three bolts securing the thermostat housing/water pump to the cylinder head.
8 Disconnect and remove the coolant and air hoses impeding access to, or connected to, the cylinder head and the manifolds. These, of course, vary with installations.
9 Disconnect the exhaust pipe from the exhaust manifold.
10 Disconnect the HT leads from the spark plugs and remove the leads together with the distributor cap.
11 Remove the rocker cover fasteners and remove the rocker cover.

27.17 Cylinder head gasket top face

27.18 Fitting cylinder head

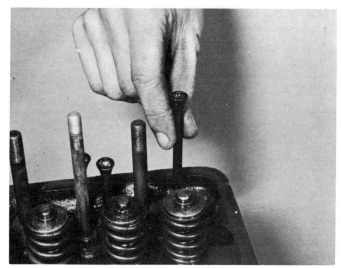

27.21 Inserting push-rods

27.22 Fitting rocker assembly

12 Undo the four nuts securing the rocker shaft assembly to the cylinder head and remove the assembly.
13 Remove the pushrods, identifying them so that they can be refitted in their original locations.
14 Remove the ten cylinder head nuts following the reverse sequence to that shown in Fig. 1.3 in Section 10. Slacken the nuts in sequence progressively before removing them.
15 Remove the engine lifting bracket and then remove the cylinder head from the engine.

Refitting
16 Thoroughly clean the cylinder block top face and check that the cylinder head joint face is also perfectly clean. Generously lubricate each cylinder with clean engine oil.
17 Fit a new cylinder head gasket to the block, easing it carefully over the studs to avoid damage. Make sure that the TOP marked face is uppermost and don't use any jointing compound (photo).
18 Lower the cylinder head into place, keeping it parallel with the block to avoid binding on the studs (photo).
19 With the head in position refit the lifting bracket and, on UK models only, the fuel pipe bracket to the right-hand rear studs. Fit the ten cylinder head nuts and washers and run them down fingertight.
20 Following the sequence shown in Fig. 1.3 in Section 10, tighten down the nuts a part of a turn at a time to the torque wrench setting quoted in the Specifications.
21 Insert the pushrods in the same positions from which they were removed, checking that the ball ends enter the tappets (photo).
22 Refit the rocker assembly to the cylinder head, checking that each rocker balljoint sits properly in the cup on its associated pushrod (photo).

28 Rocker gear – dismantling and reassembly

1 With the rocker shaft assembly removed from the cylinder head, take out the split pin from the front end of the shaft and carefully slide off the rockers, pedestals, springs and spacers noting the sequence of components for use on reassembly.
2 Remove the screw in the rear pedestal to release the shaft from the pedestal and slide the pedestal off together with the washer and rocker.

Reassembly
3 Lubricate and fit the No 8 rocker into position in the rear pedestal complete with its spring washer, and insert the shaft through the pedestal and rocker bores.
4 Apply a suitable thread-locking compound to the threads of the pedestal/shaft locating screw and fit the screw to the pedestal, making sure that it correctly engages with the shaft. Tighten the screw.
5 Now, in their original sequence and positions, slide the individual springs, rockers, pedestals, and washers onto the rocker shaft, lubricating with engine oil as the assembly is built-up. Finally fit a new split pin to secure the assembly on the shaft.

29 Engine sump – removal and refitting

The operations are as described in Part B, Section 11. Note that USA catalytic converter models have a metal pipe and banjo bolt connection at the engine oil drain plug hole.

30 Pistons, connecting rods and big-end bearings – removal and refitting (engine in car)

The operations are as described in Part B, Section 12.

31 Timing cover, sprockets and chain – removal and refitting

1 If the engine is installed in the car, some removal or repositioning of components will be necessary before the crankshaft pulley can be removed, more especially in the case of USA models. For this reason separate procedures are provided up to the removal of the pulley.

Models except N. America
2 Refer to Chapter 2 and drain the cooling system, remove the radiator and the fan. Slacken the alternator mounting bolts and remove the fanbelt.
3 Undo and remove the pinch bolt and nut from the bottom of the steering column. Remove the two bolts and washers in each clamp securing the steering rack to its mounting brackets. Then remove the three bolts and washers which secure each mounting bracket to the chassis crossmember.
4 Remove the mounting brackets, but look out for packing washers fitted under the bracket at the pinion end of the rack. Keep these washers carefully so that they can be refitted on reassembly.
5 Disconnect the steering column from the rack pinion shaft and carefully ease the rack forwards and downwards to provide access to the crankshaft pulley nut.
6 Lock the crankshaft by engaging top gear and applying the handbrake fully. Undo the pulley retaining nut and remove the pulley from the crankshaft.

N. American models
7 Refer to Chapter 2 and drain the cooling system, remove the radiator and the fan.
8 With a suitable receptacle to catch oil spillage, remove the oil filter by unscrewing it from its mounting. Disconnect the hose from the air pump, and remove the pump and its drivebelt.
9 Slacken the alternator mounting bolts and remove the fanbelt.
10 Refer to Chapter 3 and remove the carburettor.
11 Remove the pre-heater duct after pressing in its retaining clips.
12 Remove the six nuts retaining the catalytic converter to the exhaust manifold.
13 Support the engine on a jack with a piece of wood interposed to protect the sump.

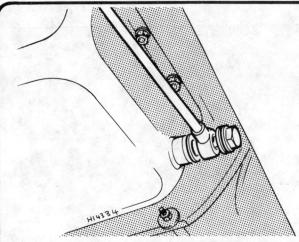

Fig. 1.19 Engine sump drain plug and banjo connection (Sec 29)

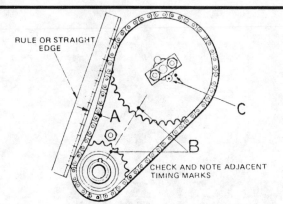

Fig. 1.20 Checking timing chain for stretch (Sec 31)

A Measure maximum deflection to assess chain wear
B Timing marks aligned and on straight line between sprocket centres
C Sprocket-to-camshaft alignment marks

Chapter 1 Engine

14 Remove the nuts, bolts and washers securing both engine front mountings to the mounting plate.
15 Carefully lift the engine assembly sufficient for removal of the crankshaft pulley.
16 Undo the pulley nut; as this will be tight, lock the crankshaft by engaging top gear and fully applying the handbrake. Remove the pulley from the crankshaft.

Engine removal
17 With the engine on the bench the only likely difficulty will be to restrain the crankshaft whilst undoing the pulley retaining nut. If the starter is removed, a large screwdriver or other suitably strong item can be inserted to jam the flywheel starter ring teeth. Alternatively, if the sump is going to be removed, do this now and interpose a block of wood between a crankshaft throw and the inner wall of the crankcase to prevent rotation. Undo the nut and remove the pulley.

All models, engine in or out of car
18 If difficulty is experienced in removing the crankshaft pulley, use two large levers to ease it off, or use a universal puller.
19 On UK models, remove the eleven setscrews and the one nut together with their washers (or on USA and Canada models, remove the eight setscrews, the one bolt, the three nuts together with their washers) and spacer securing the timing cover to the block and, on USA and Canada models, the air pump adjusting links.
20 Where fitted, remove the bottom air pump bolt and the spacer.
21 Remove the timing cover together with its gasket. It's worth checking the timing chain at this stage to see if it will need renewing on reassembly. Lay a straight edge along the slack leg of the chain run and measure the extent of chain slack from the straight edge. If it is more than 0.4 in (10 mm) the chain should be renewed.
22 To align the timing marks, turn the crankshaft, chock the front wheels and jack one of the rear wheels clear of the ground; check that the jack supporting the engine (USA and Canada cars procedure only) is stable. Release the handbrake and select top gear. By rotating the rear wheel in the forward roll direction, the engine can be turned in the normal direction of rotation. The marks on the two sprockets should be adjacent to each other on a straight line joining the two sprocket centres.
23 Remove the oil thrower from the crankshaft and bend back the locktabs on the camshaft sprocket bolts. Note that an alignment mark on the camshaft sprocket is adjacent to the one on the camshaft which can be seen through one of the holes in the camshaft sprocket. Undo and remove the two camshaft sprocket retaining bolts.
24 From this stage on, take care not to move the camshaft or crankshaft. Ease the two sprocket wheels off their shafts complete with timing chain; if necessary, use two suitable levers to do this, working on alternate sprockets until the assembly is free. Retrieve the crankshaft drive key to prevent its loss and remove the shim washers from behind the crankshaft sprocket if any are fitted.

Refitting
25 Check that the shim washers, if any, are refitted to the crankshaft. Leave the crankshaft key out of its key slot for the moment.
26 Fit the two sprockets into position on their shafts but without the timing chain; push them fully home and put a straight edge across their two faces. If one of the sprockets is out of line adjust their relative position by adding or removing shim washers behind the crankshaft sprocket. When satisfactory remove the two sprockets and fit the crankshaft key (photos).
27 Lay the sprockets on a clean surface so that the two timing marks are adjacent to each other. Slip the timing chain over them and pull the sprockets back into mesh with the chain so that the timing marks although further apart, are still adjacent to each other.
28 With the timing marks adjacent to each other, hold the sprockets above the crankshaft and camshaft. Turn the camshaft and crankshaft so that the Woodruff key will enter the slot in the crankshaft sprocket, and the camshaft sprocket is in the correct position relative to the camshaft. This will be indicated by being able to see the alignment mark on the camshaft driving flange, through the hole in the camshaft sprocket, nearest to the alignment mark near the bolt holes.
29 Fit the timing chain and sprockets assembly onto the camshaft and crankshaft, keeping the timing marks adjacent to each other (photo). Fit a new double tab washer in place on the camshaft sprocket, then fit and tighten the two retaining bolts to their specified torque.
30 Lever up the tabs on the lockwasher to lock the bolts (photo).
31 The oil seal in the front cover should be renewed before refitting the cover. Press or lever the old seal out, taking care to avoid damaging the cover.
32 Lubricate the new seal with engine oil and carefully press or tap it into position in the cover, ensuring that the seal lip will face in towards the crankshaft sprocket on assembly (photo).
33 Fit the oil deflector onto the crankshaft. Make sure that the dished periphery is towards the cover when it is fitted. Generously lubricate the chain and sprockets with engine oil and visually check that all is ready for fitting the cover.
34 Fit a new gasket to the timing cover, making sure that it is properly located on the dowels (photo).
35 It may be possible to fit the cover by holding it at an angle, so that the chain tensioner can be first located, then swinging the cover into its correct position. Alternatively, use a bent piece of stiff wire to hold the tensioner and away from the chain while the cover is being put on and, just before pushing the cover home, carefully remove the wire, making sure that the gasket is not damaged in the process.
36 Refit the setscrews, nut(s) and bolt (where appropriate) with their washers and spacer to their original positions and tighten them progressively and evenly to their specified torque.
37 Check that the crankshaft pulley is clean and undamaged and oil the seal housing surface. Carefully insert the pulley into the seal, aligning the keyway with the key, and push it fully home (photo). Fit the retaining nut.

31.25 Fitting crankshaft shim

31.26 Crankshaft Woodruff key

31.29 Timing chain and sprockets in position

31.30 Bending up sprocket lockplate tab

31.32 Timing cover oil seal correctly fitted

31.33 Crankshaft oil deflector

31.34 Timing cover gasket

31.37 Fitting crankshaft pulley

Chapter 1 Engine

38 Tighten the retaining nut to its specified torque. To hold the crankshaft whilst this is done, insert a cleanpiece of wood between the crankshaft and crankcase if the engine is on the bench or, if installed, make sure that both rear wheels are on the ground, that the handbrake is fully on, and that top gear is selected.
39 For details of timing chain tensioner, refer to Part B Section 13, Paragraphs 13 to 15.

32 Camshaft – removal and refitting

1 The operations are as described in Part B, Section 14.

33 Oil pump – removal and refitting (engine in car)

1 The operations are as described in Part B Section 15.

34 Engine – complete dismantling (engine removed)

1 Remove the following components in the sequence given as described earlier in this Section or Part B as indicated.

Rocker gear and cylinder head
Timing gears and chain
Camshaft and distributor drive gear
Sump
Piston/rod assemblies
Oil pump

2 Remove the flywheel and engine rear plate as described in Part B, Section 16, Paragraphs 8 to 11.
3 Unbolt and remove the engine front plate.
4 Remove the crankshaft Woodruff key and any shims that may have been fitted behind the crankshaft sprocket wheel.
5 Remove the two screws retaining the front sealing block which is located across the No 1 main bearing cap, and remove the sealing block and its gaskets.
6 Undo the seven bolts securing the crankshaft rear oil seal housing to the block, and remove the housing with the rear oil seal in it.
7 Note the identification marks on the big-end and main bearing caps for reference during reassembly.
8 Undo and remove the main bearing cap bolts. Remove the caps and half bearing shells. Remove the two semi-circular thrust washers from the rear main bearing and make sure that, when they need to be refitted, you will be able to fit each in the correct location.
9 Carefully lift out the crankshaft and retrieve the upper half bearing shells from the main bearings and the connecting rod big-ends. In each case identify them so that, if you do not intend to renew them, they can all be refitted in their original positions.

35 Dismantling sub-assemblies

1 The operations are similar to those described in Part B, Section 17 with the following exceptions.

Cylinder head
2 On this engine, double valve springs are used with split collet retainers.

Rocker gear
3 The detail differences in components should be noted.

Piston/connecting rods
4 The gudgeon pins are retained in the pistons by circlips.

36 Examination and renovation

Refer to Part B, Section 18.

37 Cylinder head – decarbonising

Refer to Part B, Sectin 19.

38 Lubrication and crankcase ventilation systems – description

1 A forced feed system of lubrication is fitted, with oil circulated round the engine from the sump below the block. The level of engine oil in the sump is indicated on the dipstick, which is fitted on the right-hand side of the engine. It is marked to indicate the optimum level, which is the maximum mark. An oil pressure warning switch and lamp are fitted.
2 The level of the oil in the sump, ideally, should not be above or below this line. Oil is replenished via the filler cap on the rocker cover.
3 The eccentric rotor-type oil pump is bolted in the left-hand side of the crankcase, and is driven by a short shaft from the skew gear on the camshaft which also drives the distributor shaft.
4 The pump is the non-draining variety to allow rapid pressure building-up when starting from cold.
5 Oil is drawn into the pump from the sump via the pick-up pipe. From the oil pump the lubricant passes through a non-adjustable relief valve to the full-flow filter. Filtered oil enters the main gallery which runs the length of the engine on the left-hand side. Drillings from the main gallery carry the oil to the crankshaft and camshaft journals.
6 The crankshaft is drilled so that oil under pressure reaches the crankpins from the crankshaft journals. The cylinder bores, pistons and gudgeon pins are all lubricated by splash and oil mist.
7 Oil is fed to the valve gear via the hollow rocker shaft at a reduced pressure by means of a scroll and two flats on the camshaft rear journal.
8 Drillings and grooves in the camshaft front journal lubricate the camshaft thrust plate, and the timing chain and gearwheels. Oil returns to the sump by gravity, the pushrods and cam followers being lubricated by oil returning via the pushrod drillings in the block.
9 The standard crankcase ventilation system is the closed type where air enters the oil filler cap on the rocker cover. This air and internal fumes from the engine are drawn through an oil separator/flame trap in the top of the rocker cover, then through hoses to the carburettors where the gas mixes with the induction charge to be burnt in the engine.
10 On USA and Canada cars fitted with emission control, the system is similar to the closed ventilation system, but more efficient and complex. The tube from the rocker cover is interconnected between the carburettor and the primary charcoal absorption canister. The engine normally draws air from the breather hole in the oil filler cap in the rocker cover. However, systems that have a carburettor fuel evaporative loss control have a sealed oil filler cap, and instead draw air through the two charcoal absorption canisters. The airflow from the canisters goes to the rocker cover and the feed pipe incorporates a restrictor connection to the carburettor.
11 An oil cooler is available as an optional extra, and is considered a worthwhile addition, if you intend to take full advantage of the power output, or envisage lengthy high speed runs.

38.1 Oil pressure switch

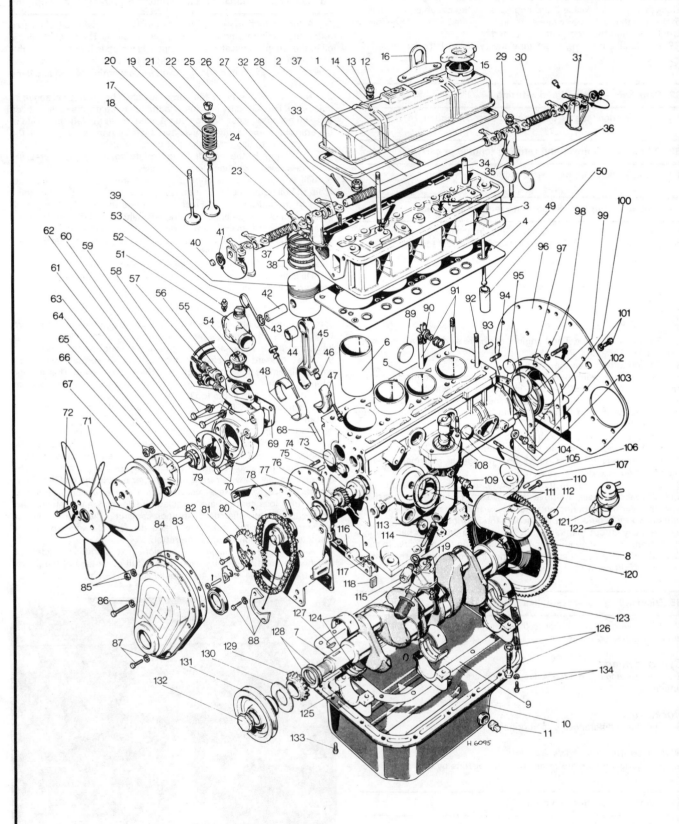

Fig. 1.21 Exploded view of 1500 engine (Sec 34)

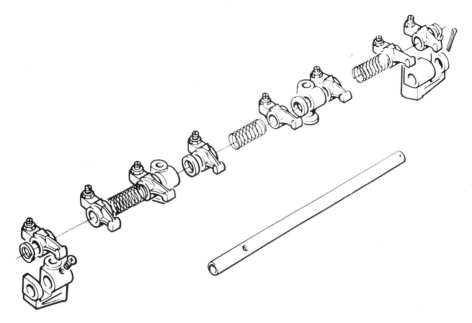

Fig. 1.22 Exploded view of rocker gear (Sec 35)

Key to Fig. 1.21

1 Rocker cover
2 Gasket
3 Cylinder head
4 Cylinder head gasket
5 Cylinder block
6 Cylinder liner
7 Crankshaft
8 Starting ring and flywheel
9 Sump gasket
10 Sump
11 Sump drain plug
12 Self-locking nut
13 Washer
14 Fibre washer
15 Oil filler cap
16 Lifting bracket
17 Exhaust valve
18 Inlet valve
19 Valve spring
20 Valve spring seating collar
21 Valve spring collet plate
22 Valve collet
23 Rocker
24 Rocker shaft bracket
25 Spacer
26 Tappet adjusting screw and locknut
27 Split-pin
28 Rocker shaft
29 Nut and washer
30 Rocker spacing spring
31 Rocker shaft bracket
32 Head securing nut and washer
33 Rocker cover securing stud
34 Valve guide
35 Rocker shaft bracket and stud
36 Core plugs
37 Inlet manifold stud
37a Exhaust manifold stud
38 Piston rings
39 Piston
40 Plug
41 Spring washer
42 Gudgeon pin
43 Circlip
44 Small-end bush
45 Connecting rod
46 Sleeve
47 Big-end cap and bolts
48 Big-end bearing shells
49 Pushrod
50 Tappet
51 Coolant water elbow
52 Elbow securing stud, nut and lockwasher
53 Dipstick
54 Thermostat
55 Gasket
56 Adaptor for temperature sensor
57 Temperature sensor unit
58 Hose adaptor
59 Locknut pump securing bolt (medium) and washer
60 Coolant pump securing bolt (long) and washer
61 Spindle and impeller
62 Spindle housing gasket
63 Spindle housing plug
64 Seal
65 Spindle and bearing housing
66 Spindle housing nut and lockwasher
67 Fan pulley and hub
68 Dipstick guide
69 Coolant pump housing gasket
70 Coolant pump securing bolt (short) and washer
71 Fan
72 Fan securing bolt and lockwasher
73 Core plugs
74 Stud
75 Dowel
76 Camshaft
77 Gasket
78 Front endplate
79 Timing chain
80 Camshaft sprocket
81 Chain tensioner
82 Sprocket bolt
83 Timing cover gasket
84 Timing cover
85 Timing cover securing nut and lockwasher
86 Timing cover securing bolt and lockwasher (long)
87 Timing cover securing bolt and lockwasher (short)
88 Camshaft locating plate
89 Core plug
90 Drain tap (washer)
91 Head securing studs (grooved)
92 Head securing stud (standard)
93 Dowel
94 Stud
95 Core plugs
96 Rear endplate
97 Rear oil seal housing
98 Seal housing bolt and lockwasher
99 Rear oil seal
100 Oil seal housing gasket
101 Bolt and lockwasher
102 Oil gallery plug
103 Drain plug and washer
104 Securing stud for fuel pump
105 Core plug
106 Securing stud for distributor housing
107 Dowel
108 Distributor driveshaft
109 Oil pressure sensing unit
110 Flywheel bolt
111 Oil filter and seal
112 Sleeve
113 Plug
114 Oil pressure relief valve
115 Relief valve nut
116 Sealing block
117 Gasket
118 Seal
119 Oil pump
120 First motion shaft bush
121 Fuel pump
122 Fuel pump securing nut and lockwasher
123 Rear main bearing thrust washer
124 Main bearing shells
125 Main bearing cap
126 Bearing cap bolt and lockwasher
127 Crankshaft key
128 Adjusting shims
129 Crankshaft sprocket
130 Oil thrower
131 Crankshaft pulley
132 Pulley retaining bolt
133 Sump securing bolt
134 Securing bolt and lockwasher

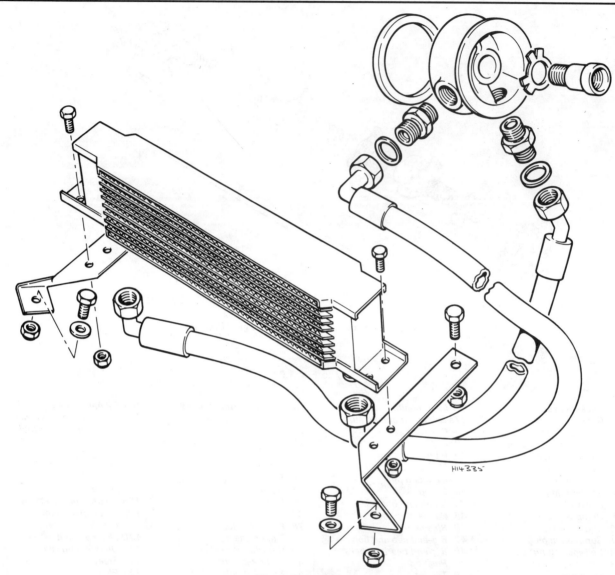

Fig. 1.23 Engine oil cooler (Sec 38)

12 In effect an oil cooler is a radiator through which the engine oil is circulated. Having said this, it must be understood that an oil cooler is more robust than a conventional cooling system radiator and has to be able to withstand far greater internal pressures.
13 The oil cooler has two pipes connected to it, a flow pipe and a return pipe. These pipes are in turn connected to an adaptor unit which is sandwiched between the engine oil filter and the cylinder block face.
14 During winter months, it may be found that the oil cooler has a tendency to over-cool the oil and in such cases it is advisable to place a cover over the cooler.
15 Removal and refitting of the various sub-assemblies making up this system is self-explanatory, making reference to Fig. 1.23. With regard to maintenance, it is advisable to periodically inspect the cooler, flexible pipes and unions for leakage. When carrying out any major engine work it is recommended to remove the oil cooler and to blow through it with a compressed air supply to remove any sludge deposits which may have built up.

39 Engine reassembly – general

Refer to Part B, Section 22.

40 Engine – complete reassembly

1   Ensure that the crankcase is thoroughly clean and that all oilways are clear. A thin twist drill or a piece of wire is useful for cleaning them out. If possible, blow them out with compressed air. Treat the crankshaft in the same fashion, and then inject engine oil into the crankshaft oilways. Commence work on rebuilding the engine by refitting the crankshaft and main bearings.
2   If the original main bearing shells are being refitted (and it is false economy not to renew them unless the originals are virtually as new), fit the three upper halves of the main bearing shells in their original positions in the crankcase after wiping their locations clean.
3   If new bearing shells are being fitted, carefully clean off all traces of the protective grease with which they are coated before fitting them to the crankcase.
4   Note that on each half shell there is a locating tag which must be fitted to the recess in either the crankcase or bearing cap housing location.
5   With the three upper bearing shells securely in place, wipe the lower bearing cap housings and fit the three lower shell bearings to their caps ensuring that the right shell goes into the right cap if the old bearings are being refitted.
6   Wipe the recesses either side of the rear main bearing which locate the thrust washers.
7   Generously lubricate the crankshaft journals and the upper and lower main bearing shells, and carefully lower the crankshaft into place (photo).
8   Fit the upper halves of the thrust washers into their grooves on each side of the rear main bearing, rotating the crankshaft in the direction towards the main bearing tabs (so that the main bearing

Chapter 1 Engine

40.7 Lowering crankshaft into position

40.9 Fitting crankshaft front main bearing cap

shells do not slide out). At the same time feed the thrust washers into their locations with their oil grooves outwards away from the bearing.
9 Fit the main bearing caps in position ensuring that they locate properly (photo). The mating surfaces must be spotlessly clean or the caps will not seat correctly. As the bearing caps were assembled to the cylinder block and then line bored during manufacture, it is essential that they are returned to the same positions from which they were removed.
10 Refit the main bearing cap bolts and washers and tighten the bolts to their specified torque (photo).
11 Test the crankshaft for freedom of rotation. Should it be very stiff to turn or exhibit high spots, that is, intermittently free and stiff, then a most careful inspection must be made, preferably by a skilled mechanic, to trace the cause of the trouble. Fortunately it is very seldom that any trouble of this kind is experienced after taking care during the fitting of the bearings and crankshaft.
12 Move the crankshaft hard in one direction along its length and using feeler gauges, measure the amount of endfloat (photo). This should be within the specified limits but, if it is not, new or oversize thrust washers will need to be fitted.
13 Next fit the sealing block and gaskets over the front main bearing cap. Smear the ends of the block with jointing compound and fit the gaskets and the block in place. Fit the securing screws but do not tighten fully. Fit new wedge seals at each end and line up the front face of the block with the front of the cylinder block (photo).
14 Do not fully tighten the securing screws until the front plate is in position. This prevents any distortion between the sealing block and plate.
15 When the plate is fitted, fully tighten the screws and cut the wedge seals flush with the crankcase flange.
16 Fit the piston/connecting rods as described in Part B, Section 12.
17 When connecting the rods to the crankshaft wipe the connecting rod half of the big-end bearing cap and the underside of the shell bearing clean, and fit the shell bearing in position with its locating tongue engaged with the corresponding rod.
18 If the old bearings are nearly new and are being refitted, then ensure that they are refitted in their respective locations on the correct rods.
19 Generously lubricate the crankpin journals with engine oil, and turn the crankshaft so that the crankpin is in the most advantageous position for the connecting rod to be drawn onto it.
20 Wipe the connecting rod bearing cap and back of the shell bearing clean and fit the shell bearing in position, ensuring that the locating tongue at the back of the bearing engages with the locating groove in the connecting rod cap (photo).
21 Generously lubricate the shell bearing and offer up the connecting rod bearing cap to the connecting rod.
22 The makers recommend that new bolts are fitted to the connecting rod big-ends on reassembly; tighten them to their specified torque noting that this varies with the type of bolt used.
23 When all the connecting rods have been fitted, rotate the crankshaft to check that everything is free, and that there are no high spots causing binding.
24 Fit a new gasket in place over the front of the cylinder block. Lightly smear it with a gasket cement to locate it.
25 Now locate the front endplate into position over the location dowel and stud in the front of the cylinder block (photo). Loosely refit the respective screws and nuts into position.
26 Tighten the two lower screws first (photo) and then tighten the crankshaft sealing block retaining screws. Now tighten the three bolts in the front of the plate.
27 Wipe the camshaft bearing journals clean and lubricate them generously with engine oil.
28 If a new camshaft is being fitted, use a centre punch to mark it in exactly the same position as the mark on the front of the drive flange on the old camshaft.
29 Gently insert the camshaft into the crankcase (photo) taking care not to damage the camshaft bearings with the cams.
30 Refit the camshaft locating plate and tighten down the two retaining bolts and washers (photo).
31 With a feeler gauge, check that the camshaft endfloat is within the specified tolerance. If it is not, renew the camshaft locating plate.
32 Fit the timing gears, chain and cover as described in Section 31.
33 Fit the pump assembly to the crankcase, and check that the driveshaft engages with the drivegear if this has not been removed.
34 Prime the pump with engine oil to reduce any possibility of oil starvation when the engine starts.
35 Refit the cover and strainer to the pump and tighten down the three securing bolts and washers (photo).
36 If this has not already been done, remove the old seal from the crankshaft rear oil seal housing and clean the housing. Lightly grease the outside of a new seal and carefully press or tap it into the housing with the lip facing in towards the crankshaft bearings.
37 Check that all traces of the old gasket has been removed from the crankcase joint face and apply a film of fresh jointing compound. Then position a new gasket on the joint face.
38 Lubricate the seal bearing surface on the crankshaft with oil and carefully fit the seal and its housing on to the crankshaft and bed the housing onto the new gasket (photo).
39 Fit the seven retaining bolts and washers, but make sure that the top bolt has a plain copper washer under its head (photo). Turn the crankshaft over several times to centralise the seal, then progressively and evenly tighten the bolts to their specified torque.
40 No gasket is fitted between the rear endplate and the block. Check that the plate and its mounting face are clean and position the plate on its locating dowels in the block. Fit and tighten the seven retaining bolts and washers (photo).
41 Clean the end of the crankshaft on which the flywheel mounts, and make sure that the spigot bush is correctly positioned in its location in the end of the shaft. Clean the mating face of the flywheel and fit it to

40.10 Tightening main bearing cap bolt

40.12 Measuring crankshaft endfloat

40.13 Front sealing block in position

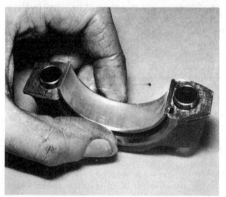
40.20 Connecting rod big-end cap and shell bearing

40.25 Fitting engine front plate

40.26 Tightening front plate screws

40.29 Installing camshaft

40.30 Fitting camshaft locating plate

40.35 Fitting oil pump

40.38 Crankshaft rear oil seal and retainer

40.39 Oil seal retainer bolt with copper washer

40.40 Engine rear plate bolted in position

Chapter 1 Engine

40.42 Tightening flywheel bolts

40.45 Fitting sump with new gasket

40.46 Sump longer bolts (arrowed)

40.48a Fitting double valve springs

the crankshaft locating it on its dowel.
42 Fit the four retaining bolts and tighten them to their specified torque (photo). Jam the starter ring gear to prevent the flywheel from turning.
43 If the starter ring gear has been renewed, a concentricity check should be made using a dial test indicator gauge (clock gauge). Mount the gauge on the rear endplate and measure the concentricity on the gear teeth crests. This should not exceed that quoted in the Specifications. After this, check the run-out of the flywheel by mounting the gauge to register at a point 3.0 in (76.2 mm) from the flywheel axis and rotate the crankshaft. Run-out, or wobble on the flywheel face, should not exceed that quoted in the Specifications.
44 Refer to Chapter 5 and refit the clutch assembly.
45 After the sump has been thoroughly cleaned, scrape all traces of the old sump gasket from the sump and crankcase flanges, fit a new gasket in place, and then refit the sump (photo).
46 Insert and tighten down the sump bolts and washers. The four longer bolts fit to the rear sump strengthening plates (photos).
47 Reassemble the cylinder head. To gain access above and below the cylinder head, rest it on its side or, alternatively, with the head joint face down on a wooden bench so that it overhangs the bench. Before starting to fit the valves, check that all components are thoroughly clean, including the cylinder head, especially if the valves have been ground in.
48 Taking each valve in turn, first fit the valve spring seating collar

40.48b Compressing valve springs

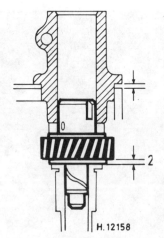

Fig. 1.24 Checking diagram for distributor drivegear endfloat (Sec 40)

1 Feeler gauge gap
2 Washer of given thickness

over the top of the valve guide. Lubricate the valve stem and fit the valve to its original guide if it has not been renewed. Fit the two springs (photo) with the close coils adjacent to the cylinder head and then fit the spring collet plate. With the base of a valve spring compressor on the valve head, compress the springs and fit the two half collets (photo). Slowly release the spring compressor and check that the half collets sit correctly in the collet plate and on the valve stem. Remove the spring compressor.

49 Repeat this procedure until all valves and springs are fitted. Then, using a hammer, strike each valve stem just hard enough to move the valve in order to settle the springs and collet.

50 Fit the tappets (cam followers), cylinder head and rocker gear as described in Section 27.

51 It is important to fit the distributor drive correctly, otherwise the ignition timing will be totally wrong. It is easy to set the driveshaft in apparently the right position but in fact exactly 180° out, by not working on the correct cylinder, which must only be at TDC but also at the end of compression and the start of the power stroke with both halves closed. The distributor driveshaft should therefore not be fitted until the cylinder head and rocker assembly are fitted and the timing chain and sprockets assembled so that the valve operations can be observed. Alternatively, if the timing cover has not been fitted, the distributor driveshaft can be refitted when the timing marks on the sprockets are adjacent to each other on a straight line through the sprocket centres.

52 Rotate the crankshaft so that No 1 piston is at TDC at the start of the power stroke (photo). This is when the inlet valve on No 4 cylinder is just opening and the exhaust valve just closing, or when the sprocket timing marks are aligned.

53 Before fitting the driveshaft in its timed position, its endfloat when installed must be checked. This endfloat is controlled by selecting an appropriate thickness gasket which fits between the flange on the distributor pedestal and the cylinder block mounting face. To determine the thickness of gasket required you will need a washer of known thickness which will fit on the driveshaft below the skew gear, and a set of feeler gauges. The washer should 0.5 in (12.7 mm) diameter. The procedure is as follows.

54 Remove the distributor pedestal/cylinder block gasket if fitted, put the washer on the lower end of the shaft assembly and fit the shaft, making sure that its drive mates with the oil pump spindle. If necessary, turn the oil pump with a screwdriver and try the driveshaft in several positions to determine which is correct.

55 Fit the distributor pedestal, without gasket of course, and holding it down on the driveshaft gear, measure the gap between its flange and the block with feeler gauges (photo).

56 Calculate the thickness of the gasket required. Three situations can arise; (i) the gap size equals the thickness of the washer; (ii) the gap is smaller than the washer; or (iii) the gap is larger than the washer. Proceed as follows.

57 If the gap equals the thickness of the washer the gasket must

40.50 Inserting cam followers into their bores

40.52 No. 1 piston at TDC

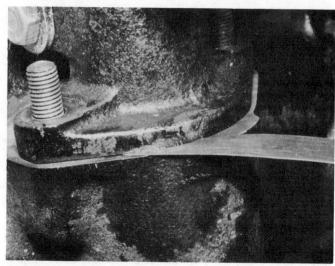

40.55 Checking gap between distributor mounting flange and cylinder block

Chapter 1 Engine

40.62 Distributor clamp plate bolt

40.67 Fitting oil filter cartridge

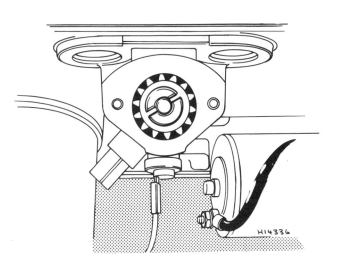

Fig. 1.25 Driveshaft slot ready to receive mechanical breaker type distributor (Sec 40)

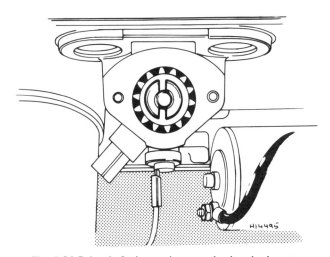

Fig. 1.26 Driveshaft slot ready to receive breakerless type distributor (Sec 40)

equal the amount of endfloat needed.

58 If the gap is smaller than the washer, without the washer there would be some endfloat, and this will equal the washer thickness minus the gap size. The gasket must therefore be thick enough to increase this difference to the required endfloat.

59 If the gap is larger than the washer, it would be found that if the washer was removed there would be an end load on the driveshaft by an amount equal to gap size minus washer thickness. Therefore the gasket must be equal to this amount plus the amount of endfloat required.

60 Remove the distributor pedestal, the driveshaft and the washer used for endfloat measurements and, with No 1 piston still at TDC at the end of the compression stroke, refit the driveshaft. This must be done so that, in addition to engaging the oil pump drive spindle, the offset slot in the top of the driveshaft is in the position shown in Figs. 1.25 and 1.26 when fully engaged with the camshaft skew gear. As the driveshaft turns every time it is fitted or removed, in order to arrive at the correct final position, several attempts at pre-positioning the shaft before fitting it may be necessary on a trial-and-error basis. On completion check that the oil pump spindle is engaged with the driveshaft.

61 Fit the gasket of the required thickness to achieve the specified driveshaft endfloat and then fit the distributor pedestal and its two nuts and washers. On USA and Canada models, two spacers are fitted to the pedestal studs together with a diverter bracket on a cylinder block stud.

62 Lubricate the distributor driveshaft with clean engine oil and then fit the distributor to the pedestal, aligning the drive with the driveshaft and securing the clamp plate with the nut, bolt and washer. In doing this, align the marks which were made during dismantling. If the clamp bolt was not disturbed the ignition timing should be as originally set. However, if the clamp bolt has been loosened, or if alignment marks were not made on dismantling, it will be necessary to re-time the ignition as explained in Chapter 4.

63 Adjust the valve clearances (Section 24).

64 Fit a new gasket to the rocker cover and carefully fit the cover in place.

65 Refit the washers over the retaining studs and ensure that the sealing washer is located under the flat steel washer. Refit the rocker cover nuts or screws as appropriate.

66 Fit the spark plugs and then refit any other ancillary components to the engine in the reverse order of removal.

67 As a basic rule, all ancillary components and fittings should be reassembled to the engine before installation, if only because access in almost every instance is easier. This means that the manifolds (using new gaskets), the carburettor(s) (again with new gaskets), the alternator, the thermostat and water pump housing, the air pump (USA and Canada models), the oil filter cartridge, and the fuel pump (using a new gasket) should all be fitted while the engine is still on the bench (photo).

68 If the engine and gearbox were removed as an assembly, refit the gearbox to the engine following the reverse of the procedure for removal. Make sure that the weight of the gearbox is not allowed to rest on the input shaft or the clutch driven plate during reassembly.

69 When the engine is installed in the car, fill up with coolant, engine oil and connect the battery.

41 Valve clearances – adjustment

The operations are as described in Section 24.

42 Starting up after major overhaul

Refer to Part B, Section 25.

43 Fault diagnosis – engine

Refer to Part B, Section 26.

Chapter 2 Cooling and heating systems

Contents

Coolant control valve – removal and refitting	16
Coolant mixture	4
Coolant pump – removal and refitting	8
Coolant pump drivebelt – adjustment	12
Coolant pump drivebelt – renewal	13
Coolant system – draining and refitting	3
Coolant temperature sender unit	11
Electric fan and switch (later N. American models) – removal and refitting	6
Fan blades – renewal and refitting	9
Fan viscous coupling – removal and refitting	10
Fault diagnosis	17
General description	1
Heater – removal and refitting	14
Heater booster motor – removal and refitting	15
Maintenance	2
Radiator – removal, repair and refitting	7
Thermostat – removal, testing and refitting	5

Specifications

System type .... 'No-loss' with expansion bottle and thermo syphon with front mounted radiator, belt-driven pump and thermostat. Most models with conventional or viscous coupling fan. Later N. American models have electric cooling fan

Radiator pressure cap .... 13 lbf in² (0.91 kgf cm²)

Thermostat
Except N. America .... 180°F (82°C)
N. America .... 190°F (88°C)

Coolant
Type/specification .... Ethylene glycol based antifreeze (Duckhams Universal Antifreeze and Summer Coolant)

Capacity:
Except N. America (1979 on) – vertical radiator .... 8 Imp pts (9.6 US pts, 4.5 l)
N. America (1979 on) – sloping radiator .... 9.3 Imp pts (11.2 US pts, 5.3 l)

Torque wrench settings

	lbf ft	Nm
Fan bolts or nuts	9	12
Thermostat housing bolts	20	27
Coolant pump housing mounting bolts	20	27
Coolant pump to housing	14	19

1 General description

1 The cooling system is of the 'no loss', thermo-syphon type which incorporates an expansion bottle to accept coolant displaced by expansion. As the coolant cools, a partial vacuum is created in the system and the coolant is drawn back to replenish that previously displaced.
2 The system has a belt-driven coolant pump, front mounted radiator, thermostat and pressurised radiator cap.
3 Later 1500 models have the cooling fan connected to a viscous coupling which limits the fan speed at high engine revolutions. This works in a similar manner to a torque converter, and provides a 'slipping clutch' effect; its aim is to reduce noise and engine loading.
4 North American models (1979 on) have an inclined radiator and an electrically driven fan, with hose connection to the automatic choke.
5 The system functions in the following fashion. Cold water in the bottom of the radiator circulates up the lower radiator hose to the water pump where it is pushed round the water passages in the cylinder block, helping to keep the cylinder bores and pistons cool.
6 The water then travels up into the cylinder head and circulates round the combustion spaces and valve seats absorbing more heat, and then, when the engine is at its proper operating temperature, travels out of the cylinder head, past the open thermostat into the upper radiator hose and so into the radiator header tank.
7 The water travels down the radiator where it is rapidly cooled by the in-rush of cold air through the radiator core, which is created by both the fan and the motion of the car. The water, now cold, reaches the bottom of the radiator, when the cycle is repeated.
8 When the engine is cold the thermostat (which is a valve which opens and closes acording to the temperature of the water) maintains the circulation of the same water in the engine. Only when the correct minimum operating temperature has been reached, as shown in the specification, does the thermostat begin to open, allowing water to return to the radiator.
9 The car interior heater is supplied with coolant from the engine cooling system and the assembly is mounted underneath the fascia panel. Air directional controls are incorporated and a variable speed booster fan is an integral feature.
10 Apart from Mk I models, the inlet manifold is heated by the engine

cooling system. On Stromberg carburettors with an automatic choke, the choke is coolant heated (see Chapter 3).

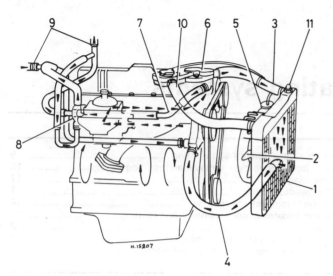

Fig. 2.1 Radiator and cooling circuit (N. America 1979 on) (Sec 1)

1 Radiator
2 Electric fan
3 Top hose
4 Bottom hose
5 Expansion bottle
6 Thermostat
7 Pipe to inlet manifold
8 Automatic choke heater pipe
9 Heater flow and return pipes
10 Air bleed hose
11 Temperature switch (electric fan)

2.1 Translucent expansion bottle

2 Maintenance

1 Every week, visibly check the coolant level through the translucent reservoir in front of the radiator (except N. America) or on the wing valance (N. America). The level should be at the half way mark, if not, top it up (photo).
2 At the intervals specified in 'Routine Maintenance' check the drivebelt tension and adjust if necessary (see Section 12).

3 Cooling system – draining and refilling

All models except N. America 1979 on

1 With the car on level ground drain the system as follows.
2 If the engine is cold remove the filler cap from the radiator by turning the cap anti-clockwise. If the engine is hot having just been run, then turn the filler cap very slightly until the pressure in the system has had time to disperse. Use a rag over the cap to protect your hand from escaping steam. If, with the engine very hot, the cap is released suddenly, the drop in pressure can result in the water boiling. With the pressure released the cap can be removed (photo).
3 If antifreeze is in the radiator drain it into a clean bucket or bowl for re-use.
4 Set the heater control lever to HOT.
5 Release the radiator drain tap or plug. On later models, a drain plug is not fitted and the radiator bottom hose will have to be disconnected to drain the coolant.
6 Unscrew and remove the drain plug from the side of the cylinder block (photo).
7 When the water has finished running, probe the drain tap orifices with a short piece of wire to dislodge any particles or rust or sediment which may be blocking the taps and preventing all the water draining out.
8 Provided the correct coolant mixture (see Section 4) has been maintained, rust and corrosion should not occur and the system may be refilled without any further action. If, however, the changing of the coolant has been neglected then a cold water hose should be placed in the radiator filler neck and the system flushed through until the water flows clear.

3.2 Radiator pressure cap

3.6 Cylinder block drain plug

Chapter 2 Cooling and heating systems

9 In severe cases, a proprietary chemical cleanser may be used strictly in accordance with the manufacturer's instructions.
10 Filling is carried out through the radiator filler cap after all drain plugs and hoses have been refitted. Fill brim-full and replace the radiator cap (photo).
11 Fill the expansion bottle with a similar mixture to that used in the rest of the system. Check that the heater control is still at HOT.
12 Start the engine and run it for three minutes at 1000 rm.
13 Switch off the engine and top-up the expansion tank, if necessary, to the half-way mark.

N. American models 1979 on

14 The operations are generally similar to those just described except that the pressure cap is not fitted to the radiator, but to the housing adjacent to the thermostat housing cover.

4 Coolant mixtures

1 Apart from the protection against freezing conditions which the use of antifreeze provides, its use is essential to minimise corrosion in the cooling system, particularly aluminium components.
2 The cooling system should contain a minimum antifreeze percentage of 30% and it is recommended that this percentage is always maintained.
3 With long-life types of antifreeze mixtures, renew the coolant every two years. With other types, drain and refill the system every twelve months. Whichever type is used, it must be of the ethylene glycol type.
4 The following table gives a guide to protection against frost and corrosion under varying climatic conditions.

Antifreeze concentration	30%	35%	50%
Protection to	-16°C	-20°C	-36°C
	3°F	-4°F	-33°F

5 Always top-up the expansion tank with antifreeze made up in a similar concentration to that used for the main quantity of coolant.
6 Do not allow antifreeze to come into contact with the paintwork as it will damage it.
7 Never use cooling system antifreeze in the windscreen washer bottle.
8 Even where climatic condition do not require the use of antifreeze, a corrosion inhibitor should be used, never plain water. This will prevent rust and corrosion developing, in particular damage to the aluminium components of the engine.

5 Thermostat removal, testing and refitting

1 Drain about four pints (2.3 l) from the cooling system in order to bring the coolant level below the thermostat housing.
2 If the particular model has the coolant temperature switch screwed into the thermostat housing, disconnect the connecting wires.
3 Unscrew the two bolts and spring washers from the thermostat housing and lift the housing and gasket away. Take out the thermostat.
4 Test the thermostat for correct functioning by dangling it by a length of string in a saucepan of cold water together with a thermometer.
5 Heat the water and note when the thermostat begins to open. This temperature is stamped on the flange of the thermostat, and is also given in the Specifications.
6 Discard the thermostat if it opens too early. Continue heating the water until the thermostat is fully open. Then let it cool down naturally. If the thermostat will not open fully in boiling water, or does not close down as the water cools, then a new unit must be fitted.
7 If the thermostat is stuck open when cold this will be apparent when removing it from the housing.
8 Replacing the thermostat is a reversal of the removal procedure. Remember to use a new gasket between the thermostat housing elbow and the thermostat.

6 Electric fan and switch (later N. American models) – removal and refitting

1 Disconnect the fan motor leads at the multi-plug connector.
2 Unbolt the fan/cowl assembly from the radiator and remove it.

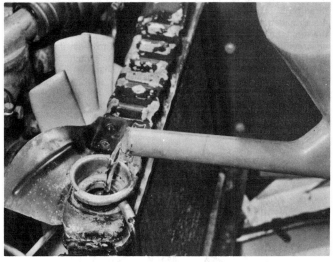

3.10 Filling the cooling system

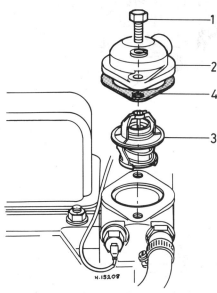

Fig. 2.2 Thermostat (except N. America 1979 on) (Sec 5)

1 Setscrew
2 Thermostat housing cover
3 Thermostat
4 Gasket

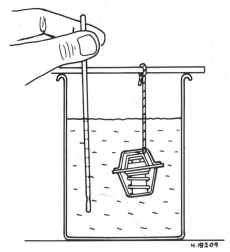

Fig. 2.3 Testing a thermostat (Sec 5)

3 The cowl and fan motor may be separated after unscrewing the three nuts.
4 Provided the cooling system is cold and pressure released, the thermostatic switch, which controls the radiator cooling fan, may be unscrewed from the radiator without the need to drain the coolant.
5 Disconnect the electrical leads from the switch, support the radiator header tank to prevent twisting and unscrew the switch.
6 Refitting is a reversal of removal, make sure that the switch sealing washer is in good order.

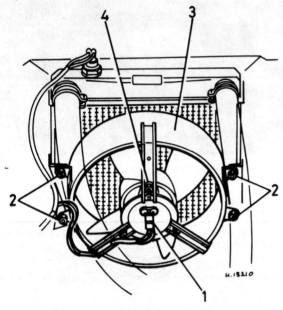

Fig. 2.4 Radiator electric fan arrangement (N. America 1979 on) (Sec 6)

1 Multi-plug
2 Fan cooling mounting bolts
3 Fan cooling
4 Fan motor mounting nuts

7.2 Releasing top hose from thermostat housing

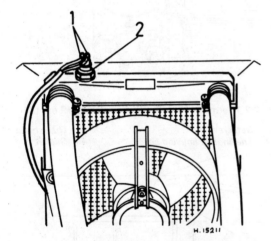

Fig. 2.5 Electric fan thermo switch (Sec 6)

1 Electrical leads
2 Switch

7.4 Releasing radiator from side supports

7 Radiator – removal, repair and refitting

All models except N. America (1979 on)

1 To remove the radiator first drain the cooling system as described in Section 3.
2 Then undo the clip which holds the top water hose to the thermostat pipe outlet. Disconnect the horn wires at the snap connector where the horns are hung on the radiator securing bolts (photo).
3 Pull the top water hose off the thermostat elbow, and then undo

7.5 Removing radiator

Chapter 2 Cooling and heating systems

the clip on the bottom hose and pull end of the hose off the radiator.
4   Undo and remove the bolts, and washers which hold the radiator in place. On later models the radiator is attached to side valances. Unhook the bonnet tension springs from their anchor brackets and undo the bolts holding each side of the radiator to the valance (photo).
5   Lift the radiator up out of the engine compartment (photo).

N. American models (1979 on)
6   Disconnect the battery and drain the cooling system.
7   Disconnect the coolant hoses from the radiator.
8   Disconnect the electrical leads which run to the cooling fan and the thermostatic switch at their multi-plug connectors.
9   Dismount the fuel evaporative control system charcoal canister (see Chapter 3).
10 Dismount the cooling system expansion bottle.
11 Release the engine side valances from the radiator.
12 Unscrew and remove the four radiator mounting bolts and lift the radiator out of the engine compartment.
13 When the radiator is out of the car it is advantageous to turn it upside down for reverse flushing. Clean the exterior of the radiator by hosing down the radiator matrix with a strong jet of water to clear away road dirt, dead flies etc.
14 Inspect the radiator hoses for cracks, internal or external perishing, and drainage caused by overtightening of the securing clips. Replace the hoses as necessary. Examine the radiator hose securing clips and renew them if they are rusted or distorted.
15 Soldering a leaking radiator is a skilled job and best left to radiator repair specialists. They will usually supply a new or reconditioned unit on an exchange basis.

8 Coolant pump – removal and refitting

1   If the water pump is badly worn normal practice is to fit an exchange reconditioned unit. Drain the cooling system as described in Section 3 and slacken the generator mounting bolts so the fan can be removed.
2   Undo the clips which hold the top and bottom water hoses to the pump body. On early models fitted with a heater, free the return pipe by undoing the union nut on the rear of the pump body. On later models with a heated inlet manifold free the small hose from the pump which leads to the manifold.
3   Pull off the temperature transmitter wire (where fitted) from its Lucar connector.
4   On later N. American models, disconnect the vacuum pipe from the front vapour trap which connects to the distributor and release the pipe union nut at the rear of the pump.
5   Undo the three bolts and washers which hold the pump body to the front of the cylinder block. Note that the bolts are all of different length. Lift the combined fan and coolant pump away from the engine.
6   The coolant pump may be separated from the housing after unscrewing and removing the securing nuts.
7   Refitting is a reversal of removal, use new gaskets and tension the drivebelt as described in Section 12.

9 Fan blades – removal and refitting

1   Remove the complete coolant pump with housing as described in the preceding Section.
2   Remove the fan mounting bolts or nuts (viscous coupling) and lift

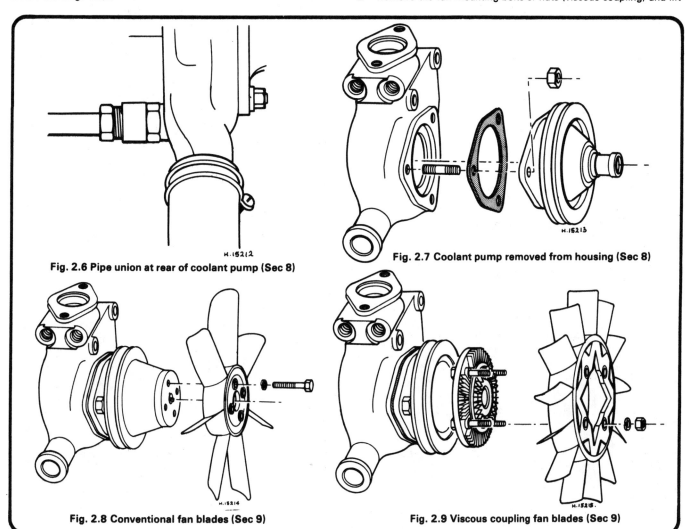

Fig. 2.6 Pipe union at rear of coolant pump (Sec 8)

Fig. 2.7 Coolant pump removed from housing (Sec 8)

Fig. 2.8 Conventional fan blades (Sec 9)

Fig. 2.9 Viscous coupling fan blades (Sec 9)

the fan from the mounting hub.
3   Refitting is a reversal of removal, but only tighten the fixing nuts or bolts to the specified torque.

10 Fan viscous coupling – removal and refitting

1   Disconnect the battery earth lead.
2   Drain the cooling system. If antifreeze is in the system then catch the coolant in a suitable receptacle for use again.
3   On North American models, slacken the air pump drivebelt and remove the drivebelt.
4   Slacken the alternator mounting bolts and remove the fan belt.
5   Remove the coolant pump housing complete with the fan assembly.
6   Remove the four nuts and spring washers which secure the fan to the unit.
7   If necessary, support the rear of the unit as close to the centre as possible, and press off the viscous coupling. Remove the tolerance ring.
8   Refitting is the reverse of the removal procedure, but always use a new tolerance ring.

11 Coolant temperature sender unit

1   If the temperature gauge fails to work either the gauge, the sender unit, the wiring or the connections are at fault.
2   It is not possible to repair the gauge or the sender unit and they must be replaced by new units if at fault.
3   First check the wiring connections and if sound check the wiring for breaks using an ohmmeter. The sender unit and gauge should be tested by substitution.
4   For details of how to remove and replace the temperature gauge see Chapter 10.
5   To remove the sender unit disconnect the battery, pull off the wire at the snap connector on the unit, and undo the unit with a spanner. On replacement renew the fibre washer to prevent the possibility of leaks developing.
6   It should be noted that with the engine at normal operating temperature, the temperature gauge needle will be three quarters of the way across the dial towards HOT.

12 Coolant pump drivebelt – adjustment

1   It is important to maintain the correct drivebelt tension at all times for two reasons (i) to ensure efficient cooling and (ii) to keep the battery fully charged.
2   The belt is correctly tensioned when its total deflection is ¾ in (19.0 mm) at the mid-point of the longest run of the belt, using moderate finger pressure.
3   To tension the belt, release the generator mounting and adjuster link bolts so that the unit pivots stiffly. Use a length of soft wood applied between the engine and the generator (drive end bracket only on alternator) as a lever to swivel the generator until belt tension is correct then tighten the mounting and adjuster link bolts.

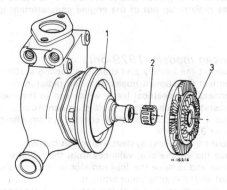

Fig. 2.10 Viscous coupling removed (Sec 10)

1   Pulley
2   Tolerance ring
3   Viscous coupling

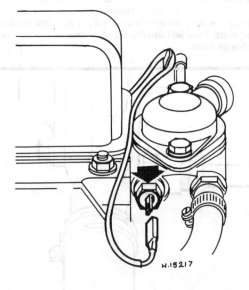

Fig. 2.11 Typical coolant temperature sender unit – arrowed (Sec 11)

10.6a Unscrewing fan bolt from viscous unit

10.6b Holding fan nut with second spanner

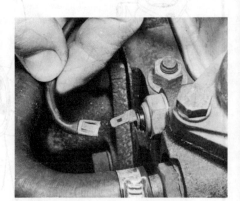

11.5 Coolant temperature sender unit

Chapter 2 Cooling and heating systems

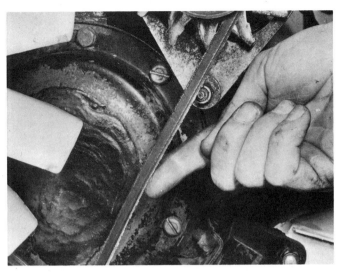

12.2 Checking coolant pump/alternator drivebelt tension

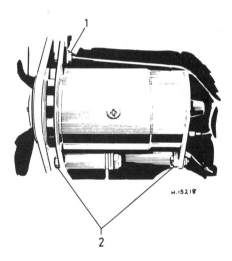

Fig. 2.12 Dynamo adjuster bolt and mounting bolts (arrowed) (Sec 12)

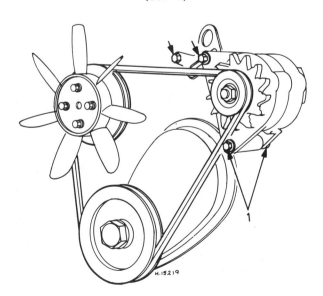

Fig. 2.13 Alternator adjuster bolt and mounting bolts (arrowed) (Sec 12)

13 Coolant pump drivebelt – renewal

Except N. America with air pump

1 Release the generator mounting and adjuster link bolts and push the generator in towards the engine as far as it will go.
2 Rotate the fan pulley and slip the belt from it by riding the belt over the pulley rims with the fingers.

N. America with air pump

3 On these models, the operations are similar to those just described, but the air pump drivebelt must be removed first as described in Chapter 3.
4 Refitting is a reversal of removal. Tension the drivebelts as described in Section 12 and Chapter 3.

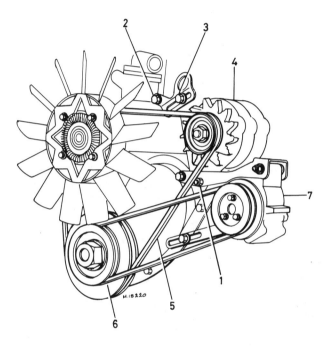

Fig. 2.14 N. American model drivebelt configuration (Sec 13)

1 Alternator mounting bolt
2 Alternator adjuster link bolt
3 Alternator adjuster link bolt
4 Alternator
5 Coolant pump drivebelt
6 Air pump drivebelt
7 Air pump

14 Heater – removal and refitting

1 Disconnect the battery negative lead.
2 Drain the cooling system as described earlier in this Chapter.
3 Disconnect both coolant hoses from the heater coolant valve on the engine compartment rear bulkhead (photo).
4 Remove the parcels shelf from the passenger side. This is simply a matter of extracting the fixing screws and bolts.
5 Extract the grub screws from the heater control lever knobs and pull off the knobs (photo).
6 Unscrew the four screws which hold the centre instrument in position and pull the parcel shelf from the fascia and allow it to remain suspended.
7 Unbolt the heater air control lever assembly (photo).
8 Disconnect the lead from the fan switch and the cable from the trunnion and clamp, but not before marking its setting with quick-drying paint.
9 Disconnect the electrical lead from the left-hand side of the heater blower motor.
10 Unclip and disconnect the two demister hoses from the heater casing.

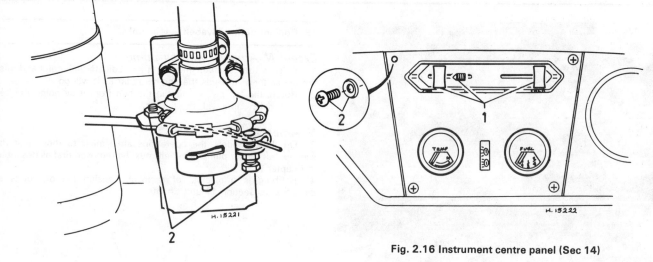

Fig. 2.15 Heater coolant temperature control valve (Sec 14)

Fig. 2.16 Instrument centre panel (Sec 14)

1 Knob grub screws 2 Panel fixing screws

14.3 Heater coolant valve

14.5 Releasing heater control knob grub screw with an Allen key

14.7 Heater control lever assembly

15.3 Heater booster motor mounting screws

Chapter 2 Cooling and heating systems

11 Working within the engine compartment unscrew and remove the four bolts which retain the heater assembly to the rear bulkhead.
12 Working inside the car, pull the heater towards you and remove it from under the fascia on the passenger side. Be prepared for some spillage of coolant, so protect the carpets accordingly.
13 Refitting is a reversal of removal, but all sealing flanges should be sealed with new gaskets and a bead of suitable mastic.

15 Heater booster motor – removal and refitting

1 The booster motor may be removed from the main heater assembly independently.
2 Lower the centre instrument panel as described in the preceding Section and reach through the aperture and pull off the wiring connection from the heater motor terminals.
3 Unscrew and remove the heater motor mounting screws (photo).
4 Withdraw the motor from the heater casing and out from under the fascia panel on the driver's side.
5 The fan wheel may be removed from the motor shaft by releasing

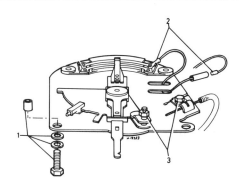

Fig. 2.17 Air control lever (Sec 14)

1 Mounting screw
2 Booster fan switch leads
3 Cable clamp and trunnion

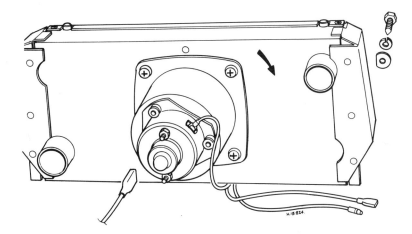

Fig. 2.18 Heater unit (Sec 14)

the retaining clip with a screwdriver.
6 Reassemble the fan wheel to the shaft so that a $\frac{1}{4}$ in (6.0 mm) clearance exists between the wheel and the motor casing. Otherwise refitting is a reversal of removal.

16 Coolant control valve – removal and refitting

1 Drain the cooling system and then disconnect the coolant hoses from the control valve.

2 Mark the setting of the cable in relation to the clamp and trunnion using quick drying paint. Disconnect the cable from the valve.
3 Unbolt the valve from its bracket.
4 Sometimes a leaking valve can be repaired by fitting a new stem seal.
5 Refitting is a reversal of removal, but in spite of the cable positioning marks made before removal, check that full movement of the fascia panel control lever does in fact move the valve from the fully closed to the fully open positions. If not, adjust the cable clamp or trunnion screws as necessary to achieve the correct movement.

17 Fault diagnosis – cooling system

Symptom	Reason(s)
Overheating Heat generated in cylinder, not being successfully disposed of by radiator	Insufficient coolant in cooling system Drivebelt slipping (accompanied by a shrieking noise on rapid engine acceleration) Radiator core blocked or radiator grille restricted Bottom coolant hose collapsed, impeding flow Thermostat not opening properly Ignition advance and retard incorrectly set (accompanied by

Chapter 2 Cooling and heating systems

loss of power, and perhaps, misfiring)
Carburettor incorrectly adjusted (mixture too weak)
Exhaust system partially blocked
Oil level sump too low
Blown cylinder head gasket (water/steam being forced down the radiator overflow pipe under pressure)
Engine not yet run-in
Brakes binding

Engine runs cool
Too much heat being dispersed by radiator

Thermostat jammed open
Incorrect grade of thermostat fitted allowing premature opening of valve
Thermostat missing

Loss of cooling water
Leaks in system

Loose clips on coolant hoses
Top, bottom, or by-pass coolant hoses perished and leaking
Radiator core leaking
Thermostat gasket leaking
Radiator cap spring worn or seal ineffective
Blown cylinder head gasket (pressure in system forcing water/steam down overflow pipe)
Cylinder wall or head cracked

Chapter 3 Fuel, carburation and emission control systems

Contents

Air cleaner (gauze element type) – servicing	4
Air cleaner (paper element type with temperature control) – servicing	6
Air cleaner (paper element type without temperature control) – servicing	5
Air cleaners – general	3
Carburettors – description	15
Carburettor (SU HS2) – overhaul	18
Carburettor (SU HS2) – removal and refitting	17
Carburettor (SU HS2) – tuning	16
Carburettors (SU HS2E, HS4) – overhaul	22
Carburettors (SU HS2E, HS4) – removal and refitting	21
Carburettors (SU HS2E) – tuning	19
Carburettors (SU HS4) – tuning	20
Carburettor (Stromberg 150 CDSE) – overhaul	25
Carburettor (Stromberg 150 CDSE) – removal and refitting	24
Carburettor (Stromberg CD4) – overhaul	31
Carburettor (Stromberg CD4T) – overhaul	33
Carburettor (Stromberg CD4) – removal and refitting	30
Carburettor (Stromberg CD4T) – removal and refitting	32
Carburettor (Stromberg CD4 and CD4T) – tuning	26
Carburettor (Stromberg 150 CDSE) – tuning	23
Emission control systems – description	34
Emission control system components – servicing, removal and refitting	35
Fault diagnosis – emission control systems	39
Fault diagnosis – fuel and exhaust systems	38
Fuel line filter – renewal	12
Fuel pump – cleaning	8
Fuel pump – description	7
Fuel pump – overhaul	11
Fuel pump – removal and refitting	10
Fuel pump – testing	9
Fuel tank – removal and refitting	13
Fuel tank transmitter – removal and refitting	14
General description	1
Maintenance	2
Manifolds and exhaust system	37
Stromberg CD4T automatic choke – removal, refitting and adjustment	29
Stromberg deceleration and by-pass valve – checking and adjusting	27
Stromberg CD4 manual choke – removal and refitting	28
Throttle and choke controls	36

Specifications

System type .... Rear fuel tank located in luggage boot. Mechanically-operated fuel pump, single or twin variable venturi carburettors. Full emission control on North American models

Fuel tank capacity
Mk II, III, IV .... 8.25 Imp gal, 9.9 US gal, 37.5 l
1500 .... 7.25 Imp gal, 8.7 US gal, 32.9 l

Fuel octane
High compression engines (8.5:1 – 9.0:1) .... 97 octane – 4 star
Low compression engines (7.0:1 – 7.5:1) .... 91 octane – 2 star
With catalytic converter .... Unleaded 91 octane

Fuel pump
Type .... Mechanically-operated

Carburettors
Mk I, II and III .... Twin SU HS2
Mk IV (except N. America) .... Twin SU HS2E
Mk IV (N. America) .... Single Stromberg 1.50 CDSE
1500 (except N. America) .... Twin SU HS4
1500 (N. America):
To 1977 (without catalytic converter) .... Single Stromberg 150 CD4
       (with catalytic converter – California) .... Single Stromberg 150 CD4T
1978 to 1981 .... Single Stromberg 150 CD4T

Carburettor calibration and setting data

SU HS2
Jet needle	Mk II AN, Mk III BO, Mk IV AN or AAN
Main jet	0.090 in
Idle speed	750 to 800 rpm
Fast idle speed	1100 to 1300 rpm

SU HS2E
Jet needle	AAN
Main jet	0.090 in
Idle speed	800 to 850 rpm
Fast idle speed	1100 to 1300 rpm

Stromberg 1.50 CDSE
Jet needle (to engine 25 000)	B5AV
(from engine 25 001)	B5CH
Main jet	0.090 in
Idle speed	800 to 850 rpm
Fast idle speed	1100 to 1300 rpm
Float height	16.0 to 17.0 mm

SU HS4
Jet needle	ABT
Idle speed	650 to 850 rpm
Fast idle speed	1100 to 1300 rpm

Stromberg 150 CD4
Jet needle (to 1978)	BIDL
(from 1978)	45P
Idle speed	700 to 900 rpm
Fast idle speed	1700 to 1900 rpm
CO at idle (nominal)	
To 1976	1.5%
1977	3.0%
1978 on	5.0%

Stromberg 150 CD4T
Jet needle (to 1977)	BIDL
(1977)	45L
(1978 on)	45N
Idle speed	700 to 900 rpm
Fast idle speed	1700 to 1900 rpm
CO at idle (nominal)	
To 1976	1.5%
1977	3.0%
1978 on	5.0%

Torque wrench settings
	lbf ft	Nm
Fuel pump to cylinder block nuts	14	19
Inlet manifold to exhaust manifold	14	19
Manifold to cylinder head nuts	24	33
Air cleaner to baseplate	8	11
Air cleaner to carburettor	8	11
Carburettor mounting nuts	14	19

1 General description

1 The fuel system on all models of the Spitfire consists of a fuel tank mounted behind the seats and at the front of the boot with a central filler; a mechanical fuel pump mounted on the left-hand side of the engine and operated by the camshaft; the necessary fuel lines between the tank and the pump, and the pump and the carburettor; single or twin variable venturi carburettors with air cleaner.

2 On North American versions, emission control systems are fitted and are covered in later Sections of this Chapter.

2 Maintenance

1 At the intervals specified, unscrew the carburettor dashpot damper caps and withdraw the cap with damper piston.

2 Top-up the damper guide with suitable oil to within a quarter of an inch of its rim (photo).

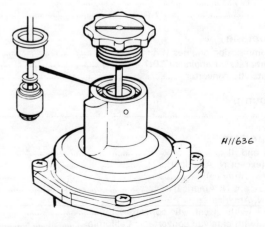

Fig. 3.1 Carburettor dashpot damper (typical) (Sec 2)

Chapter 3 Fuel, carburation and emission control systems

2.2 Topping-up carburettor damper

5.1a Air cleaner intake hoses

5.1b Removing air cleaner from carburettor flanges

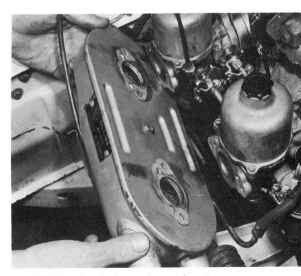

5.1c Withdrawing air cleaner from carburettors

3 Refit the damper. Some pressure (resistance) may be felt as this is done. This is normal.
4 Periodically, check the hoses and unions of the fuel lines for condition and security.

3 Air cleaners – general

The type of air cleaner fitted will depend upon the date of production of the car, the number of carburettors fitted and the market for which it was made.

4 Air cleaner (gauze element type) – servicing

1 The carburettors on early models of the Spitfire were fitted with expanded gauze air cleaners.
2 Unsrew the air cleaner cover bolts, remove the cover and extract the gauze element.
3 Wash the element in petrol, allow it to dry and then dip it in engine oil.
4 Allow the oil to drain from it and refit to the air cleaner casing.
5 Refit the cover and bolts.

5 Air cleaner (paper element type without temperature control) – servicing

1 At the intervals specified in Routine Maintenance, pull the convoluted air intake hoses from the air cleaner. Unscrew and remove the air cleaner cover bolts and take off the air cleaner complete. Unscrew the bolt which holds the cover and baseplate together (photos).
2 Take out the element(s) and tap them on a hard surface to remove adhering dust (photo).
3 If a compressed supply is available, blow the elements through from the centre hole outwards. Refit the element(s).
4 Wipe out the casing and refit, making sure that the sealing gasket is in good order and the cover alignment peg engages in its baseplate notch.
5 Check that the carburettor flange gaskets are unbroken. Renew them if necessary.
6 The operations are identical when renewing the element, as are those at the service intervals specified in Routine Maintenance at the beginning of this Chapter.

5.1d Removing air cleaner cover screw

5.2 Air cleaner elements exposed

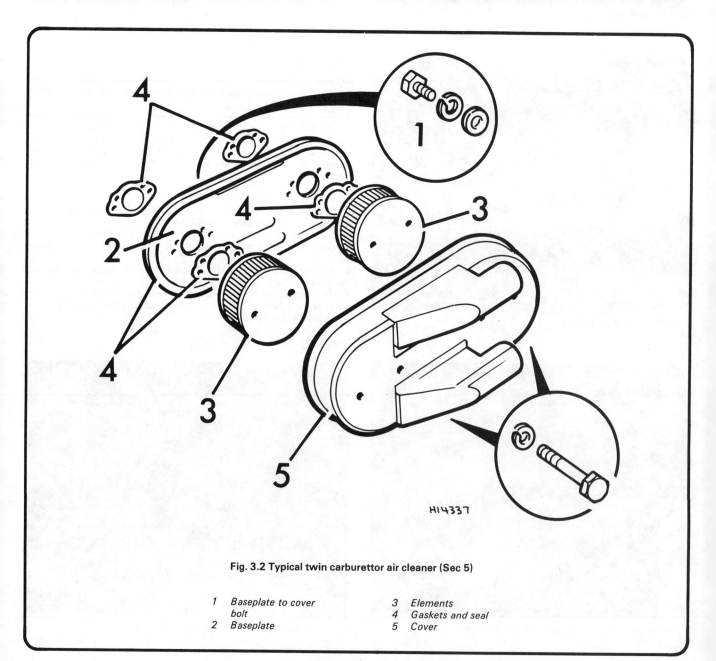

Fig. 3.2 Typical twin carburettor air cleaner (Sec 5)

1 Baseplate to cover bolt
2 Baseplate
3 Elements
4 Gaskets and seal
5 Cover

Chapter 3 Fuel, carburation and emission control systems

6 Air cleaner (paper element type with temperature control) – servicing

1 The air cleaner fitted in conjunction with a single Stromberg carburettor to North American models incorporates an air temperature control unit which regulates the air flow from hot and cold sources to maintain a set air temperature at the carburettor intake.
2 To gain access to the element for cleaning or renewal, disconnect the emission control hoses from the rear face of the air cleaner baseplate.
3 Disconnect the hot air hose from the air cleaner intake.
4 Unscrew the two bolts which hold the air cleaner to the carburettor flange.
5 Withdraw the air cleaner and separate the baseplate from the cover by inserting a screwdriver at its joint.
6 Take out the element. Either clean it (see Section 5, paragraphs 2 and 3) or renew it.
7 Wipe out the casing and check that the gaskets are in good order.
8 Reassembly and refitting are reversals of removal and dismantling, but make sure that the baseplate alignment peg engages in the notch in the cover. The temperature control valve is accessible in the following way.
9 Pull off the hot air hose and elbow connector.
10 Slacken the hose clip securing the air intake control valve to the air cleaner joining hose. Pull the air intake control valve assembly out.
11 Refitting is the reverse of the dismantling procedure.

7 Fuel pump – description

1 The mechanically operated fuel pump is actuated through a spring loaded rocker arm. One arm of the rocker bears against an eccentric on the camshaft and the other arm operates a diaphragm pull rod.
2 As the engine camshaft rotates, the eccentric moves the pivoted rocker arm octwards which in turn pulls the diaphragm pull rod and the diaphragm down against the pressure of the diaphragm spring.
3 This creates sufficient vacuum in the pump chamber to draw in fuel from the tank through the fuel filter gauze and non-return valve.
4 The rocker arm is held in constant contact with the eccentric by an anti-rattle spring and as the engine camshaft continues to rotate the eccentric allows the rocker arm to move inwards. The diaphragm spring is thus free to push the diaphragm upwards forcing the fuel in the pump chamber out to the carburettor through the non-return outlet valve.
5 When the float chamber in the carburettor is full the float chamber needle valve will close so preventing further flow from the fuel pump.
6 The pressure in the delivery line will hold the diaphragm downwards against the pressure of the diaphragm spring, and it will remain in this position until the needle valve in the float chamber opens to admit more petrol.

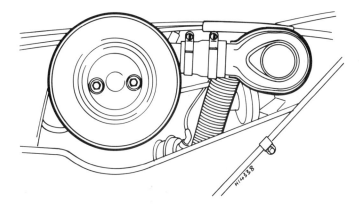

Fig. 3.3 Single Stromberg type air cleaner with temperature control (Sec 6)

10.1 Fuel pump hose

8 Fuel pump – cleaning

1 At the intervals specified in Routine Maintenance, undo the bolt in the centre of the cover and lift off the cover.
2 Inspect the filter gauze for sediment and lift it out and clean it with petrol and a soft brush if dirty.
3 Check the condition of the gasket and renew if it has hardened or broken. Replacement is a straightforward reversal of the removal sequence – do not forget to refit the fibre washer under the head of the retaining bolt and tighten the bolt just enough for the cover to make a leak-proof joint with the gasket.

9 Fuel pump – testing

Assuming that the fuel lines and unions are in good condition and that there are no leaks anywhere, check the performance of the fuel pump in the following manner. Disconnect the fuel pipe at the carburettor inlet union, and the high tension lead to the coil, and with a suitable container or a large rag in position to catch the ejected fuel, turn the engine over on the starter motor solenoid. Good regulator spurts of petrol should be ejected.

10 Fuel pump – removal and refitting

1 Remove the fuel inlet and outlet pipes by unscrewing the union nuts or by pulling off the hoses according to type (photo).
2 Undo the two nuts and spring washers which hold the pump to the crankcase. Note the special nut at the rear of the pump with a slotted head. Ensure it is replaced on the correct stud, ie the stud nearest the rear of the engine.
3 Lift the pump together with the gasket away from the crankcase.
4 Refitting of the pump is a reversal of the above process. Remember to use a new crankcase to fuel pump gasket to ensure no oil leaks, ensure that both faces of the flange are perfectly clean, and check that the rocker arm lies on top of the camshaft eccentric and not underneath it.

11 Fuel pump – overhaul

1 The various fuel pumps on earlier models can be dismantled and repaired, but later pumps are of sealed type and any dismantling is limited to removal of the cover for cleaning the filter (refer to Section 8).
2 Unscrew the securing bolt from the centre of the cover and lift the cover away. Note the fibre washer under the head of the bolt.
3 Remove the sealing washer and the fine mesh filter gauze.

Chapter 3 Fuel, carburation and emission control systems

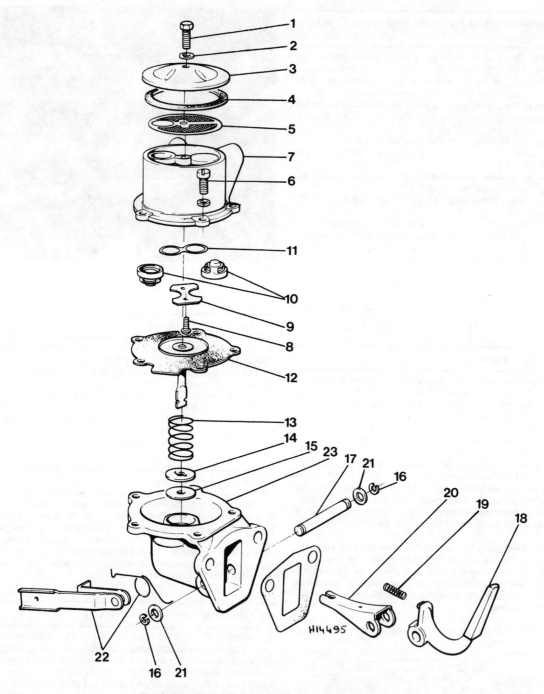

Fig. 3.4 Exploded view of early type fuel pump (Sec 11)

1 Cover screw	7 Pump upper body	13 Spring	19 Rocker arm spring
2 Washer	8 Screws	14 Washer	20 Operating fork
3 Cover	9 Retainer	15 Gland washer	21 Spacer
4 Gasket	10 Valves	16 Pivot pin retainer	22 Hand priming lever
5 Filter gauze	11 Upper retainer	17 Pivot pin	23 Pump lower body
6 Screws	12 Diaphragm	18 Rocker arm	

4 If the condition of the diaphragm is suspect or for any other reason it is wished to dismantle the pump fully, proceed as follows. Mark the upper and lower flanges of the pump that are adjacent to each other. Unscrew the five screws and spring washers which hold the two halves of the pump body together. Separate the two halves with great care, ensuring that the diaphragm does not stick to either of the two flanges.

5 Unscrew the screws which retain the valve plate and remove the plate and gasket together with the inlet and outlet valves. (Some later pumps have a simplified valve plate arrangement which is released by one screw).

6 Press down and rotate the diaphragm a quarter of a turn (in either direction) to release the pull rod from the operating lever, and lift away the diaphragm and pull rod which is securely fixed to the diaphragm and cannot be removed from it. Remove the diaphragm spring and the metal and fibre washer underneath it.

7 If it is necessary to dismantle the rocker arm assembly, remove the retaining circlips and washer from the rocker arm pivot and slide out

Chapter 3 Fuel, carburation and emission control systems

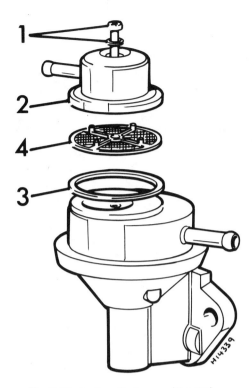

Fig. 3.5 Later type fuel pump (Sec 11)

1 Cover screw and washer
2 Cover
3 Seal
4 Filter screen

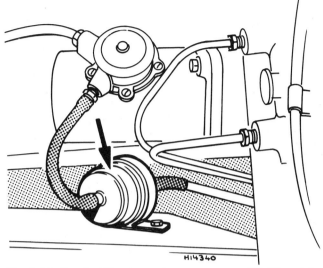

Fig. 3.6 In-line fuel filter located next to pump on Mk III models (Sec 12)

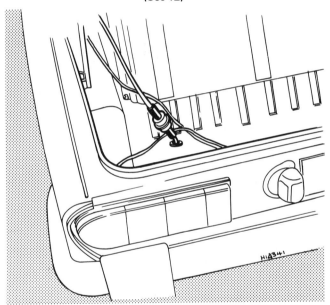

Fig. 3.7 In-line filter located near fuel tank (Sec 12)

the rod which will then free the rocker arm, operating rod, and anti-rattle spring.

8 Check the condition of the cover sealing washer, and if it is hardened or broken it must be replaced. The diaphragm should be checked similarly and replaced if faulty. Clean the pump thoroughly and agitate the valves in paraffin to clean them out. This will also improve the contact between the valve seat and the valve. It is unlikely that the pump body will be damaged, but check for fractures and cracks. Renew the cover if distorted.

9 To reassemble the pump proceed as follows. Replace the rocker arm assembly comprising the operating link, rocker arm, anti-rattle spring and washer in their relative positions in the pump body, and washers with the holes in the body and insert the pivot pin.

10 Refit the circlips to the grooves in each end of the pivot pin.

11 Earlier pumps used valves which had to be built up, while later versions used ready assembled valves which are merely dropped into place in the inlet and outlet ports. Ensure that the correct valve is dropped into each port.

12 Reassemble the earlier type of valve as follows. Position the delivery valve in place on its spring. Place the inlet valve in position in the pump body and then fit the spring. Place the small four legged inlet valve spring retainer over the spring with the legs positioned towards the spring.

13 Place the valve retaining gasket in position, replace the plate, and tighten down the three securing screws, or single screw in the case of later models. Check that the valves are working properly with a suitable piece of wire.

14 Position the fibre and steel washer in that order in the base of the pump and place the diaphragm spring over them.

15 Replace the diaphragm and pull rod assembly with the pull rod downwards and the small tab on the diaphragm adjacent to the centre of the flange and rocker arm.

16 With the body of the pump held so that the rocker arm is facing away from one, press down the diaphragm, turning it a quarter of a turn to the left at the same time. This engages the slot on the pull rod with the operating lever. The small tab on the diaphragm should now be at an angle of 90° to the rocker arm and the diaphragm should be firmly located.

17 Move the rocker arm until the diaphragm is level with the body flanges and hold the arm in this position. Reassemble the two halves of the pump ensuring that the previously made marks on the flanges are adjacent to each other.

18 Insert the five screws and lockwashers and tighten them down finger tight.

19 Move the rocker arm up and down several times to centralise the diaphragm, and then with the arm held down, tighten the screws securely in a diagonal sequence.

20 Replace the gauze filter in position. Fit the cover sealing washer, fit the cover, and insert the bolt with the fibre washer under its head. Do not over-tighten the bolt but ensure that it is tight enough to prevent any leaks.

12 Fuel line filter – renewal

1 Certain Mk III models are fitted with a disposable fuel line filter which must be renewed at the intervals specified in Routine Maintenance.

2 To remove the filter loosen the bracket bolt clamp and then pull off the flexible inlet (lower) hose and pull off the top hose from the other side of the filter unit, which can now be removed and thrown away.

Chapter 3 Fuel, carburation and emission control systems

3 Refitting is a straightforward reversal of the removal sequence. Ensure the arrow marked on the filter points to the outlet hose.
4 A similar filter is fitted to certain Mk IV and 1500 models. However, later cars have the filter fitted near the fuel tank. It is reached from the boot. Remove three screws holding the top of the trim pad hiding the tank and pull the pad out.
5 Prise the fuel pipes off the filter. Replace the filter with the end marked IN uppermost.

13 Fuel tank – removal and refitting

1 Disconnect the battery.
2 If the tank contains more than a small amount of fuel, syphon it out into a metal container which can be sealed.
3 Remove the trim panel from the luggage boot to expose the fuel tank (photo).
4 Disconnect the leads from the terminals of the tank transmitter unit. Identify the leads for ease of reconnection (photo).
5 Disconnect the filler hose by releasing the hose clips.
6 On North American vehicles, disconnect the hoses from the vapour separator (see Section 35).
7 Disconnect the fuel feed pipe from the top of the fuel tank.
8 Unscrew and remove the tank mounting bolts and withdraw the tank from the car (photo).
9 If the tank contains sediment, remove the transmitter unit (see Section 14) and then pour in some paraffin. Shake the tank vigorously and then let it drain. Repeat as necessary.
10 If the tank is leaking, a temporary repair may be carried out using one of the proprietary products available, but if a more permanent repair is necessary, leave it to the specialists (often radiator repairers). Never be tempted to weld or solder a fuel tank yourself owing to the danger from explosion. The tank will have to be steamed out for some time to remove fumes.
11 Refitting is a reversal of removal.

14 Fuel tank transmitter – removal and refitting

1 Disconnect the earth lead from the battery and the two wires from their Lucar blade terminals on the sender unit.
2 Undo the six screws which hold the sender unit to the tank, or on models where the sender unit is held in place with a retaining ring use a screwdriver to turn the ring anti-clockwise.
3 Carefully lift the complete unit away making sure that the float lever is not bent or damaged in the process.
4 Refitting of the unit is a reversal of the above process. To ensure a fuel-tight joint, clean the tank and sender gauge mating flanges, and always use a new joint gasket, together with sealing compound in the case of the screw-retained unit.

15 Carburettors – description

SU carburettors

1 The SU carburettor is a relatively simple instrument and is basically the same irrespective of its size and type. It differs from most other carburettors in that instead of having a number of various sized fixed jets for different conditions, only one variable jet is fitted to deal with all possible conditions.
2 Air passing rapidly through the carburettor venturi draws petrol from the jet so forming the petrol/air mixture. The amount of petrol drawn from the jet depends on the position of the tapered carburettor needle, which moves up and down the jet orifice according to engine load and throttle opening, thus effectively altering the size of the jet so that exactly the right amount of fuel is metered for the prevailing road conditions.
3 The position of the tapered needle in the jet is determined by engine vacuum. The shank of the needle is held at its top end in a piston which slides up and down the dashpot in response to the degree of manifold vacuum. This is directly controlled by the position of the throttle.
4 With the throttle fully open, the full effect of inlet manifold vacuum is felt by the piston which has an air bleed into the choke tube on the

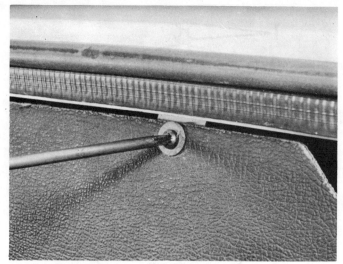
13.3 Fuel tank cover panel

13.4 Fuel tank sender unit

13.8 Fuel tank mounting bolt

Chapter 3 Fuel, carburation and emission control systems

outside of the throttle. This causes the piston to rise fully, bringing the needle with it. With the accelerator partially closed only slight inlet manifold vacuum is felt by the piston, and the piston only rises a little, blocking most of the jet orifice with the metering needle.

5 To prevent the piston fluttering, and to give a richer mixture when the accelerator is suddenly depressed, an oil damper and light spring are fitted inside the dashpot.

6 The only portion of the piston assembly to come into contact with the piston chamber or dashpot is the actual central piston rod. All the other parts of the piston assembly, including the lower portion, have sufficient clearances to prevent any direct metal to metal contact which is essential if the carburettor is to work properly.

7 The correct level of the petrol in the carburettor is determined by the level of the float in the float chamber. When the level is correct the float rises and by means of a lever resting on top of it closes the needle valve in the cover of the float chamber. This closes off the supply of fuel from the pump. When the level in the float chamber drops as fuel is used in the carburettor the float sinks. As it does, the float needle comes away from its seat so allowing more fuel to enter the float chamber and restore the correct level.

8 In order to reduce exhaust emission levels, later HS2E and HS4SU carburettors have been modified.

9 The changes were not all introduced at the same time, but basically consist of the spring biased needle described earlier, over-run valves in the throttle, and jet adjustment restrictors.

10 The spring biased needle is supplied complete with shouldered spring seats. No attempt should be made to alter the position of the spring or to convert a fixed type needle to spring-loaded applications. The raised pip form in the needle guide ensures the needle is correctly centralised. This pip should not be removed or repositioned. The needle with its spring and in its guide should be assembled into the piston so that the lower edge of the guide is flush with the face of the piston. The needle must be at the correct axis in the piston. Some guides have a line etched on their face. This must be between the two piston transfer holes. Alternative needle guides have a flat machined on them which must be positioned so that the locking screw tightens down onto it. If the guide is incorrectly positioned the locking screw will stick out from the piston. Guide locking screws for spring loaded needles are shorter than the ones for normal fixed needles.

11 The spring over-run valve in the throttle opens under the influence of the great manifold depression that occurs when the engine is slowing down on over-run. Otherwise such conditions provoke excessive emissions. The valve should limit manifold depression to about 21 in Hg compared with 24-25 in Hg without the valve. If the valve is not seating correctly the idle speed will be faster than normal. A proper check can be made using a vacuum gauge, noting the reading as the engine slows from high speed after snapping the throttle shut.

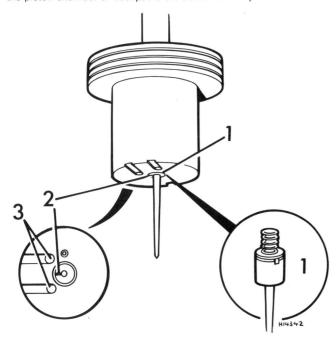

Fig. 3.8 Spring biased needle on SU carburettor (Sec 15)

1 Flush fitting of needle
2 Etch mark
3 Transfer holes

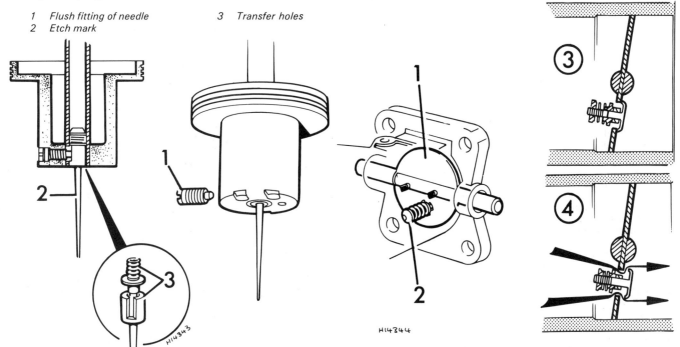

Fig. 3.9 SU needle securing arrangement (Sec 15)

1 Grub screw
2 Metering needle
3 Spring and guide

Fig. 3.10 Throttle over run valve (Sec 15)

1 Throttle butterfly valve
2 Over run valve
3 Over run valve closed
4 Over run valve open

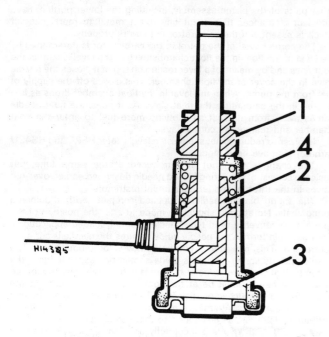

Fig. 3.11 Sectional view of temperature compensator (Sec 15)

1 Jet adjusting nut
2 Jet head
3 Thermal element
4 Control spring

12 On HS4 carburettors, a temperature compensator is incorporated on both carburettors at the jet heads. The compensating device comprises a thermal element and control spring contained in a housing around the die-cast jet head. The thermal element is a sealed unit containing heat-sensitive wax which expands as temperature rises and through a piston, reduces the effective area of the jet.

Stromberg carburettors

13 The carburettors are fitted to N. American versions in order to meet emission control regulations.
14 Basically the Stromberg type of carburettor works on the same principal as the SU, a piston acting as an air valve, and withdrawing a tapered needle from the fixed jet to vary the fuel. However, there are a number of important differences. The Stromberg piston has a flexible diaphragm to seal the depression in the air chamber of the dashpot. The jet is fixed, but the needle is movable for height in the piston to adjust the mixture.

16 Carburettor (SU HS2) – tuning

1 Before carrying out any adjustment to the carburettors, make sure that the valve clearances and ignition settings are correct.
2 Have the engine at normal operating temperature and remove the air cleaners.
3 Slacken the clamp bolts on the throttle spindle connections.
4 Close the throttle valve plates by unscrewing the throttle speed screws until the ends of the screws are just in contact with the cam levers. Now turn each screw in one and a half turns.
5 Remove the retaining screws and withdraw the dashpots and pistons.
6 Screw up the jet adjusting nut on each carburettor until each jet is flush with the carburettor bridge.
7 Refit the piston/dashpot assemblies and check that the pistons fall freely after the lift pin has been pressed upward with the finger. If the piston tends to stick, then the jet will require centring. To do this remove the union holding the nylon feed tube to the base of the jet, together with the jet and jet adjusting nut securing spring, after removing the link between the jet head and lever.
8 Replace the jet and nylon feed tube and press them up under the

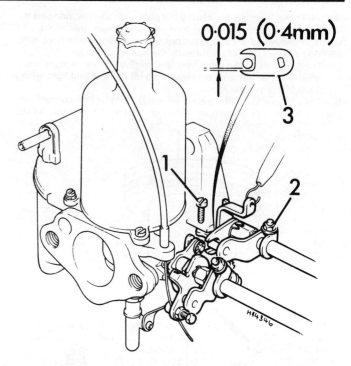

Fig. 3.12 SU carburettor adjustment points (Sec 16)

1 Throttle speed screw
2 Throttle spindle clamp bolts
3 Throttle lever interconnecting pin setting

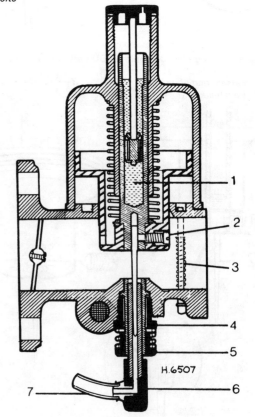

Fig. 3.13 Sectional view of SU carburettor (Sec 16)

1 Oil damper reservoir
2 Needle securing screw
3 Piston lifting pin
4 Jet locking nut
5 Jet adjusting nut
6 Jet head
7 Feed tube

head of the large hexagonal jet locking nut. Unscrew this nut slightly until the jet bearing can be turned.

9 Remove the damper securing nut and damper from the top of the dashpot and push the piston assembly right down so that the metering needle enters fully into the jet.

10 Tighten the jet locking nut and test the piston assembly to check that the needle is still quite free to slide in the jet orifice. On lifting the piston and then releasing it the piston should hit the inside jet bridge with a soft metallic click, and the intensity of the click should be the same whether the jet is in its normal position or is fully lowered.

11 If the sound is different when the jet is fully lowered then the jet is not yet properly centralised and the process must be repeated.

12 When all is correct, remove the jet, replace the jet adjusting nut securing spring, the adjusting nut and jet, and the link between the jet head and the lever.

13 Set the jet flush with the bridge and re-check the piston fall as previously described.

14 Now unscrew (turn downwards) the jet adjusting nuts two complete turns.

15 Start the engine and turn the throttle speed screws an equal amount until the idle speed conforms to that specified.

16 Now set the throttle lever interconnecting lever pins so that there is a gap of 0.015 in (0.4 mm) between the pin and the lower prong of the fork. Also make sure that there is a side clearance of $\frac{1}{32}$ in (0.8 mm) between the interconnecting levers and the throttle nuts.

17 The jet adjusting nuts should now be screwed up or down by an equal amount until the fastest idle speed is obtained consistent with even running. While the nuts are being turned, keep the jets pressed upwards to ensure that they are in contact with the nuts.

18 Reduce the idle speed to the specified level by unscrewing the throttle speed screws equally.

19 Now raise the piston lifting pin on the front carburettor, but only through a distance of $\frac{1}{32}$ in (0.8 mm).

20 If the engine speed increases, the mixture strength of the front carburettor is too rich; if the speed decreases the mixture is too weak.

21 When it is found that the speed increases very slightly, but momentarily, and then returns to normal, the mixture is correctly adjusted.

22 Check the mixture in the same way for the rear carburettor.

23 If adjustment has been carried out correctly, the exhaust note should be even and regular. The appearance inside the end of the exhaust tail pipe will give an indication of tune after a few miles running. A black sooty appearance indicates a rich mixture, while a light greyish colour indicates a correct mixture. White indicates a weak mixture.

24 A further tuning refinement is to synchronise the twin carburettors. With the mixture correctly set, the idling suction must be equal on both. It is best to use a vacuum synchronising device available from motor parts shops. If this is not available, it is possible to obtain fairly accurate synchronisation by listening to the hiss made by the air flow into the intake throats of each carburettor.

25 The aim is to adjust the throttle butterfly disc so that an equal amount of air enters each carburettor. Loosen the clamping bolts on the throttle spindle connections. Listen to the hiss from each carburettor and if a difference in intensity is noticed between them, then unscrew the throttle adjusting screw on the other carburettor until the hiss from both the carburettors are the same.

26 With a vacuum synchronisation device all that it is necessary to do is to place the instrument over the mouth of each carburettor in turn and adjust the adjusting screws until the reading on the gauge is identical for both carburettors.

27 Tighten the clamping bolts on the throttle spindle connections which connect the throttle disc of the two carburettors together, at the same time holding down the throttle adjusting screws against their idling stops. Synchronisation of the two carburettors is now complete.

28 Finally with the engine idling and the choke fully off, adjust the fast idle screw to give a clearance of 0.015 in (0.4 mm) between the end of the screw and the rocker lever.

17 Carburettor (SU HS2) – removal and refitting

1 Referring to Fig. 3.15 take off the air cleaners and free the breather pipe (where fitted).

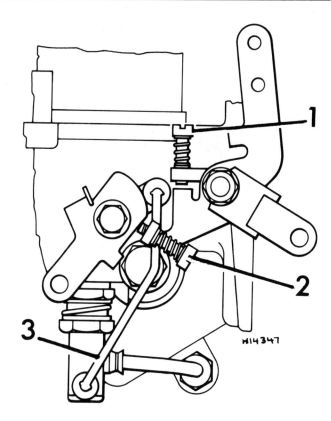

Fig. 3.14 SU carburettor interconnection lever screws (Sec 16)

1 Throttle speed screw
2 Fast idle screw
3 Link rod

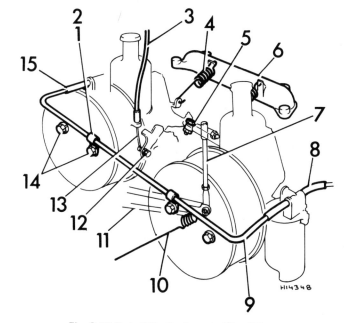

Fig. 3.15 Twin SU attachments (Sec 17)

1 Pipe clamp
2 Fuel pipe
3 Choke outer cable
4 Pull off spring
5 Throttle spindle clamp
6 Balance pipe
7 Throttle control rod
8 Connecting hose
9 Fuel pipe
10 Return spring
11 Throttle actuating lever
12 Choke cable trunnion
13 Choke inner cable
14 Air cleaner casing bolts
15 Connecting hose

Chapter 3 Fuel, carburation and emission control systems

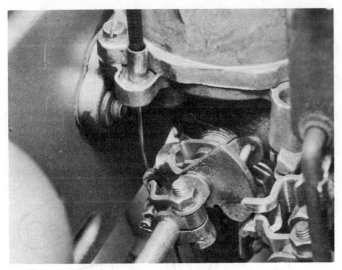

17.2 Typical choke cable attachment to carburettor

17.3 Throttle return spring

2 Disconnect the choke cable from the front carburettor by loosening the nut which clamps the inner cable in place on the short actuating arm (photo).
3 Unhook the throttle return springs, (the front return spring is shown in the photo), and then disconnect the accelerator control rod.
4 Carefully pull off the rubber tubes which carry the fuel to disconnect the petrol feed pipes.
5 Undo the nuts which hold the inlet manifold and lift off the twin carburettors complete with their linkages.
6 To replace the carburettors reverse the above procedure using new gaskets throughout. Make sure that with the throttles closed a gap of 0.015 in (0.8 mm) exists between the pins on the end of each of the short levers attached to the carburettor interconnecting rod, and the forked lever attached to each of the throttle spindles. Adjust if necessary as described in Section 16.

18 Carburettor (SU HS2) – overhaul

1 The SU carburettor is a straightforward instrument to service, but at the same time it is a delicate unit and clumsy handling can cause much damage. In particular it is easy to knock the finely tapering needle out of true, and the greatest care should be taken to keep all the parts associated with the dashpot scrupulously clean.
2 Remove the oil dashpot plunger nut from the top of the dashpot.
3 Unscrew the set screws holding the dashpot to the carburettor body, and lift away the dashpot, light spring, and piston and needle assembly.
4 To remove the metering needle from the choke portion of the piston unscrew the sunken retaining screw from the side of the piston choke and pull out the needle. When replacing the needle ensure that the shoulder is flush with the underside of the piston.
5 Release the float chamber from the carburettor by releasing the clamping bolt and sealing washers from the side of the carburettor base.
6 Normally, it is not necessary to dismantle the carburettor further, but if, because of wear or for some other reason, it is wished to remove the jet, this is easily accomplished by removing the clevis pin holding the jet operating lever to the jet head, and then just removing the jet by extracting it from the base of the carburettor. The jet adjusting screw can then be unscrewed together with the jet adjusting screw locking spring.
7 If the larger locking screw above the jet adjusting screw is removed, then the jet will have to be recentred when the carburettor is reassembled. With the jet screws removed it is a simple matter to release the jet bearing.
8 To remove the throttle and actuating spindle release the two screws holding the throttle in position in the slot in the spindle, slide the throttle out of the spindle and then remove the spindle.

9 To dismantle the float chamber, first disconnect the inlet pipe from the fuel pump at the top of the float chamber cover, if this has not already been done.
10 Undo the three screws which hold the float chamber cover in position, and lift off the cover.
11 If it is not wished to remove the float chamber completely and the carburettor is still attached to the engine, carefully insert a thin piece of bent wire under the float and lift the float out.
12 To remove the float chamber from the carburettor body undo the bolt which runs horizontally through the carburettor.
13 Make a careful note of the rubber grommets and washers and on reassembly ensure that they are replaced in the correct order. If the float chamber is removed completely it is a simple matter to turn it upside down to drop the float out. Check that the float is not cracked or leaking. If it is, it must be repaired or renewed.
14 The float chamber cover contains the needle valve assembly which regulates the amount of fuel which is fed into the float chamber.
15 One end of the float lever rests on top of the float, rising and falling with it, while the other end pivots on a hinge pin which is held by two lugs. On the float cover side of the float lever is a needle which rises and falls in its brass seating according to the movement of the lever.
16 With the cover in place the hinge pin is held in position by the walls of the float chamber. With the cover removed the pin is easily pushed out so freeing the float lever and the needle.
17 Examine the tip of the needle and the needle seating for wear. Wear is present when there is a discernible ridge in the chamfer of the needle. If this is evident then the needle and seating must be renewed. This is a simple operation and the hexagon head of the needle housing is easily screwed out.
18 Never renew either the needle or the seating without renewing the other part as otherwise it will not be possible to get a fuel tight joint.
19 Clean the fuel chamber out thoroughly.
20 Examine all components for wear or damage especially the following:

The carburettor needle

21 If this has been incorrectly assembled at some time so that it is not centrally located in the jet orifice, then the metering needle will have a tiny ridge worn on it. If a ridge can be seen then the needle must be renewed. SU carburettor needles are made to very fine tolerances and should a ridge be apparent no attempt should be made to rub the needle down with fine emery paper. If it is wished to clean the needle it can be polished lightly with metal polish.

The carburettor jet

22 If the jet is worn it is likely that the rim of the jet will be damaged where the needle has been striking it. It should be renewed as otherwise fuel consumption will suffer. The jet can also be badly worn or ridged on the outside from where it has been sliding up and

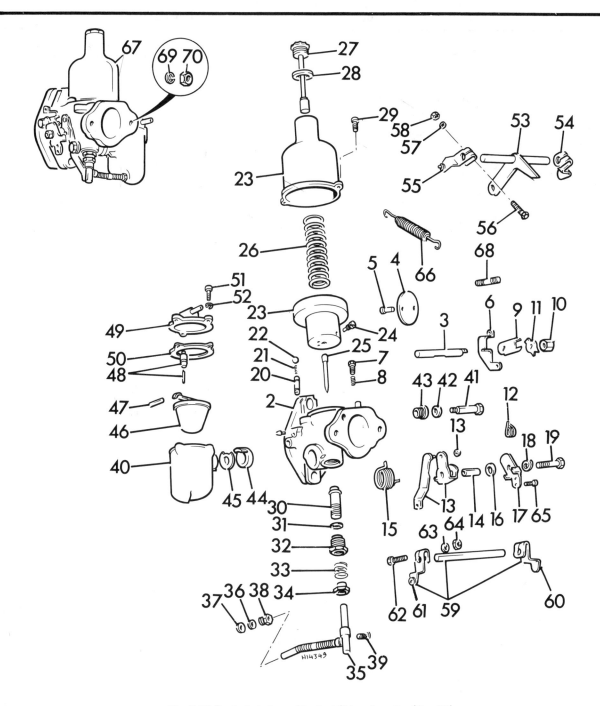

Fig. 3.16 Exploded view of typical SU carburettor (Sec 18)

2 Body	19 Pivot bolt	37 Brass washer	54 Lever
3 Throttle spindle	20 Piston lifting pin	38 Nut	55 Lever
4 Throttle valve plate	21 Spring	39 Screw	56 Bolt
5 Screw	22 Circlip	40 Float chamber	57 Washer
6 Lever	23 Dashpot/piston assembly	41 Bolt	58 Nut
7 Throttle speed screw	24 Screw	42 Steel washer	59 Rod
8 Spring	25 Metering needle	43 Flexible bush	60 Lever
9 Lost motion lever	26 Piston spring	44 Grommet	61 Lever
10 Nut	27 Damper	45 Washer	62 Bolt
11 Lockplate	28 Sealing washer	46 Float	63 Washer
12 Return spring	29 Screw	47 Float pivot pin	64 Nut
13 Pick-up lever	30 Jet bearing	48 Fuel inlet needle valve	65 Pivot screw
14 Sleeve	31 Washer	49 Float chamber cover	66 Throttle return spring
15 Return spring	32 Screw	50 Gasket	67 Gasket
16 Spacer	33 Spring	51 Screw	68 Stud
17 Cam	34 Screw	52 Spring washer	69 Lock washer
18 Spring washer	35 Jet/flexible pipe	53 Throttle lever	70 Nut
	36 Rubber washer		

Chapter 3 Fuel, carburation and emission control systems

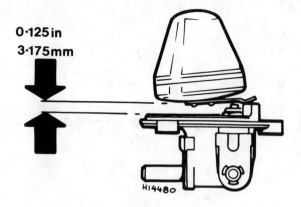

Fig. 3.17 SU float setting diagram (Sec 18)

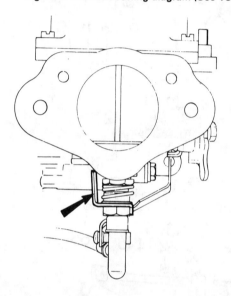

Fig. 3.18 Jet restrictor (SU HS2E carburettor) (Sec 19)

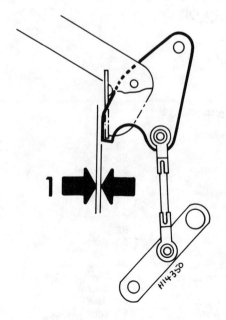

Fig. 3.19 Gap (throttle relay lever to manifold machined surface) (1) = 0.030 in (0.76 mm) (Sec 19)

down between the jet bearings every time the choke has been pulled out. Removal and renewal is the only answer here as well.

General

23 Check the edges of the throttle and the choke tube for wear. Renew if worn. The washers fitted to the base of the jet, to the float chamber, and to the petrol inlet union may all leak after a time and can cause much fuel wastage. It is wisest to renew them automatically when the carburettor is stripped down.

24 Reassembly is a straightforward reversal of the dismantling procedure. It will be necessary to centre the jet. How to do this correctly is described in Section 16.

25 It is essential that the fuel level in the float chamber is always correct as otherwise excessive fuel consumption may occur. On reassembly of the float chamber check the fuel level before replacing the float chamber cover, in the following manner.

26 Invert the float chamber so that the needle valve is closed. It should now be just possible to place a $\frac{1}{8}$ in (3.175 mm) diameter bar between the machining float chamber lip parallel to the float lever hinge, so that face of the float lever just rests on the bar, when the float needle is held fully on its seating.

27 If the bar lifts the lever or if the lever stands proud of the bar then it is necessary to bend the lever at the bifurcation point between the shank and the curved portion until the clearance is correct. Never bend the flat portion of the lever.

19 Carburettors (SU HS2E) – tuning

1 These carburettors are fitted with a tamperproof jet adjuster nut.
2 A jet restrictor is fitted between the jet adjuster nut and the carburettor body and takes the form of a bent tag. This tag limits the amount of adjustment on the nut unless the tag on the restrictor is bent up. Provided the needle is in the correct position the amount of adjustment available, without bending up the tag, should be sufficient to correct the mixture. If it is not, then a full carburettor tuning service will be needed, which involves undoing the tag. Before contemplating doing this make sure that you are not causing a breach of local regulations which may prevent such adjustments. It is essential that the person making such adjustments has the necessary skill, and test equipment, such as an exhaust gas analyser and a vacuum gauge, otherwise the adjustments made might cause the emissions to exceed the allowed amount.
3 The mixture of each carburettor is set as described in Section 16. The difference is that in this case extreme accuracy is needed. The throttles must be synchronised first (Section 16), and set to give the idle speed given in the Specifications Section. The correct mixture setting is halfway between the two points where the engine is noticeably affected by wrong mixture; one way too rich, the other too weak. Ideally an exhaust gas analyser should be used, but at the least a vacuum gauge is required. This gauge will be well worth its modest expense. It should be plumbed into the inlet manifold using its own adaptor, and not temporarily onto one of the others, as the absence of the normal connections may affect the setting. The idling should be set to give the highest steady reading on the gauge. A mixture slightly too weak may have a high reading, but interrupted by fluctuations. Having set the mixture hold the jet adjusting nut to prevent it from turning, and move the restrictor round the nut till the upright tag contacts the carburettor body on the left as seen from the air cleaner. In this position bend down the tag to engage it with a flat on the nut. Now the mixture can be weakened, but not richened, the latter being the alteration most likely to give excessive emissions.
4 When idling the engine for long periods when making adjustments, it may run raggedly, and not seem to respond properly to adjustments. To prevent this, after every minute or so at idling speed, open the throttle and run the engine at 3000 rpm, for half a minute.
5 Using an exhaust gas analyser the twin SU carburettors should be set to give a CO percentage as given in the Specification Section.
6 The throttle linkage in the twin SU carburettors must be set so that there is a small amount of lost motion before the connection from the pedal starts to move the levers on the carburettors. This is to ensure that the throttle stop screws are controlling the idle speed, and not the pedal linkage. The throttle linkage, when released, should rest against a machined surface on the inlet manifold balance pipe. Put a

0.030 in (0.76 mm) feeler gauge between this surface and the throttle relay lever. Slacken the screws clamping the throttle interconnecting levers, and press each lever until it is lightly resting on the carburettor throttle lever. Carefully tighten the clamps so as not to disturb the levers. Check that the interconnecting spindle has about $\frac{1}{32}$ in (1 mm) endfloat so that the throttles cannot bind. Remove the feeler gauge. Move the throttle lever by hand and check that the free play on both throttles is taken up at the same instant.

20 Carburettors (SU HS4) – tuning

1 These carburettors are fitted with tamperproof seals on the mixture jet adjusting nut and the throttle speed adjusting screw. The following operations will normally only be required after major dismantling and overhaul.
2 Remove the air cleaner and check the throttle control for ease of operation.
3 Remove the tamperproof seals from the jet adjusting nut and the throttle speed screw.
4 Unscrew the fast idle screws until they are well clear of the cams and then disconnect the choke control cable from the trunnion on the rear carburettor.
5 Now unscrew the throttle speed screws until they are just clear of the throttle levers with the throttles closed and then screw them in $1\frac{1}{2}$ turns.
6 Raise the piston of each carburettor and check that it falls with a distinct 'click' onto the carburettor bridge. Should the pistons stick, then the piston/dashpot must be removed for cleaning in the following way.
7 Mark the position of the dashpot in relation to the carburettor body. Remove the damper and the dashpot.
8 Take out the coil spring and the piston assembly and empty the oil from the piston rod.
9 Clean deposits and discoloration from the piston and dashpot interior with fuel or methylated spirit.
10 Reassemble to the carburettor and top-up the damper with engine oil until it is $\frac{1}{2}$ in (12.5 mm) above the top of the hollow piston rod.
11 With the piston now falling freely, continue with the tuning operations by lifting and retaining the pistons clear of the carburettor bridges so that the jets are visible.
12 Turn the jet adjusting nuts until the jets are flush with the carburettor bridges. Check that the needle shank is flush with the lower face of the piston.
13 Turn the jet adjusting nuts down two turns.
14 Make sure that the piston damper is topped-up with oil.
15 Start the engine and run it at a fast idle until normal operating temperature is reached, and then continue to run it for a further five minutes.
16 Increase the engine speed to 2500 rpm for a thirty second period.
17 Push the probe of an exhaust gas analyser into the exhaust tailpipe and connect the analyser in accordance with the manufacturers instructions.
18 Slacken the clamp nuts and bolts on the throttle spindle interconnections.
19 Slacken both nuts and bolts on the jet control interconnections.
20 Using a carburettor balancing device (see Section 16, paragraph 24) synchronise the carburettors by turning the throttle speed screws.
21 Turn the jet adjusting nuts up (weaken(or down (enrich) until the fastest idle speed is obtained as indicated on the car tachometer.
22 Now screw each jet adjusting nut up one flat at a time until the engine speed just starts to drop and then screw the nuts downs by the least amount necessary to regain the fastest engine speed.
23 If adjustment continues beyond a three minute period, rev up the engine to 2500 rpm momentarily to clear the fuel/air intake passages then continue at idling speed.
24 Turn each throttle speed screw equally to regain the specified idle speed.
25 Check the CO level indicated on the exhaust gas analyser, this should be within specified tolerance.
26 Set the throttle interconnection clamp levers until the lever pins rest on the lower prongs of the forks.
27 Set the clearance between the fulcrum plate and machined face on the intake manifold to between 0.030 and 0.035 in (0.75 and 0.89 mm). Tighten the clamp nuts and bolts, but make sure that a slight

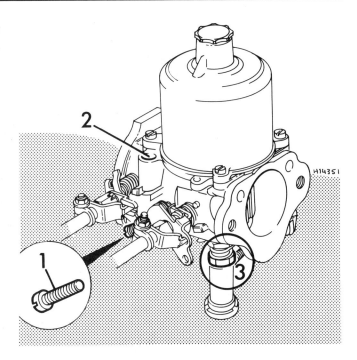

Fig. 3.20 SU HS2E carburettor adjusting screws (Sec 20)

1 Fast idle screw
2 Throttle speed screw
3 Jet adjusting nut

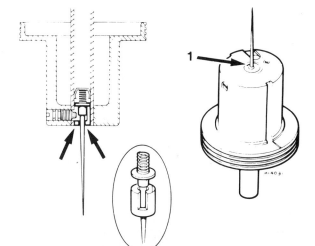

Fig. 3.21 SU metering needle set flush with lower face of piston (Sec 20)

1 Needle shank

clearance exists between the pins and the fork lower prongs (see Fig. 3.12).
28 Recheck the carburettor synchronisation.
29 Push the fast idle cams against their stops and tighten the jet control interconnection clamps.
30 Connect the choke control cable to the trunnion on the rear carburettor, allowing a free movement of $\frac{1}{16}$ in (1.5 mm).
31 With the engine still idling, pull out the choke control until the linkage is just about to actuate the jet. Again using the balancing device to retain synchronisation, turn each fast idle screw until the fast idle speed is at specified level.
32 Fit new seals to the carburettor adjustment screws, refit the air cleaner, remove the exhaust gas analyser and switch off the engine.

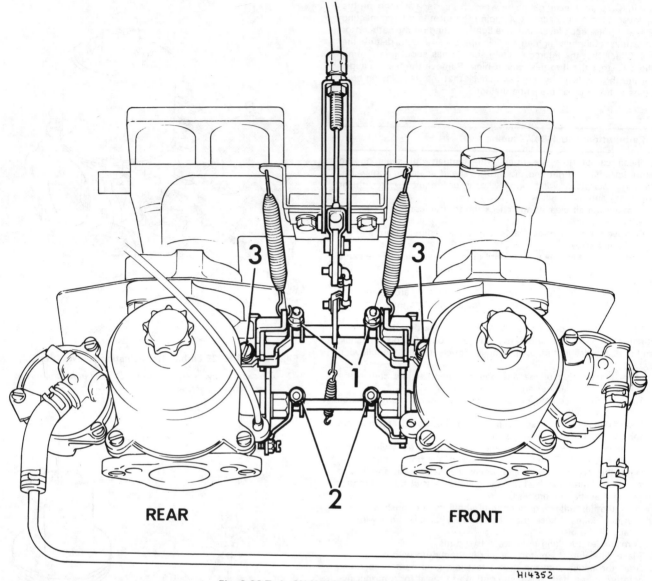

Fig. 3.22 Twin SU HS2E installation (Sec 20)

1 Throttle spindle clamp bolts
2 Jet control interconnections
3 Throttle speed screws

21.1 Cable type throttle control

21 Carburettors (SU HS2E, HS4) – removal and refitting

The procedure is similar to that described for the HS2 units in Section 17, but note the cable type throttle control.

22 Carburettors (SU HS2E, HS4) – overhaul

1 The operations are similar to those described for the HS2 unit, but observe the differences itemized in Section 15, paragraph 8 onward with particular reference to the needle retaining details.

2 On HS4 units, the inclusion of the throttle overrun valve and the temperature compensator device will complicate dismantling to a small extent.

23 Carburettor (Stromberg 150 CDSE) – tuning

1 Before making any carburettor adjustments it is essential to ensure that such items as the spark plugs, valve clearances, contact breaker points and ignition timing have all been attended to. It is also

Chapter 3 Fuel, carburation and emission control systems

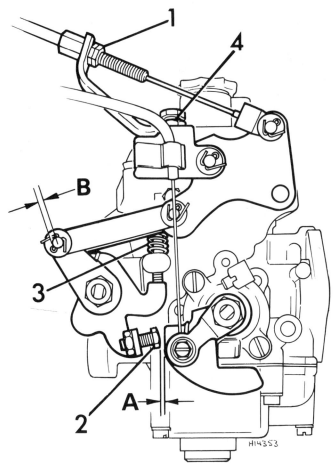

Fig. 3.23 Stromberg 150 CDSE carburettor adjusting diagram for idle speed and throttle linkage (Sec 23)

1. Throttle cable adjuster nut
2. Fast idle screw
3. Throttle speed screw
4. Throttle spindle bracket bolt

A. Fast idle screw to cam clearance
B. Clevis pin free play in slot

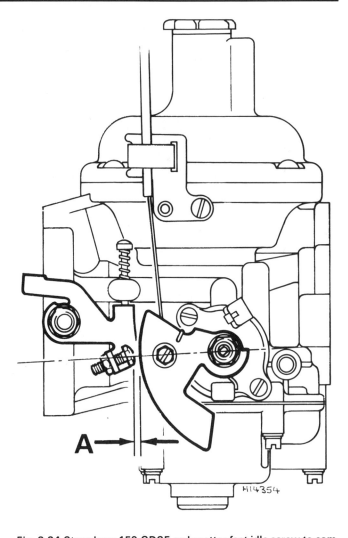

Fig. 3.24 Stromberg 150 CDSE carburettor fast idle screw to cam clearance (Sec 23)

A = 0.020 to 0.025 in (0.5 to 0.6 mm)
w9

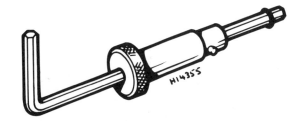

Fig. 3.25 Stromberg 150 CDSE carburettor adjusting tool (Sec 23)

essential to ensure that the engine is operating at its normal temperature. Obviously, if you have just completed a carburettor overhaul this condition cannot be achieved. In such cases the best policy is to achieve an approximately correct setting, take the car for a short run to warm the engine up, and then repeat the necessary adjustments.

2 Remove the air cleaner and slacken the throttle cable abutment nuts right back.

3 Adjust the fast idle screw to obtain maximum cam clearance.

4 Turn the throttle speed screw until the throttle butterfly is just closed them turn the screw in a clockwise direction by $1\frac{1}{2}$ turns to obtain an initial setting.

5 Make sure that the choke control knob is pushed fully home and then start the engine and attain normal running temperature.

6 Set the engine idling speed to that given in the Specifications Section by turning the throttle speed screw. Now stop the engine.

7 Slacken the locknut and set the linkage adjusting screw until the clevis pin is moved to the engine side of the slots in the linkage straps, then tighten the locknut.

8 Now retighten the throttle cable abutment nuts to remove all the slack.

9 Ensure that the choke cable is correctly adjusted and does not have too much slackness or tightness.

10 Now adjust the clearance gap between the fast idle screw and fast idle cam (A in Fig. 3.24) until it is between 0.020 in and 0.025 in (0.51 and 0.64 mm). Remember to retighten the locknut after making any necessary adjustment.

11 Restart the engine and operate the choke control until the cam is turned to a position where the cam pivot, cable clamp screw and fast idle screw are in alignment. Adjust the fast idle screw, if necessary, to give the fast idle speed in the Specifications. Tighten the locknut afterwards.

12 Push the choke control knob fully home.

13 On the adjustable needle (biased type) carburettor adjustment of the mixture is as follows.

14 Remove the carburettor damper and insert the special service tool Triumph number S353 or Zenith B20379Z (Stromberg carburettors being made by the Zenith Company).

15 Insert the tool slowly until the outer part of the tool engages in the

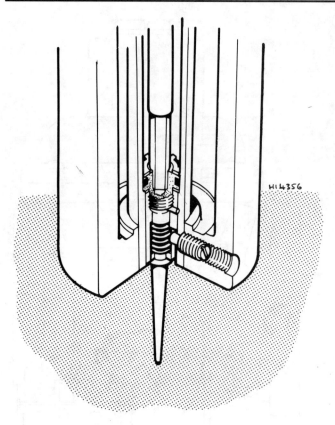

air valve and the inner tool engages with the hexagon in the needle adjuster tool. To use the special tool, the outer mandrel is held by one hand to prevent tearing the diaphragm, whilst the other turns the needle.

16 Turning the adjuster clockwise pulls up the needle, and enriches the mixture. If the adjuster is turned too far anticlockwise the needle assembly may come off the bottom of its thread. If this happens, take off the dashpot (suction chamber cover), take out the piston, and turn the screw whilst holding the needle up into position. Note that the grub screw in the side of the piston does not clamp the needle housing; it merely prevents it from rotating.

17 Try the effect of turning the needle adjusting tool a quarter of a turn. The principles are the same as with other carburettors described earlier. The correct setting is that which gives the fastest idle, and, if you have access to an exhaust gas analyser, the recommended CO reading. If the mixture is wrong lift the piston with a thin screwdriver about $\frac{1}{16}$ in; if the engine speeds up the mixture is rich. If it slows down it is weak. The final setting, theoretically, will be for the needle to be halfway between the two points where there is a noticeable effect of the mixture being wrong; one way too rich, the other too weak.

18 If the mixture was originally a long way out, the engine idle speed will increase as adjustment is corrected, so the throttle stop screw will need readjustment.

19 During the adjustment the idle will become ragged or rough so about every minute open the throttle and run the engine at about 3000 rpm for half a minute to clear the inlet manifold and spark plugs.

24 Carburettor (Stromberg 150 CDSE) – removal and refitting

1 Remove the air cleaner.
2 Disconnect the fuel feed pipe.
3 Disconnect the rocker cover vent pipe.
4 Unhook the throttle return spring.

Fig. 3.26 Cut-away view of Stromberg 150 CDSE adjusting tool being used (Sec 23)

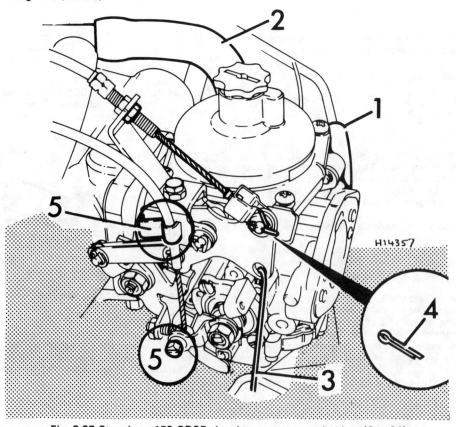

Fig. 3.27 Stromberg 150 CDSE showing some removal points (Sec 24)

1 Fuel pipe flexible connector
2 Rocker cover vent hose
3 Throttle return spring
4 Throttle cable and adjuster
5 Choke control cable clip

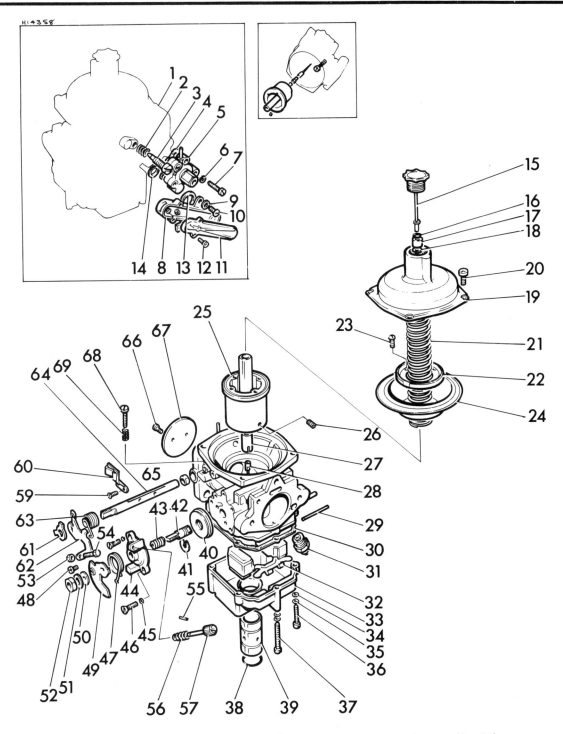

Fig. 3.28 Exploded view of Stromberg 150 CDSE non-biased needle carburettor (Sec 24)

1 Carburettor
2 Spring
3 Idle trimming screw
4 Gasket
5 Bypass valve
6 Lockwasher
7 Screw
8 Temperature compensator unit
9 Lockwasher
10 Screw
11 Cover
12 Screw
13 Seal
14 Seal
15-17 Damper assembly
18 Circlip
19 Dashpot
20 Screws
21 Spring
22 Ring
23 Screw
24 Diaphragm
25 Air valve piston
26 Screw
27 Needle holder assembly
28 Needle assembly
29 Float pivot pin
30 Gasket
31 Needle valve
32 Float assembly
33 Float chamber cover
34-35 Washers
36-37 Screws
38 Rubber O-ring
39 Plug
40 Valve plate
41-44 Cold start valve parts
45-52 Cold start assembly
53 Locknut
54 Screw – fast idle
55 Choke stop pin
56 Stop spring
57 Choke stop
58 Cable abutment bracket
59 Screw
60 Spring clip
61-64 Throttle parts
65 Seal
66 Screw
67 Throttle disc
68 Throttle speed screw
69 Spring

Inset A biased type needle

Chapter 3 Fuel, carburation and emission control systems

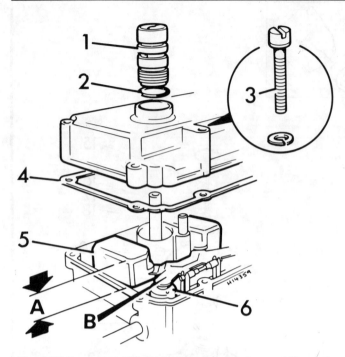

Fig. 3.29 Stromberg 150 CDSE float chamber inverted (Sec 24)

1. Base plug
2. O-ring seal
3. Screw
4. Gasket
5. Float
6. Fuel inlet needle valve
A. Float height 16.0 to 17.0 mm

5 Disconnect the throttle inner and outer cables.
6 Pull off the distributor vacuum pipe.
7 Unscrew the carburettor mounting flange nuts.
8 Remove the carburettor.
9 Refitting is a reversal of removal, but use a new flange gasket.

25 Carburettor (Stromberg 150 CDSE) – overhaul

1 Remove the carburettor from the inlet manifold and carefully clean off any dirt with a paint brush and a cleaning solvent.
2 Undo and remove the damper assembly and drain out the oil from the air valve piston tube by inverting the carburettor.
3 Undo the six screws securing the float chamber cover and remove it.
4 The float assembly can now be removed by gently prising the spindle from the clips.
5 Unscrew the fuel needle valve using a suitably sized box or socket spanner. **Note**: *There is a washer beneath the needle valve.*
6 Remove the four screws retaining the dashpot to the main body. Lift the dashpot away, followed by the air valve return spring and the air valve piston/diaphragm assembly.
7 Further dismantling of the piston/diaphragm assembly can now take place by undoing the four screws securing the diaphragm and retaining the ring to the air valve piston.
8 Note that early versions of the CDSE carburettor have fixed jet and unbiased needle. The needle assembly of this type is fitted into a socket in the air valve guide rod in a pre-set position and locked by a conventional grub screw. The position of the needle should not be altered. The jet assembly is pressed into the body and is non-adjustable and non-removable.
9 If the carburettor has an adjustable biased needle it will be necessary to use Triumph tool number S353 to remove the needle.
10 Insert the tool in the stem of the air valve and turn it in an anti-clockwise direction approximately two turns. The needle and housing can now be pulled out from the lower face of the air valve piston with the fingers.
11 Undo the screws retaining the starter box and remove it.
12 From the other side of the carburettor and undo the temperature compensator screws and remove the compensator noting the two rubber washers are of different diameters.
13 Now remove the screws retaining the bypass valve and gasket.
14 Undo the two screws securing the throttle butterfly to the spindle. Turn the spindle until the butterfly is on the horizontal plane and then pull the butterfly out from the slot in the spindle.
15 The spindle and spring can now be withdrawn.
16 The spindle seals can be removed from the carburettor body using a small screwdriver.
17 Now clean all the components in clean petrol or a suitable solvent and then allow to air dry. Lay all the parts out on a clean surface and discard all seals and gaskets.
18 Examine all moving components for wear or deterioration, paying particular attention to the check valve needle and seat and the air valve and diaphragm. Unless these components are in first class condition they should be renewed.
19 If a compressed air supply is available then the various ports, needle valve and starter box can be blown through to remove any possible obstructions.
20 Having obtained the necessary replacement parts or repair kit, reassembly can now proceed.
21 First fit the spindle seals to the carburettor body. These should be carefully tapped in until the metal casing of the seals is level with the carburettor body.

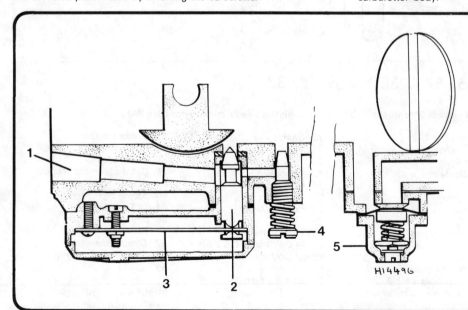

Fig. 3.30 Sectional view of temperature compensator on Stromberg 150 CDSE carburettor (Sec 25)

1. Air intake
2. Temperature compensating valve
3. Tempeature compensator spring
4. Idle trim screw
5. Throttle by-pass

22 Insert the spindle and at the same time load and locate the return spring.
23 Turn the throttle spindle and insert the butterfly. To ensure that the butterfly is correctly positioned there are two pips cast onto it. These pips should face towards the rear of the carburettor (manifold face) and be lying beneath the throttle spindle. Refit the two retaining screws and tighten them evenly.
24 Refit the starter box and tighten the retaining screws evenly.
25 The bypass valve can now be refitted but remember that a new gasket must be used.
26 Now refit the temperature compensator noting that the two rubber washers of different diameters should be renewed.
27 The needle valve assembly can now be refitted into the base of the air valve. On biased needle type carburettors insert the needle assembly into the base of the air valve piston, fit special tool number S353 and turn it in a clockwise direction to engage the threads of the needle valve assembly with the adjusting screw; continue turning until the slotted portion of the needle valve is in line with the grub screw and the needle is flush with the end of the guide rod. Note: *The locating grub screw does not tighten down onto the needle housing but locates into the slot. This makes sure that during adjustments the needle will remain in its correct operating position, ie biased by the spring in the needle housing.*
28 Fit the rubber diaphragm unit to the air valve piston securing it using the retaining ring and screws. Note that there is a locating tag formed on the inner part of the diaphragm which locates in a similar recess in the air valve.
29 Refit the air valve assembly taking care to locate the tag formed on the outer ring of the diaphragm in the recess in the carburettor body.
30 Place the damper spring over the stem of the air valve piston and refit the top cover. The top cover must be refitted so that the bulge on its housing neck faces towards the air intake side of the carburettor. Remember to tighten the top cover retaining screws evenly.
31 The float chamber needle valve can now be refitted, but remember to fit a new sealing washer beneath it.
32 The float assembly can be refitted by carefully levering the pivot pin gently into position.
33 With the carburettor still inverted check the float heights by measuring the distance between the carburettor gasket face and the highest point of the floats. If this dimension is not 16 to 17 mm (0.625 to 0.672 in) it can be adjusted by bending the tabs. It is essential to ensure that the tab sits on the needle valve at right angles when making adjustments.
34 Fit a new float chamber gasket, refit the float chamber and secure it with the six screws.
35 Now fit a new O-ring to the bottom plug and screw the plug into the base of the carburettor.
36 Refill the damper with engine oil. The level should be such that when the damper is pushed in, resistance is felt when the damper's threaded plug is $\frac{1}{4}$ in (6 mm) above the dashpot.
37 The carburettor can now be refitted to the car and tuned.

26 Carburettor (Stromberg CD4 and CD4T) – tuning

1 With the engine at normal operating temperature, have it idling at the specified idling speed. If necessary, turn the throttle speed screw to achieve the correct idling speed.

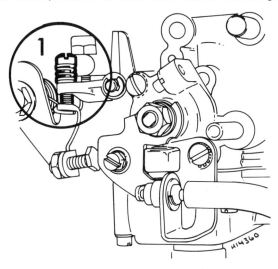

Fig. 3.31 Stromberg CD4 and CD4T throttle speed screw (1) (Sec 26)

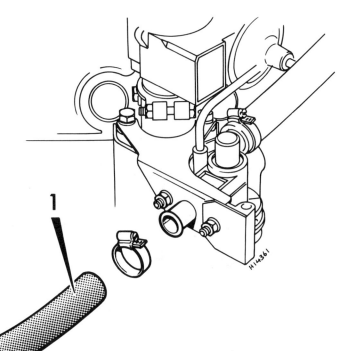

Fig. 3.32 Air pump outlet hose connection (1) at diverter/relief valve on Stromberg 150 CDSE (Sec 26)

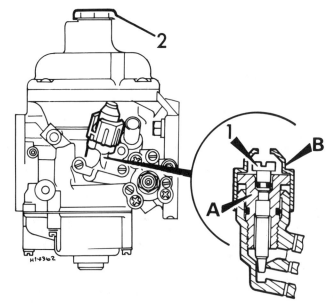

Fig. 3.33 Idle control screws on Stromberg CD4 and CD4T carburettors (Sec 26)

A Coarse idle control (do not disturb)
B Plastic cap
1 Fine idle control screw
2 Damper piston cap

2 Disconnect the air pump outlet hose at the diverter/relief valve to vent the pump to atmosphere.
3 Insert the probe of an exhaust gas analyser into the tail pipe of the exhaust system and check the CO level against the specified tolerance shown on the vehicle decal. If marginal adjustment is required, turn the fine idle control screw and the recheck. On no account break off the tamperproof cap to alter the coarse idle screw which is pre-set during production.
4 If the mixture is grossly out of adjustment, then the procedure for correcting it using the special tool is as described in Section 23 for the Stromberg 150CDSE.

27 Stromberg deceleration and by-pass valve – checking and adjusting

1 Run the engine at a fast idle speed at its operating temperature for a period of five minutes.
2 From the distributor remove the vacuum pipe and seal the pipe end with a finger. The idle speed should increase to 1300 rpm. If the idle speed rises sharply to between 2000 and 2500 rpm, the bypass valve is floating and needs adjustment.
3 With the vacuum pipe plugged, the adjustment screw is turned anti-clockwise to reduce the engine speed to 1300 rpm.
4 Sharply increase and decrease the engine speed with the throttle and ensure that the engine speed always drops back to 1300 rpm. Adjust, if necessary, and then turn the by-pass valve adjustment screw anti-clockwise half a turn to seat the valve. Unplug the vacuum pipe which can be reconnected to the distributor.

28 Stromberg CD4 manual choke – removal and refitting

1 Disconnect the choke operating cable from the carburettor.
2 Disconnect the vacuum pipe from the EGR control valve.
3 Extract the two screws which hold the choke assembly to the carburettor body.
4 Remove the EGA cut-out valve and withdraw the choke assembly.
5 Refitting is a reversal of removal.

29 Stromberg CD4T automatic choke – removal, refitting and adjustment

1 Remove the air cleaner and the carburettor from the engine.
2 Open the throttle valve plate and hold it open by inserting a piece of soft wood.
3 Extract the three screws which hold the automatic choke in position noting that the lower one is shorter than the other two.
4 Remove the choke assembly and its gasket.
5 Dismantle the choke housing cover by removing the central bolt, sealing ring, screws and clamp ring.
6 Take out the finned aluminium heat mass without straining the bi-metallic coil.
7 Remove the heat insulator.
8 Commence refitting by locating a new gasket and screwing on the choke lower housing.
9 Adjust the fast idle screw to give a clearance between the base circle of the cam and the end of the pin of 0.020 in (0.50 mm).
10 Locate the heat insulator so that the bi-metallic lever protrudes through the slot in the insulator.
11 Fit the aluminium heat insulator with fins facing outwards and engaging the rectangular loop on the end of the bi-metallic coil carefully over the bi-metallic lever. Check for correct assembly by rotating the heat mass in both directions through 30° to 40° and note that it springs back to its static position.
12 Fit the clamp ring, but leave its fixing screws slack.
13 Rotate the heat mass in an anti-clockwise direction until the scribed line on its outer edge is aligned with the datum line on the insulator and the autochoke cover. Hold it in this position and fully tighten the clamp ring screws.
14 Fit the sealing ring.
15 Fit the choke housing cover, but leave the centre bolt slack.
16 Refit the carburettor, connect the coolant pipes to the choke housing cover and then tighten the centre bolt.

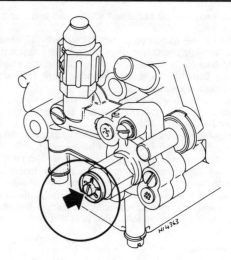

Fig. 3.34 Deceleration/by-pass valve adjusting screw on Stromberg carburettor (Sec 27)

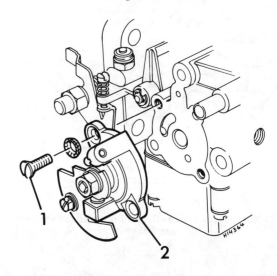

Fig. 3.35 Removing manually-operated choke from Stromberg carburettor (Sec 28)

1 Retaining screw 2 Choke housing

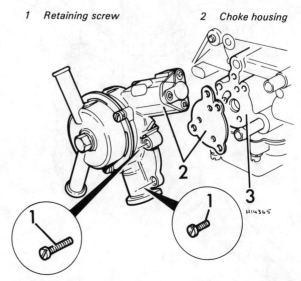

Fig. 3.36 Removing automatic choke from Stromberg carburettor (Sec 29)

1 Securing screws 3 Choke mounting face
2 Choke housing and gasket

Chapter 3 Fuel, carburation and emission control systems

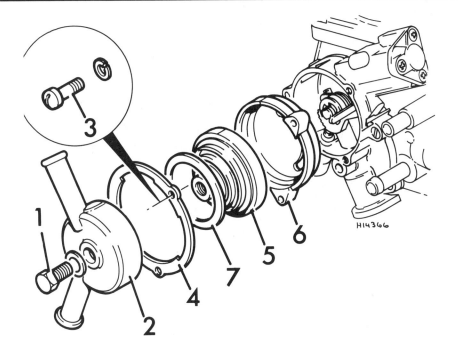

Fig. 3.37 Exploded view of Stromberg automatic choke (Sec 29)

1. Centre bolt
2. Choke housing cover
3. Clamp ring screw
4. Clamp ring
5. Aluminium mass
6.
7. Sealing ring

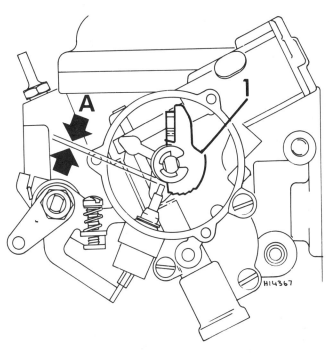

Fig. 3.38 Automatic choke adjustment (Sec 19)

A 0.020 in (0.50 mm)   1 Cam

17 Top-up the damper with engine oil.
18 Depress the accelerator to reset the automatic choke.
19 Fit the air cleaner.
20 Start the engine and run it to normal operating temperature and then adjust the idle speed. After the engine has cooled down check the coolant level and top-up if necessary.

30 Carburettor (Stromberg CD4) – removal and refitting

1 Remove the air cleaner assembly (see beginning of this Chapter).
2 Disconnect the float chamber vent pipe by squeezing together the ends of the retaining clips and pulling the pipe from its connection.
3 Disconnect the fuel feed pipe by following the instructions given in the preceding paragraph.
4 Pull off the vacuum pipe from the EGR air bleed valve.
5 Pull off the hose from the engine breather where it joins the carburettor.
6 Pull off the vacuum hose from the carburettor to the front vapour trap.
7 Loosen the retaining screw of the mixture control cable at the starter box lever. Withdraw the cable after releasing the spring fastening clip.
8 To free the throttle cable yoke from the carburettor, withdraw the split-pin from the spring retaining rod and remove the rod and spring.
9 The throttle valve is now released from its support bracket by unscrewing the outer cable nut.
10 Pull off the vacuum hose from the carburettor to the rear vapour trap.
11 Undo the two nuts which secure the carburettor to the inlet manifold and lift away the two spring washers. Withdraw the carburettor complete with insulating block.
12 To refit the carburettor reverse the sequence of dismantling but the throttle cable abutment nuts will have to be adjusted to take up the inner cable slack. **Note:** *When refitting the carburettor it is essential that the mating faces are clean, new gaskets are used and the retaining nuts are tightened evenly.*

31 Carburettor (Stromberg CD4) – overhaul

Note: *Before dismantling, check the availability of spare parts. If the needle is to be adjusted, special tool number S353 (Zenith B20379Z) will be required.*

1 Refer to Fig. 3.39. With the carburettor removed from the car, clean off the exterior.
2 Remove the damper cap by unscrewing, then carefully lift the piston whilst simultaneously withdrawing the retaining cap from the air valve unit.
3 From the float chamber, pull out the bottom plug, and drain the oil and petrol from the carburettor.
4 Now unscrew the screws retaining the float chamber to the body and remove it complete with gasket.
5 Prise the spindle from the clip at opposing ends of the float assembly and remove it.
6 Unscrew the needle valve and washer.
7 Remove the top cover by unscrewing the four retaining screws. Lift off the cover complete with spring and the air valve unit.
8 The diaphragm can now be carefully removed after unscrewing the four securing screws, and lifting off the retaining ring.

Chapter 3 Fuel, carburation and emission control systems

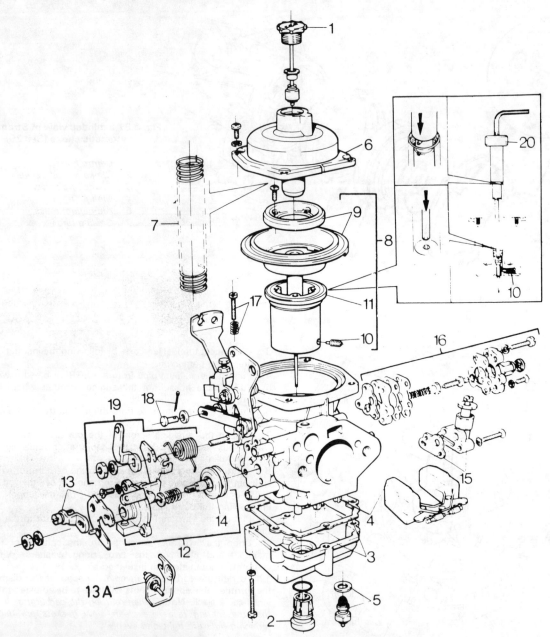

Fig. 3.39 Exploded view of Stromberg 150 CD4 carburettor (Sec 31)

1 Damper cap
2 Bottom plug
3 Float chamber and gasket
4 Float
5 Needle valve and washer
6 Top cover
7 Spring
8 Air valve unit
9 Retaining ring and diaphragm
10 Grub screw
11 Piston
12 Starter box unit
13 Choke cable lever
14 Starter disc and spindle
15 Air regulator and gasket
16 Deceleration bypass valve unit and gasket
17 Idle adjustment screw and spring
18 Throttle lever clevis pin and split pin
19 Throttle lever and spring unit
20 Special tool No. S353

9 At this stage, to remove the needle and housing special tool S353 (Zenith B20379Z) will be required. Insert it to the stem of the air valve and turn anti-clockwise about two turns. Loosen the grub screw in the side of the piston and carefully withdraw the needle and housing by pulling directly out, using only finger pressure. **Note:** *The needle adjuster is a fixed assembly in the air valve and it must not be removed.*
10 Remove the starter box retaining screws and withdraw the box unit.
11 Now disconnect the EGR air bleed valve and bracket.
12 Dismantle the starter box assembly by unscrewing the nut and lockwasher from the spindle, disconnect the choke cable lever, and withdraw the starter disc and spindle.
13 Remove the C-clip from the spindle and withdraw the spring.
14 Remove the air regulator and gasket by unscrewing the securing screws.
15 Withdraw the deceleration by-pass valve unit and gasket by removing its three retaining screws. To dismantle this unit, release the baseplate from the valve housing by unscrewing the retaining screws and then disconnect the spring, valve and gaskets from the body of the deceleration valve. To release the adjustment screw, remove the star washer. Now remove the adjustment screw O-ring and unscrew the locknut.
16 Unscrew the idle adjustment screw and spring.
17 Withdraw the split-pin and clevis pin with washers from the throttle spindle lever and disconnect it.

Chapter 3 Fuel, carburation and emission control systems

18 The throttle lever retaining nut can now be removed from the spindle to disconnect the lever with retainer and spring.
19 With the carburettor fully dismantled the various parts can be cleaned off in clean petrol and allowed to drain dry. The ports, needle valve and starter box can be blown dry with compressed air.
20 Carefully inspect all parts for signs of wear, in particular the needle and seat, and the air valve and diaphragm. Renew them if required. If the starter box unit, deceleration bypass valve or idle and air regulator are damaged or worn, the carburettor must be scrapped and a new replacement unit obtained.
21 Before assembly, obtain a repair kit for the carburettor, which contains the relevant seals and gaskets which must be renewed.
22 Reassemble the components of the carburettor in the reverse order of dismantling paying particular attention to the following:

(a) Ensure that all components are spotlessly clean. Do not overtighten retaining screws and use only hand-pressure to fit the components
(b) When fitting the deceleration bypass valve between the baseplate gaskets, ensure that the spring register is towards the valve body
(c) When refitting the disc unit to the starter body, ensure that the lug detent ball is located in the disc slot and the largest of the series of holes
(d) When refitting the cam and choke cable levers, ensure that the cam lever is located with the detent ball
(e) When using the special tool to engage the needle valve threads with those of the adjusting valve screw, turn the tool clockwise until the needle housing slot is in line with the grub screw. Then tighten the grub screw. The grub screw does not tighten onto the needle housing – it only locates into the slot, which ensures that during adjustment, the needle remains in its operating position
(f) When fitting the diaphragm it locates to the inner tag in the recess of the air valve
(g) The air valve is fitted so that the outer tag of the diaphragm is located in the recesses of the carburettor body
(h) The top cover is located in position with the bulge of the housing neck to the air intake. Tighten the top cover retaining screws evenly
(i) When the float assembly has been refitted, the float height is checked by measuring between the carburettor gasket face and the high point of the floats. It is essential that the float height is equal and set to 0.625 in to 0.672 in (16 mm to 17 mm). If adjustment is required bend the tabs accordingly, but the tab must seat on the needle valve at right-angles
(j) When the carburettor is refitted to the car, the damper dashpot is lubricated with clean light engine oil. Use the damper as a dipstick, and when the threaded plug is 0.25 in (6 mm) above the dashpot, resistance should be felt. Then lift the piston and press the damper securing cap into the air valve taking great care. Screw the damper cap into the cover
(k) The carburettor is now ready for tuning as described in Section 26

32 Carburettor (Stromberg CD4T) – removal and refitting

1 Disconnect the battery earth lead and drain the cooling system.
2 Remove the air cleaner assembly.
3 Disconnect the inlet and outlet water hoses at the auto-choke.
4 Pull off the following hoses:

(a) Vacuum pipe to front vapour trap
(b) Vacuum pipe to rear vapour trap
(c) Engine breather pipe at carburettor connection
(d) Vacuum pipe from EGR control valve

5 Disconnect the fuel inlet pipe and float chamber vent pipe by squeezing together the ends of the retaining clips and pulling the pipes off.
6 From the roller lever, disconnect the carburettor link rod balljoint.
7 Remove the two carburettor securing nuts and washers. Lift the carburettor away together with the insulating block.
8 Refitting the carburettor is the reverse of removal, but the automatic fuel enrichment unit must be adjusted then tuned as described in Section 26.

33 Carburettor (Stromberg CD4T) – overhaul

Note: *Before dismantling check the availability of spare parts. If the mixture needle is to be adjusted or removed, use special tool number S353 (Zenith B20379Z).*

1 Repeat the operations described in Section 31, paragraphs 1 to 10 and 14 and 15.
2 Remove the auto-choke operating lever linkage rod by unscrewing the retaining nut and washer. Then disconnect the auto-choke operating lever bush and spring, and also the stop lever and spring.
3 The auto-choke is now removed completely with gasket by unscrewing the three retaining screws.
4 To dismantle the auto-choke, remove the retaining bolt and washer from the water jacket. Withdraw the water jacket and sealing ring. The heat mass is retained by three screws. Remove them and withdraw the heat mass with its insulator, but be careful not to strain or damage the extra sensitive coil (refer also to Section 29).
5 Finally, remove the vacuum kick piston cover and gasket which is retained by three screws.
6 Having dismantled the carburettor, carefully clean the various components in clean petrol and where possible allow to drain dry. The ports, needle valve and starter box can be blown dry with compressed air.
7 Carefully inspect all the components for wear, in particular the needle and seat, and the air valve and diaphragm, all of which must be in good condition. Renew them if they show any signs of wear or damage. If the auto-choke idle unit, air regulator or deceleration bypass valve are damaged in any way, renew the carburettor.
8 Before reassembling the carburettor components, obtain a repair kit which will contain the relevant seals and gaskets which must be renewed.
9 Reassembly is the reverse of dismantling. Refer to Section 31, paragraph 22 but ignoring sub-paragraphs (c) and (d) and turning instead to Section 29 for automatic choke details.
10 The carburettor can now be refitted to the car and the damper dashpot lubricated with clean light grade engine oil. Use the damper as a dipstick. When the threaded plug is $\frac{1}{4}$ in (6 mm) above the dashpot, resistance should be felt. Then lift the piston and press the damper securing cap into the air valve, taking great care. Screw the damper cap into the cover.
11 The carburettor is now ready for tuning, see Section 26.

34 Emission control systems – description

1 All models are fitted with some form of emission control equipment.
2 Cars not destined for North America have a basic system which includes the carburettor itself (designed for lean mixture, low exhaust emission operation) and the crankcase ventilation system (refer to Chapter 1, Sections 21 or 38).
3 Vehicles destined for operation in North America however are equipped with sophisticated systems and on models sold in California, a catalytic converter is also incorporated into the exhaust system.

Anti run-on (dieseling) valve
4 This valve prevents the engine from running-on after the ignition has been switched off. This is caused by engine heat creating a compression-ignition situation.
5 When the ignition is switched off, a solenoid valve is activated and seals of the inlet to the bottom of the charcoal absorption canister. A connection to the inlet manifold thus applies a partial vacuum to the canister and consequently to the float chamber(s) of the carburettor(s). This vacuum is sufficient to prevent fuel being drawn into the engine. When the engine has stopped and the oil pressure has dropped to zero, the solenoid valve is de-activated and the engine is thus capable of being restarted.

Air temperature control air cleaner
6 This type of air cleaner maintains the air being drawn into the carburettor intake at a constant temperature (see Section 6).
7 A sponge rubber valve attached to a bi-metal strip is positioned in the air cleaner intake spout. When the engine is cold the valve closes the air intake spout from the atmosphere and the air supplied to the

carburettor is drawn via a hot air pipe from a box attached to the exhaust system. As the temperature of the air drawn from around the exhaust system rises, so the bi-metal strip deflects and gradually shuts off the hot air supply, and the air is subsequently drawn from the atmosphere.

Air injection system (AIS)
8 This system is used to reduce the hydrocarbons, nitric oxide and carbon monoxide in the exhaust gases, and comprises an air pump, a combined diverter and relief valve, a check valve and an air manifold.
9 The rotary vane type pump is belt driven from the engine and delivers air to each of the four exhaust ports.
10 The diverter and relief valve diverts air from the pump to the atmosphere during deceleration, being controlled in this mode by manifold vacuum. Excessive pressure is discharged to the atmosphere by operation of the relief valve.
11 The check valve is a diaphragm spring operated non-return valve. Its purpose is to protect the pump from exhaust gas pressures both under normal operations and in the event of the drivebelt failing.
12 The air manifold is used to direct the air into the engine exhaust ports.

Exhaust gas recirculation (EGR) system
13 To minimise nitric oxide exhaust emissions, the peak combustion temperatures are lowered by circulating a metered quantity of exhaust gas through the inlet manifold.
14 A control signal is taken from the throttle edge tapping of the carburettor. At idle or full load no recirculation is provided, but under part load conditions a controlled amount of recirculation is provided according to the vacuum signal of the metering valve. The EGR valve is mounted on the exhaust manifold.
15 An EGR control valve (if fitted) cuts the vacuum signal when the choke is in operation by opening an air bleed into the vacuum line.

Fuel evaporative control system
16 This system uses an activated absorption canister through which the fuel tank is vented, and incorporates the following features:

(a) The carburettor float chamber is vented to the engine when the throttle is open, and to the absorption canister when the throttle is closed
(b) The carburettor constant depression is used to induce a purge condition through the canister via the anti-run-on valve. The crankcase breathing is also coupled to this system
(c) A separator tank is used to prevent fuel surges from reaching the canister which could otherwise saturate the system
(d) A sealed filler cap is used to prevent loss by evaporation
(e) The fuel filler tube extends into the fuel tank to prevent complete filling, this permits the fuel to expand in hot weather

Catalytic converter
17 To further reduce the emission of carbon monoxide and hydrocarbons a catalytic converter is installed in the exhaust system of Californian models.
18 The catalytic converter is basically a steel casing holding a catalytic carrier in the form of ceramic beads cotated with platinum or palladium. As exhaust gases pass through the converter, the hydrocarbons and carbon monoxide content of the gases is oxidised and changed to water and carbon dioxide.
19 Overheating of the catalytic converter can occur during descent of long gradients or by operating the car with faulty carburettor or ignition settings. The following precautions should be taken with cars equipped with a catalytic converter:

(a) Avoid heavy impacts to the casing which can cause damage to the internal ceramic material
(b) Never use anything but unleaded fuel, or the emission control system efficiency will be seriously impaired
(c) The catalytic converter becomes extremely hot during operation of the car so allow the converter to cool before touching it during any repair or maintenance operations
(d) Do not push or tow start the car
(e) Do not use the type of pump which is screwed into a spark plug hole and is used for paint spraying or for tyre inflation

35 Emission control system components – servicing, removal and refitting

The following operations are included to facilitate routine maintenance (see specified intervals in Routine Maintenance at the beginning of this Manual) and to enable faulty components to be renewed.

Air injection system air pump drivebelt – removal, refitting and tensioning
1 Slacken the adjusting link bolt and nut, adjusting link pivot bolt and nut, and the air pump pivot nut and bolt.
2 Push the air pump towards the engine and slip the drivebelt off the pulleys.
3 Refitting of the belt is the reverse of the removal procedure. Do not tighten the adjusting and pivot bolt nuts until the belt has been tensioned.
4 The nuts should be sufficiently tightened to allow the air pump to be moved stiffly on its mounting pivots.
5 Carefully lever against the pump to obtain a total deflection of $\frac{1}{2}$ in (12.7 mm) at the mid-point of the longest run of the belt.
6 Tighten the adjusting and pivot bolt nuts whilst maintaining the tension. Recheck the belt tension after about 150 miles (250 km) of travelling.

Air distribution manifold – removal and refitting
7 Disconnect the air hose at the diverter and relief valve.
8 Remove the rocker cover nut retaining the air distribution manifold support bracket.

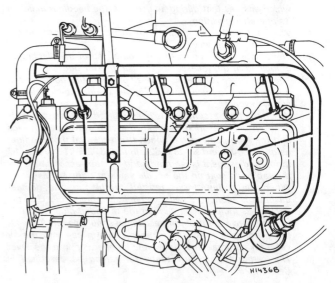

Fig. 3.40 AIS air distribution manifold (Sec 35)

1 Pipe union nuts
2 Manifold and check valve

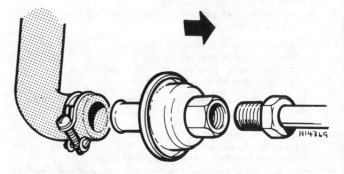

Fig. 3.41 AIS check valve direction of installation (Sec 35)

Chapter 3 Fuel, carburation and emission control systems   111

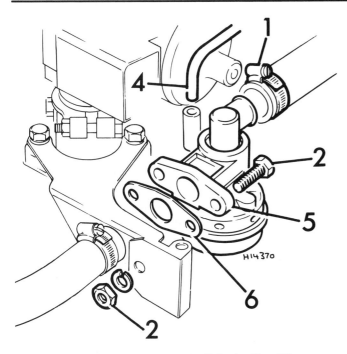

Fig. 3.42 AIS diverter and relief valve (Sec 35)

1 Hose from check valve
2 Fixing nuts and bolts
4 Vacuum pipe
5 Diverter and relief valve
6 Gasket

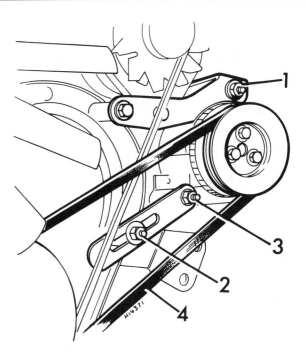

Fig. 3.43 AIS air pump and mountings (Sec 35)

1 Pump pivot bolt
2 Adjuster link bolt
3 Adjuster link pivot bolt
4 Drivebelt

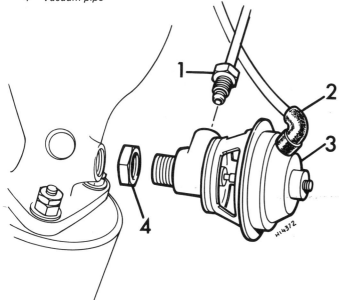

Fig. 3.44 EGR valve (catalytic converter model) (Sec 35)

1 Pipe to inlet manifold
2 Vacuum pipe
3 EGR valve
4 Locknut

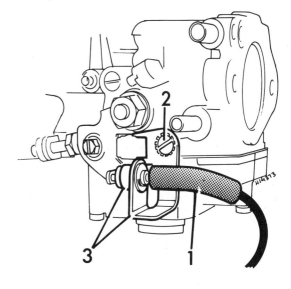

Fig. 3.45 EGR control valve (Sec 35)

1 Vacuum pipe to EGR valve
2 Support bracket
3 Valve and bracket abutment

9 Unscrew the four union nuts, and withdraw the complete manifold (and check valve) from the exhaust ports.
10 If necessary, hold the manifold in a vice, and unscrew the check valve.
11 Refitting is the reverse of the removal procedure.

Check valve – removal, testing and refitting

12 Disconnect the air hose at the check valve.
13 Using two open-ended spanners, unscrew the check valve whilst preventing strain on the air manifold.
14 If necessary the valve can be checked by blowing air (by mouth only) through the valve. Air should pass through from the hose connection end but not from the manifold end. Renew a defective valve.
15 Refitting is the reverse of the removal procedure.

Diverter and relief valve – removal and refitting

16 Slacken the hose clip and pull off the hose from the diverter and relief valve to the check valve.
17 Pull out the vacuum sensing pipe from the rubber sleeve connector.
18 Remove the nut and bolt retaining the drive resistor. Place the

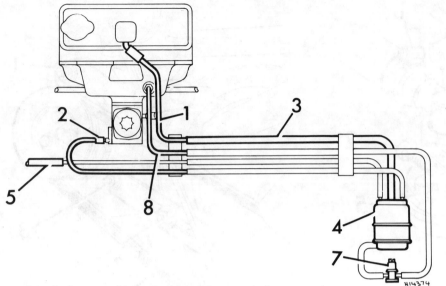

Fig. 3.46 Fuel evaporative control system (Sec 35)

1 Crankcase purge pipe
2 Float chamber vent pipe
3 Canister purge line
4 Charcoal canister
5 Fuel tank vent pipe
7 Anti-run-on valve
8 Manifold vacuum line

drive resistor back to one side but leave the wires attached.
19 Remove the other nut and bolt and lift the diverter and relief valve away. Separate the gasket from the face of the valve.
20 Refitting is the reverse of the removal procedure, but note that it is recommended that a new gasket is used between the valve and bracket.

Air pump – removal and refitting

21 Remove the left-hand engine valance and oil filter element to gain better access to the air pump.
22 Disconnect the air hose from the pump.
23 Slacken the link pivot bolt and remove the pump pivot and adjusting bolt.
24 Slip off the air pump drivebelt and remove the pump.
25 Refitting is the reverse of the removal procedure. Refer to paragraphs 1 to 6 for belt tensioning.

Exhaust gas recirculation (EGR) system – removal, maintenance and refitting of valve

26 Remove the air cleaner assembly.
27 Disconnect the vacuum pipe T-piece from the EGR valve (non-catalytic converter USA models only). On catalytic converter models there is only a single vacuum pipe to be disconnected from the EGR valve.
28 Undo the banjo bolt retaining the metal pipe to the inlet manifold.
29 Undo the union nut on the EGR valve and remove the metal pipe.
30 Loosen the locknut and turn the EGR valve anti-clockwise to remove it from the exhaust manifold.
31 Clean the joint faces of the EGR valve with a wire brush. The valve and seat are best cleaned using a spark plug cleaning machine. The valve opening is inserted into the machine, the diaphragm carefully and evenly raised, and then the valve blasted for a short burst of about 30 seconds. Repeat this operation until all the carbon deposits are removed. The steel pipe can be cleaned out using a flexible wire brush. All parts, when cleaned, should be blown clear with compressed air to remove all traces of carbon grit. Pipes or fittings showing signs of damage or deterioration must be renewed.
32 Refit the components in the reverse order to dismantling.

Exhaust gas recirculation (EGR) control valve – removal and refitting

33 Disconnect the pipe from the valve, then remove the screw and shakeproof washer securing the valve bracket to the carburettor.
34 Remove the valve and bracket together.
35 Refitting is the reverse of the removal procedure.

Fuel evaporative control system – vapour separator – removal and refitting

36 Remove the rear compartment trim pad which is retained by six screws and cup washers.

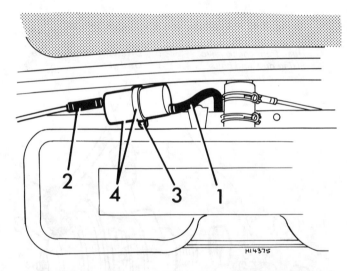

Fig. 3.47 Vapour separator unit near fuel tank filler (Sec 35)

1 Hose from tank
2 Hose to carbon canister
3 Fixing bolts
4 Vapour separator and clip

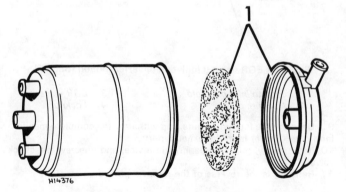

Fig. 3.48 Early type charcoal canister (Sec 35)

1 Screw base and disposable filter

Chapter 3 Fuel, carburation and emission control systems

37 Remove the boot trim pad and disconnect the leads to the boot illumination lamp.
38 Squeeze together the ends of the hose clips and pull off the fuel tank to vapour separator hose and vapour separator to absorption canister hose.
39 Remove the nut and bolt securing the vapour separator to the fuel tank and lift away the vapour separator.
40 Reverse the order of dismantling to refit the separator.

Absorption canister (early USA models) – removal, servicing and refitting

41 Early systems had the vapour line from the tank only, without the float chamber vent.
42 Maintenance at 25 000 mile (40 000 km) intervals is to renew the filter pad housed in the base of the unit.
43 Remove the vapour and purge pipes from the top of the unit and slacken the screw clamping the canister in the bracket.
44 Remove the air intake pipe and unscrew the base of the canister. The filter can now be changed.
45 Reassembly of the unit is the reverse of the dismantling sequence. The complete absorption canister should be renewed at 50 000 mile (80 000 km) intervals.

Absorption canister (later USA models) – removal, servicing and refitting

46 This later system does not have a renewable filter pad and the complete absorption canister should be renewed at 50 000 mile (80 000 km) intervals.
47 Pull off the canister (centre) purge pipe.
48 Squeeze together the ends of the retaining clips and pull off the canister to fuel tank pipe and carburettor vent pipe at the top of the absorption canister.
49 Slacken the screw and nut of the clamping band and lift out the absorption canister.
50 Reverse the removal instructions for refitting.

Anti-run on valve – removal and refitting

51 Detach the solenoid electrical leads.
52 Detach the vacuum pipe and the hose leading from the absorption canister.
53 Twist the valve out of its retaining bracket.
54 Refitting is the reverse of the removal procedure.

Catalytic converter – removal and refitting

55 Make sure that the exhaust pipe and catalytic converter are cool.
56 Remove the air cleaner assembly and two rear screws securing the right-hand engine valance.
57 Remove the three nuts and locking nuts securing the converter to the exhaust manifold flange.
58 Raise the car up at the front and support it on axle stands or strong packing blocks.
59 From underneath the car undo the three nuts and bolts which secure the converter to the intermediate pipe. Extract the olive.
60 From within the engine compartment pull back the right-hand engine valance and withdraw the converter and downpipe assembly.
61 Refitting the converter and downpipe assembly is the reverse of the dismantling procedure, but fit a new gasket between the converter and manifold flange and also apply grease to both faces of the olive. Tighten the flare bolts and nuts evenly.

36 Throttle and choke controls

Throttle control – early models

1 Early cars use a rod linkage with one ball-jointed section.
2 Dismantling is carried out by removing the roll pin and circlips.
3 When refitting, adjust the length of the ball-jointed link rod and the pedal stop bolt to give a travel from the fully closed to fully open positions of the throttle valve butterflies.

Throttle control – later models

4 Later models have a cable controlled throttle linkage.
5 The accelerator pedal is removed by unscrewing the pivot bracket

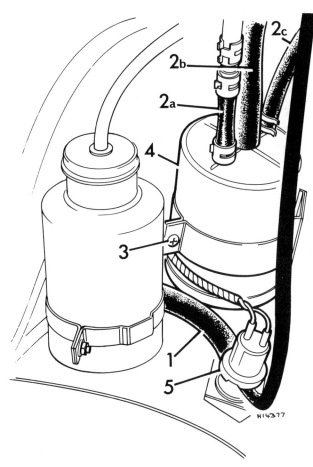

Fig. 3.49 Later type charcoal canister (Sec 35)

1 Hose to charcoal canister
2a Hose to separator
2b Purge line
2c Carburettor vent pipe
3 Clamp screw
4 Charcoal canister
5 Anti-run-on valve

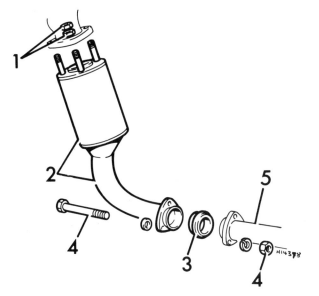

Fig. 3.50 Catalytic converter (Sec 35)

1 Flange nuts
2 Converter
3 Olive
4 Flange connecting bolts
5 Exhaust intermediate pipe

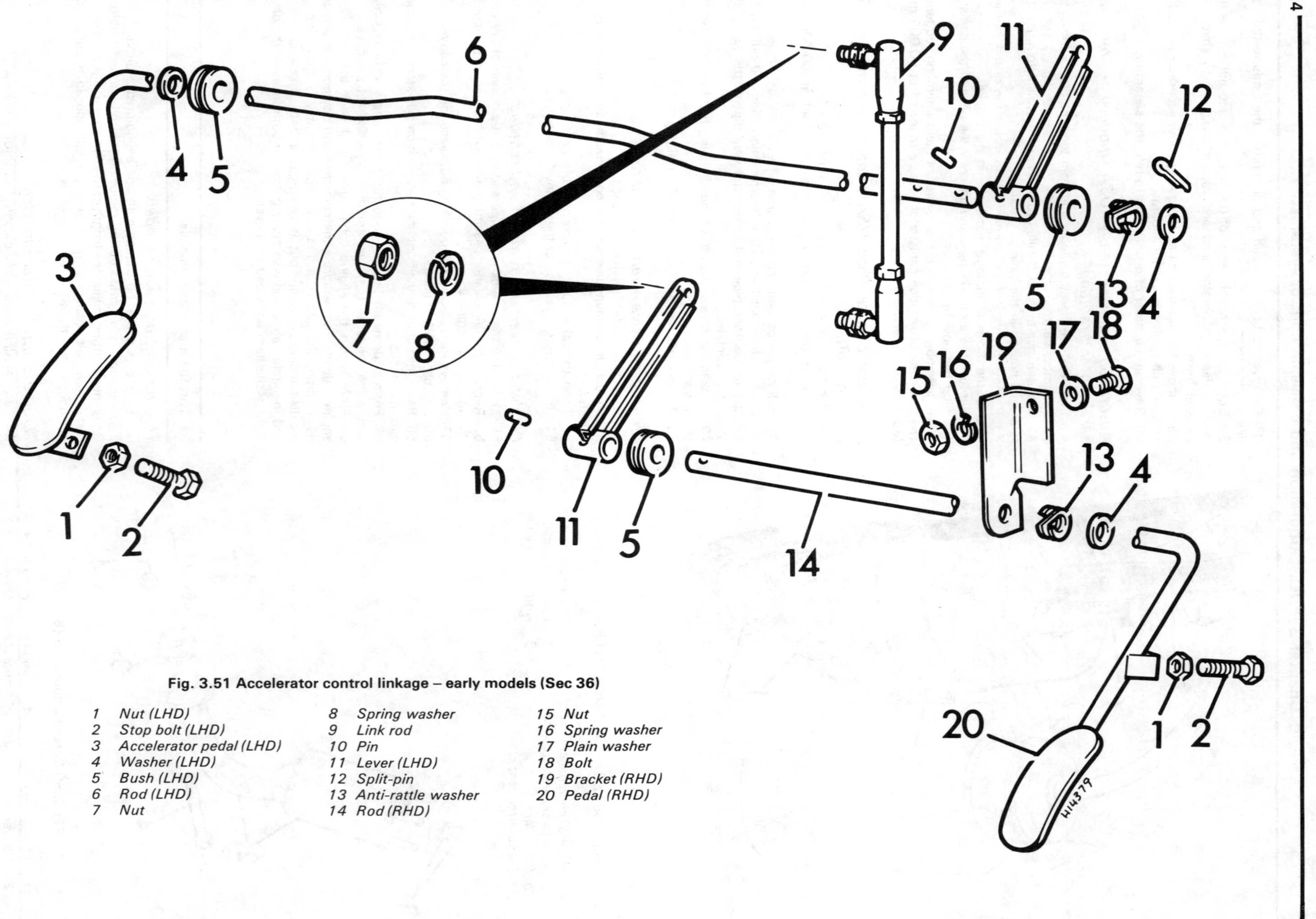

Fig. 3.51 Accelerator control linkage – early models (Sec 36)

1 Nut (LHD)
2 Stop bolt (LHD)
3 Accelerator pedal (LHD)
4 Washer (LHD)
5 Bush (LHD)
6 Rod (LHD)
7 Nut
8 Spring washer
9 Link rod
10 Pin
11 Lever (LHD)
12 Split-pin
13 Anti-rattle washer
14 Rod (RHD)
15 Nut
16 Spring washer
17 Plain washer
18 Bolt
19 Bracket (RHD)
20 Pedal (RHD)

Chapter 3 Fuel, carburation and emission control systems

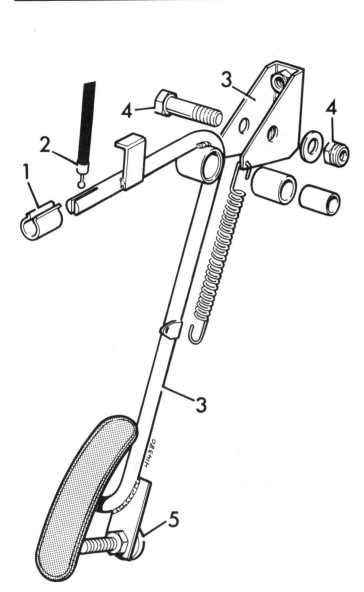

Fig. 3.52 Later type accelerator pedal (Sec 36)

1 Cable clip
2 Cable end nipple
3 Bracket and pedal
4 Pivot bolt
5 Pedal stop bolt

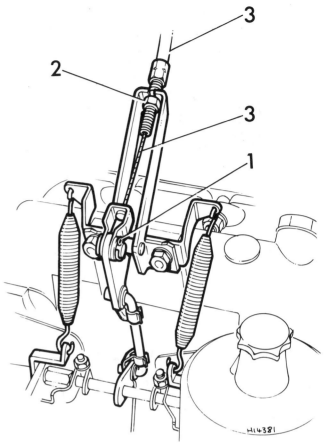

Fig. 3.53 Throttle cable attachment (Twin SU carburettors) (Sec 36)

1 Split-pin and clevis pin
2 Cable adjuster
3 Throttle inner and outer cables

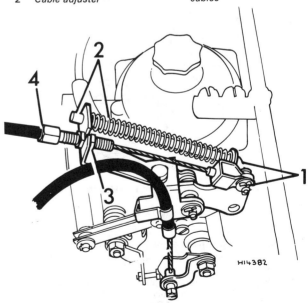

Fig. 3.54 Throttle cable attachment (Stromberg manual choke) (Sec 36)

1 Split-pin and spring-loaded rod
2 Spring and rod
3 Locknut
4 Adjuster

from the bulkhead after having disconnected the cable from the pedal arm by sliding the retaining clip off.

6 The throttle valve attachment to the carburettor varies according to the carburettor used. In all cases cable removal is as follows. Release the outer cable locknuts and extract the split-pins and clevis pins from the cable end fittings. Slide the cable out of the slot in the support bracket and withdraw the complete cable through the grommets in the engine compartment rear bulkhead.

7 Refitting is a reversal of removal, but note the different methods of adjustment according to carburettor type.

Twin SU carburettors

8 Tension the cable by means of the outer cable end fitting nuts until the fulcrum plate is just in contact with the machined face on the inlet manifold.

Single Stromberg carburettor (manual choke)

9 Adjust the cable end fitting nuts to provide $\frac{1}{16}$ in (1.6 mm) free movement at the carburettor linkage.

Chapter 3 Fuel, carburation and emission control systems

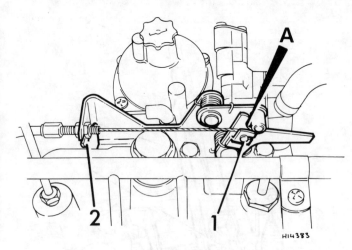

Fig. 3.55 Throttle cable attachment (Stromberg automatic choke) (Sec 36)

A   Roller to progression lever gap
1   Throttle cable nipple
2   Adjuster

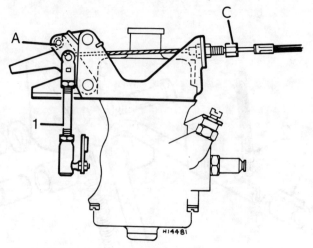

Fig. 3.56 Throttle cable adjustment (Stromberg automatic choke/carburettor) (Sec 36)

A   Roller to progression lever gap
C   Adjuster
1   Link rod

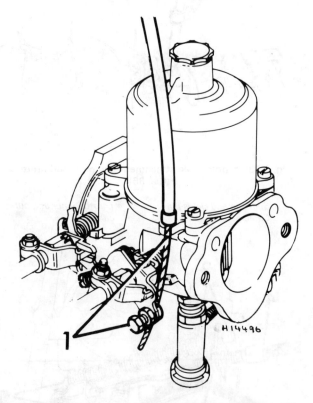

Fig. 3.57 Choke cable attachment to SU carburettor (Sec 36)

1   Cable and trunnion

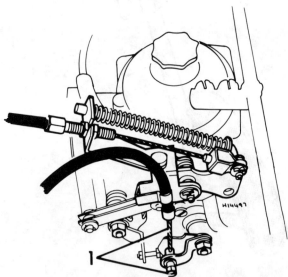

Fig. 3.58 Choke cable attachment to Stromberg carburettor (Sec 36)

1   Cable and trunnion

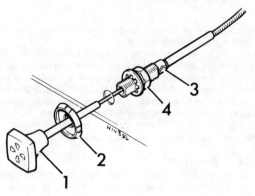

Fig. 3.59 Choke control knob components (Sec 36)

1   Knob
2   Bezel
3   Position locking clip
4   Backnut

Single Stromberg carburettor (automatic choke)
10 The adjustment should be carried out with the engine at normal running temperature.
11 Disconnect the link rod and adjust its length until at reconnection there is a gap (A) between the roller and progression lever of between 0.010 and 0.190 in (0.25 and 4.82 mm).
12 Eliminate all slack from the cable by adjusting the outer cable end fitting nuts.

All carburettors
13 Adjust the pedal stop bolt to give wide open throttle without straining the cable by over-depressing the pedal past full throttle.

Chapter 3 Fuel, carburation and emission control systems

Choke control cable

14 The inner cable on all models is connected to the carburettor operating arm by means of a small trunnion and pinch screw.
15 The outer cable is secured either by the engagement of the end fitting in the support bracket (SU) or by means of a spring clip (Stromberg).
16 The choke control knob is secured to the fascia panel by means of a bezel and back nut.
17 To remove the cable, release the cable from the carburettor and the control knob from the fascia panel and withdraw the cable through the engine compartment rear bulkhead grommets.
18 Refitting is a reversal of removal, but allow a slight free movement in the cable before tightening the pinch screw.

37 Manifolds and exhaust system

1 As the engine cylinder head is not of the crossflow type, both the

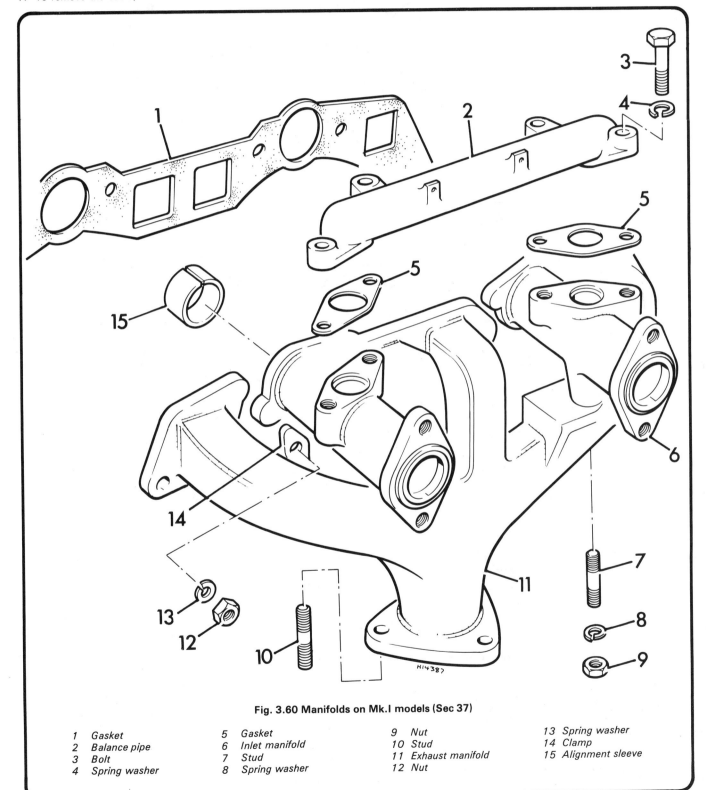

Fig. 3.60 Manifolds on Mk.I models (Sec 37)

1	Gasket	5	Gasket	9	Nut	13	Spring washer
2	Balance pipe	6	Inlet manifold	10	Stud	14	Clamp
3	Bolt	7	Stud	11	Exhaust manifold	15	Alignment sleeve
4	Spring washer	8	Spring washer	12	Nut		

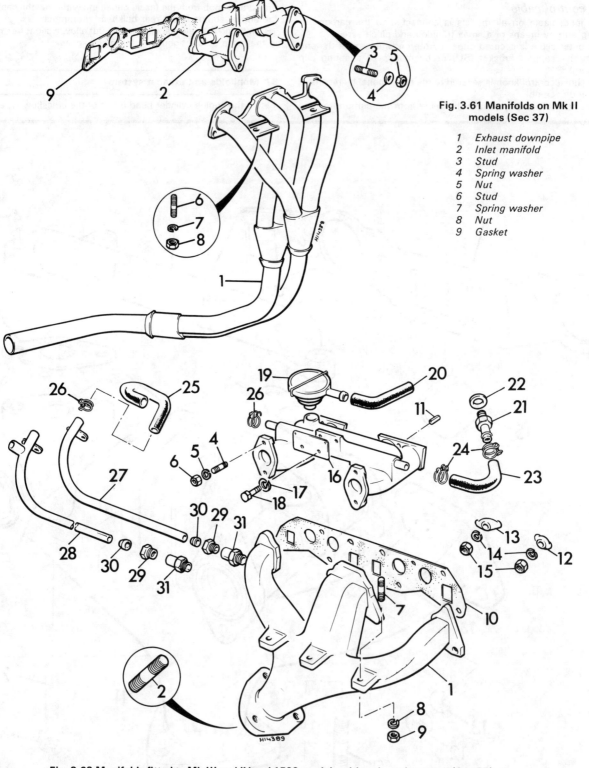

Fig. 3.61 Manifolds on Mk II models (Sec 37)

1 Exhaust downpipe
2 Inlet manifold
3 Stud
4 Spring washer
5 Nut
6 Stud
7 Spring washer
8 Nut
9 Gasket

Fig. 3.62 Manifolds fitted to Mk III and IV and 1500 models with twin carburettors (Sec 37)

1 Exhaust manifold
2 Stud
3 Inlet manifold
4 Stud
5 Spring washer
6 Nut
7 Stud
8 Spring washer
9 Nut
10 Gasket
11 Positioning dowel
12 Clamp
13 Clamp
14 Spring washer
15 Nut
16 Throttle return spring plate
17 Spring washer
18 Bolt
19 Crankcase breather valve (early models)
20 Hose
21 Adaptor
22 Fibre washer
23 Coolant hose
24 Clips
25 Coolant return hose
26 Clip
27 Return pipe (only if heater not fitted)
28 Return pipe (with heater fitted)
29 Nut
30 Adaptor
31 Return pipe

Chapter 3 Fuel, carburation and emission control systems

inlet and exhaust manifold are located on the same side of the engine.
2 The inlet manifold on Mk II and later models is coolant-heated and the cooling system will therefore have to be partially drained before the manifold can be removed.
3 Removal of the manifolds is otherwise straightforward. Once the attachments are disconnected, unscrew the retaining nuts and the exhaust downpipe flange nuts and remove them.
4 When refitting manifolds, always use new flange gaskets and always tighten the manifolds to the cylinder head before tightening the nuts which connect the inlet and exhaust manifolds to each other. Tighten all nuts to the specified torque.
5 The exhaust system may be one of several designs depending upon the year of production and model. The system may be of multi-section construction incorporating an expansion box and silencer or a silencer alone with a single or dual downpipe arrangement.
6 Removal of the exhaust system is simply a matter of disconnecting the downpipes from the manifold and then releasing the system mountings. Withdraw the complete system from beneath the car, jacking-up, if necessary, to obtain better access (photos).
7 It is not recommended that corroded or damaged sections of the exhaust system are removed while the system is still in position as further damage can be caused to good sections and the mountings distorted. It is better to remove the complete system where more purchase can be applied to disengage the sections.
8 Reassemble the new components but do not tighten any pipe clamps until the system has been installed on its mountings and the silencer and expansion box checked for correct alignment.

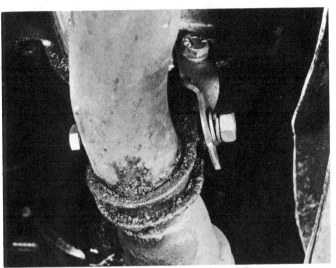

37.6a Exhaust bracket at gearbox

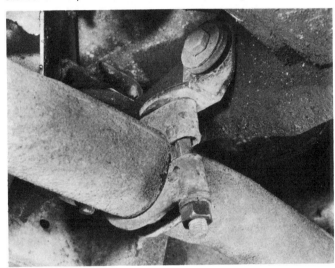

37.6b Exhaust bracket on rear exhaust section

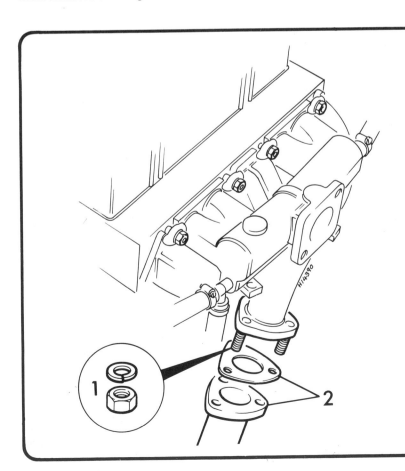

Fig. 3.63 Manifold on 1500 single carburettor models (Sec 37)

1 Nut
2 Flange and gasket

38 Fault diagnosis – fuel and exhaust systems

Symptom	Reason(s)
Fuel consumption excessive	Air cleaner choked giving rich mixture Leak from tank, pump or fuel lines Float chamber flooding due to incorrect level or worn needle valve Carburettor incorrectly adjusted Idling speed too high Incorrect valve clearances
Lack of power, stalling or difficult starting	Faulty fuel pump Leak on suction side of pump or in fuel line Intake manifold or carburettor flange gaskets leaking Carburettor incorrectly adjusted
Poor or erratic idling	Weak mixture Leak in intake manifold Leak in distributor vacuum pipe Leak in crankcase extractor hose

39 Fault diagnosis – emission control systems

Symptom	Reason(s)
Low CO content of exhaust gases (weak or lean mixture)	Fuel level incorrect in carburettor Incorrectly adjusted carburettor
High CO content of exhaust gases (rich mixture)	Incorrectly adjusted carburettor Choke sticking Absorption canister blocked Fuel level incorrect in carburettor Air injection system faulty
Noisy air injection pump	Belt tension incorrect Relief valve faulty Diverter valve faulty Check valve faulty

Chapter 4 Ignition system

Contents

Breakerless distributor – air gap adjustment	6
Capacitor (mechanical breaker distributor) – testing and renewal	12
Description – electronic ignition system	2
Description – mechanical breaker ignition system	1
Distributor – removal and refitting	7
Distributor (breakerless type) – overhaul	9
Distributor (mechanical breaker type) – overhaul	8
Drive resistor (breakerless electronic ignition)	15
Dwell angle – checking	5
Fault diagnosis – breakerless (electronic) ignition	17
Fault diagnosis – mechanical breaker ignition	16
Ignition coil	13
Ignition timing (breakerless type distibutor)	11
Ignition timing (mechanical breaker type distributor)	10
Mechanical contact breaker – points adjustment	3
Mechanical contact breaker – points renewal	4
Sparking plugs and leads	14

Specifications

System type
Except N. America	12V, battery, coil and mechanical breaker distributor
N. America	12V, Opus electronic breakerless system

Sparking plugs
Mk I and II	Champion L87Y
Mk III and IV	Champion N9Y
1500	Champion N12Y
Sparking plug gap	0.025 in (0.64 mm)

Ignition coil
Mk I, II and III	Lucas 12V
Mk IV and 1500	Lucas 6V with ballast resistor

Distributors
Mk I, II, III and IV	Delco Remy
1500 (except N. America)	Lucas 45D4
(N. America)	Lucas 45DE4

Delco Remy distributor
Contact points gap	0.014 to 0.016 in (0.35 to 0.40 mm)
Dwell angle	38 to 40°
Rotor rotation	Anti-clockwise
Capacitor capacity	0.18 to 0.23 μF

Lucas 45D4 distributor
Contact points gap	0.015 in (0.4 mm)
Dwell angle	46 to 56°
Rotor rotation	Anti-clockwise
Capacitor capacity	0.18 to 0.25 μF

Lucas 45DE4 distributor
Pick-up air gap .... 0.015 in (0.4 mm)
Rotor rotation .... Anti-clockwise

Ignition timing

	Static	Dynamic
Mk I and II models	17° BTDC	17° BTDC
Mk III models	6° BTDC	6° BTDC
Mk IV models:		
To engine No. FH/FK 25 000	6° BTDC	6° BTDC
From engine No. FH/FK 25 001	8° BTDC	6° BTDC
N. America	–	2° ATDC at 800/850 rpm
1500 models:		
Except N. America	10° BTDC	10° BTDC
N. America:		
To 1976	–	2° ATDC*
1977 to 1949 States	–	10° BTDC*
1977 to 1981 California	–	2° ATDC

* at idle speed 700 to 900 rpm, distributor vacuum pipe connected.

Torque wrench settings

	lbf ft	Nm
Distributor clamp plate bolt	20	27
Sparking plug	20	27

1 Description – mechanical breaker ignition system

1 In order that the engine can run correctly it is necessary for an electrical spark to ignite the fuel/air mixture in the combustion chamber at excatly the right moment in relation to engine speed and load. The ignition system is based on feeding low tension voltage from the battery to the coil where it is converted to high tension voltage. The high tension voltage is powerful enough to jump the sparking plug gap in the cylinders many times a second under high compression pressures, providing that the system is in good condition and that all adjustments are correct.
2 The ignition system is divided into two circuits. The low tension circuit and the high tension circuit.
3 The low tension (sometimes known as the primary) circuit consists of the battery, lead to the control box, lead to the ignition switch, lead from the ignition switch to the low tension or primary coil windings (terminal +), and the lead from the low tension coil windings (coil terminal +) to the contact breaker points and capacitor in the distributor.
4 The high tension circuit consists of the high tension or secondary coil windings, the heavy ignition lead from the centre of the coil to the centre of the distributor cap, the rotor arm, and the sparking plug leads and sparking plugs.
5 The system functions in the following manner. Low tension voltage is changed in the coil into high tension voltage by the opening and closing of the contact breaker points in the low tension circuit. High tension voltage is then fed via the carbon brush in the centre of the distributor cap to the rotor arm of the distributor.
6 The rotor arm revolves anti-clockwise at half engine speed inside the distributor cap, and each time it comes in line with one of the four metal segments in the cap, which are connected to the sparking plug leads, the opening and closing of the contact breaker points causes the high tension voltage to build up, jump the gap from the rotor arm to the appropriate metal segment and so via the sparking plug lead to the sparking plug, where it finally jumps the spark plug gap before going to earth.
7 The ignition is advanced and retarded automatically, to ensure the spark occurs at just the right instant for the particular load at the prevailing engine speed.
8 The ignition advance is controlled both mechanically and by a vacuum operated system. The mechanical governor mechanism comprises two lead weights, which move out from the distributor shaft as the engine speed rises due to centrifugal force. As they move outwards they rotate the cam relative to the distributor shaft, and so advance the spark. The weights are held in position by two light springs and it is the tension of the springs which is largely responsible for correct spark advancement.
9 The vacuum control consists of a diaphragm, one side of which is connected via a small bore tube to the carburettor, and the other side to the contact breaker plate. Depression in the inlet manifold and carburettor, which varies with engine speed and throttle opening, causes the diaphragm to move, so moving the contact breaker plate, and advancing or retarding the spark. A fine degree of control is achieved by a spring in the vacuum assembly.
10 One of two different makes of distributor may be fitted. Slight detail differences will be observed from the illustrations.

2 Description – electronic ignition system

1 This type of ignition is fitted to North American 1500 models from 1975 on.
2 The conventional cam, contact breaker points and capacitor are replaced by an oscillator, timing rotor, pick-up, amplifier and power transistor. These components are housed in a standard Lucas distributor body.
3 During operation, the oscillator supplies pulses to the pick-up in a continuous sequence. As the timing rotor, which is driven by the distributor shaft and has a ferrite rod for each hole, passes the pick-up, one of its ferrite rods catches one of the pulses and applies it to the amplifier. This amplified pulse causes the power transistor to switch off; this results in a collapse of the ignition coil primary current and a high secondary voltage is induced as in a conventional system.
4 The conventional centrifugal advance and retard unit is retained, although the vacuum advance unit is no longer used. However, there is a vacuum retard unit which comes into operation as described in the early paragraphs of this Section.
5 A separately mounted drive resistor is used as part of the control circuit for the amplifier. This is mounted on the diverter and relief valve bracket (see Section 15).

3 Mechanical contact breaker – points adjustment

1 To adjust the contact breaker points to the correct gap, first pull off the two clips securing the distributor cap to the distributor body, and lift away the cap. Clean the cap inside and out with a dry cloth. It is unlikely that the four segments will be badly burned or scored, but if they are the cap will have to be renewed.
2 Push in the carbon brush located in the top of the cap once or twice to make sure that it moves freely.
3 Gently prise the contact breaker points open to examine the condition of their faces. If they are rough, pitted or dirty, it will be necessary to remove them for resurfacing, or for replacement points to be fitted.
4 Assuming the points are satisfactory, or that they have been cleaned and replaced, measure the gap between the points by turning the engine over until the contact breaker arm is on the peak of one of the four cam lobes.
5 A 0.015 in (0.38 mm) feeler gauge should now just fit between the points.
6 If the gap varies from this amount, slacken the contact plate securing screw.

Chapter 4 Ignition system

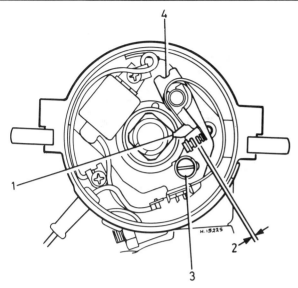

Fig. 4.1 Lucas contact points adjustment (Quickafit) shown (Sec 3)

1 Heel on high point of cam
2 Points gap
3 Adjuster/clamp screw
4 Adjuster slot

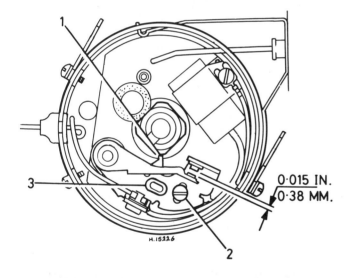

Fig. 4.2 Delco contact points adjustment (Sec 3)

1 Heel on high point of cam
2 Adjuster clamp screw
3 Fixed contact

Fig. 4.3 Lucas Quickafit contact arm and terminal plate (Sec 4)

7 Adjust the contact gap by releasing the contact arm fixing screw and insert a screwdriver in the contact screw driver slot and turn to vary the gap as required until the feeler gauge just slips in. Tighten the securing screw.
8 Replace the rotor arm and distributor cap and clip the spring blade cap retainer into place.
9 This method of adjusting the points should be regarded as a means of getting the engine started. More precise setting should be carried out by checking and adjusting the dwell angle as described in Section 5.

4 Mechanical contact breaker – points renewal, Delco Remy type

1 If the contact breaker points are burned, pitted or badly worn, starting is likely to be difficult and running erratic. Examine the faces of the points every 6000 miles (9600 km) and either resurface or fit new ones as wished.
2 Lift off the distributor cap and remove the rotor arm. Undo the fixed contact screw and lift the contact breaker assembly just enough to get at the terminal nut. Undo the nut and washer, and take off the LT cable and the capacitor from the terminal stud. Undo the nut from the terminal stud and lift off the contacts.
3 To reface the points, rub their faces on a fine carborundum stone, or on fine emery paper. It is important that the faces are rubbed flat and parallel to each other so that there will be complete face to face contact when the points are closed. One of the points will be pitted and the other will have deposits on it.
4 It is necessary to completely remove the built-up deposits, but not necessary to rub the pitted point right down to the stage where all the pitting has disappeared, though obviously if this is done it will prolong the time before the operation of refacing the points has to be repeated.
5 Thoroughly clean the points before refitting them. Place the fixed point in the distributor housing and loosely fit the locking screw and stud unit.
6 Apply one drop of oil to the arm pivot and install the moving point arm with the spring fitted between the insulator and the low tension wire terminal. Tighten the stud nut and set the gap between the points as described in Section 3 and 5.

Lucas type

7 Unscrew the retaining screw and remove complete with spring and flat washer. Raise the contact points and press the spring to release the terminal plate.
8 Carry out the operations described in Paragraphs 3 and 4.
9 Before refitting new or refaced points, clean the contacts with methylated sprit or petrol and lightly smear the pivot post with grease.
10 To refit the points, first position the adjustable contact breaker plate, and secure it with its screw spring and flat washer. On the early models, fit the fibre washer to the terminal pin, and fit the contact breaker arm over it. Insert the flanged nylon bush with the condenser lead immediately under its head, and low tension lead under that, over the terminal pin. Fit the steel washer and screw on the securing nut. On later models using Quickafit points check that the contact breaker spring is correctly located in the insulation.
11 The points are now reassembled and the gap should set as detailed in Sections 3 and 5.
12 It is worth remembering that poor earthing of the engine/transmission or the distributor baseplate can cause severe burning of the points. A faulty capacitor can also cause similar results (see Section 12).

5 Dwell angle – checking

1 This operation applies only to cars equipped with a mechanical breaker type distributor.
2 The dwell angle is the number of degrees through which the distributor cam turns during the period between the instants of closure and opening of the contact breaker points. Checking the dwell angle not only gives a more accurate setting of the points gap, but this method also evens out any variations in the gap which could be caused by wear in the distributor shaft or its bushes, or difference in height of any of the cam peaks.

Chapter 4 Ignition system

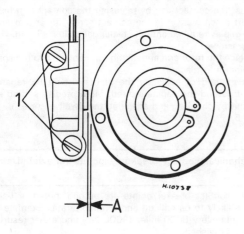

Fig. 4.4 Air gap adjustment diagram (Sec 6)

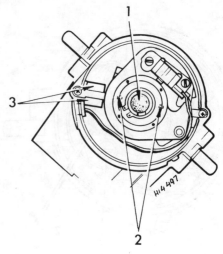

Fig. 4.5 Lubrication points on breakerless distributor (Sec 6)

1. Felt pad
2. Centre bearing
3. Lubrication holes

3 The angle should be checked with a dwell meter connected in accordance with the maker's instructions. Refer to the Specifications for the correct dwell angle.
4 If the dwell angle is too large, increase the points gap, if it is too small, reduce the points gap.
5 The dwell angle should always be adjusted before checking and adjusting the ignition timing. The ignition timing must always be checked after adjusting the dwell angle.

6 Breakerless distributor – air gap adjustment

1 This operation will normally only be required after overhaul of the distributor or renewal of its components.
2 With the battery disconnected, and the distributor cap, anti-flash shield and rotor arm removed, insert a 0.015 in (0.375 mm) feeler gauge into the gap and adjust so that it is a slight interference fit.
3 Adjust by slackening the two pick-up locking screws, and then move the pick-up to suit. Retighten the screws to lock the pick-up with the correct gap setting.
4 Re-check the gap after the pick-up screws have been tightened to ensure that the correct gap has been retained.
5 Lubricate with light oil all mechanical moving parts.
Note: *Do not insert the feeler gauge into the gap with the ignition circuit switched on!*

7 Distributor – removal and refitting

1 Check that the ignition is switched off.
2 Unclip the distributor cap and place it to one side without disconnecting the HT lead.
3 On mechanical breaker type distributors, disconnect the distributor LT wire.
4 On breakerless distributors, disconnect the the wire connectors.
5 Pull off the vacuum pipe from the vacuum capsule and the tachometer lead (later models) (photo).
6 Remove the bolt which holds the clamp plate to the cylinder block, but not release the clamp plate pinch bolt or the timing will be upset (see Section 10 or 11).
7 Withdraw the distributor.
8 Refitting is simply a matter of aligning the distributor shaft offsets and pushing it into position.
9 If the clamp plate pinch-bolt was slackened then turn the crankshaft until No. 1 piston is rising on its compression stroke. This is easily ascertained by removing No. 1 sparking plug and placing a finger over the plug hole to feed the compression being generated as the crankshaft is turned.
10 Continue to turn the crankshaft until the timing notch on the crankshaft pulley is aligned with the correct timing mark for your particular engine (see Specifications).

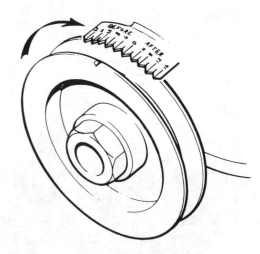

Fig. 4.6 Timing marks aligned at TDC (Sec 7)

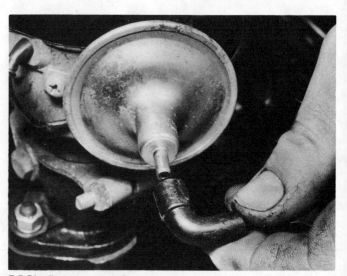

7.5 Distributor vacuum pipe

Chapter 4 Ignition system

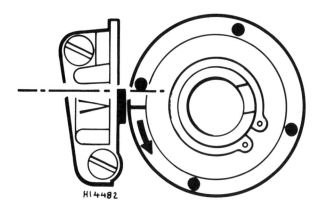

Fig. 4.7 Pick-up to ferrite rod relationship (breakerless distributor – static timing) (Sec 7)

11 Now turn the distributor (clamp pinch-bolt released) until the contact points are just about to open (mechanical breaker) or the pick-up to ferrite rod relationship is as indicated by the dotted line in Fig. 4.7 on breakerless distributors.
12 Tighten the clamp bolt.
13 Reconnect the distributor cap and wiring.
14 Time the ignition using a stroboscope as described in Section 10 or 11 according to type of distributor.
15 It should be realised that correct setting of the distributor is dependent upon the correct setting of the offset dogs on the driveshaft gear. If the gear has been disturbed, refer to Chapter 1, Section 14 or 40 according to engine type for the correct alignment procedure.

8 Distributor (mechanical breaker type) – overhaul

Delco Remy

1 The only time when the distributor should be dismantled is when it is wished to recondition it, and certain parts should always be discarded as a matter of course. Ensure that these parts, described in the following text, are available before taking the distributor down.
2 With the distributor removed from the car and on the bench, remove the distributor cap and lift off the rotor arm. If very tight, lever it off gently with a screwdriver.
3 Remove the points from the distributor as described.
4 Carefully remove the vacuum unit assembly after undoing the two screws from the side of the distributor body which also serves to partially hold the contact breaker plate in place.
5 Tap back the stakings which hold the end plug in place and drift out the tachometer drive gear, thrust washer, and the end cover using a 0.15 (3.81 mm) diameter rod.
6 Note particularly that the teeth on the drive dog are offset to the left when facing the slot which engages the rotor arm at the top of the shaft.
7 Drift out the rivet, and remove the drive dog, and bottom washer.
8 Pull out the compete mainshaft and cam assembly from the distributor body and then take off the upper washer. Pull out the clip and lift out the oil retaining felt.
9 The top and bottom washers and the advance weight springs are bound to have worn and stretched respectively and should be renewed.
10 Check the points as described in Section 3. Check the distributor cap for signs of tracking indicated by a thin black line between the segments. Replace the cap if any signs of tracking are found.
11 If the metal portion of the rotor arm is badly burned or loose, renew the arm. If slightly burnt clean the arm with a fine file. Check that the rotor contact spring conforms to setting shown in Fig. 4.9.
12 Check that the carbon brush moves freely in the centre of the distributor cover.
13 Examine the fit of the breaker plate on the bearing plate and also check the breaker arm pivot for looseness or wear and renew as necessary.

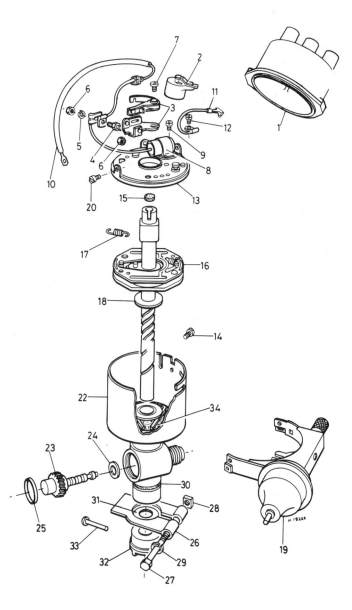

Fig. 4.8 Exploded view of Delco Remy distributor (Sec 8)

1 Cap
2 Rotor arm
3 Contact set
4 LT terminal stud
5 Lockwasher
6 Nut
7 Points adjusting lockscrew
8 Capacitor
9 Capacitor fixing screw
10 LT lead and grommet
11 Earth lead
12 Earth screw
13 Breaker plate
14 Screw
15 Felt pad
16 Shaft/cam assembly
17 Counterweight springs
18 Washer
19 Vacuum unit
20 Screw
21 Cap clip
22 Body
23 Tachometer drivegear
24 Thrust washer
25 Blanking plug
26 Clamp plate
27 Pinch bolt
28 Nut
29 Washer
30 O-ring seal
31 Thrust washer
32 Drive dog
33 Coupling pin
34 Clip for felt lubrication pad

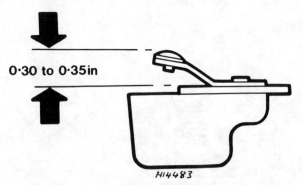

Fig. 4.9 Delco Remy distributor rotor arm spring setting (Sec 8)

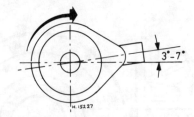

Fig. 4.10 Drive dog to rotor arm relationship (Delco Remy) (Sec 8)

14 Examine the balance weights and pivot pins for wear, and renew the weights or cam assembly if a degree of wear is found.
15 Examine the length of the balance weight springs and compare them with new springs. If they have stretched they must be renewed. It is almost inevitable that they will have stretched and it is best to fit new springs as a matter of course.
16 Examine the gear teeth on the drivegear and mainshaft for wear or damage.
17 Check the fit of the mainshaft in the housing bushes and if very loose renew.
18 Check that the vacuum unit is working properly and that there are no holes in the diaphragm.
19 Reassembly is a straightforward reversal of the dismantling process, but there are several points which should be noted.
20 Lubricate with SAE 20 engine oil the balance weights and other parts of the mechanical advance mechanism, the cam, the mainshaft, and the felts during assembly.
21 Always use a new upper and lower thrust washer and if fitting the original mainshaft and sleeve ensure they line up correctly. Check the mainshaft endfloat between the bottom thrust washer and the housing with a feeler gauge. The dimension should be between 0.002 and 0.005 in (0.02 and 0.13 mm), if using new thrust washers.
22 If a new undrilled distributor shaft is being fitted, the hole for the drive dog pin must be drilled so that the relationship between the slot for the rotor arm and the offset drive dog is as shown in Fig. 4.10 when viewed from the drive dog.
23 Ensure the rivet holes are aligned, and fit a new rivet. Also check that the endfloat is correct. Drill a rivet hole in the mainshaft using a No. 12 drill.
24 Always grease the tachometer drivegear and shaft and remember to stake over the housing plug in five or six places.
25 Finally set the contact breaker gap to the correct clearance.

Lucas
26 With the distributor removed from the car take off the rotor and remove the contact breaker assembly.
27 Pick out the felt pad from the recess in the top of the shaft.
28 Extract the capacitor screw and remove it with its connecting wire.
29 Extract the two fixing screws and remove the vacuum unit.
30 Extract the baseplate screw and take out the moving plate earth lead.
31 Using a small screwdriver, prise the expandable tag of the fixed baseplate inwards to release it from the cut-out in the distributor body.
32 Withdraw the fixed baseplate assembly.
33 Tap out the pin from the drivegear, noting the relationship of offset dogs to the shaft cutout for the rotor, remove the gear and the thrust washer.
34 Withdraw the shaft from the top of the distributor body.
35 Remove the distance collar from the shaft.
36 If the counterweights and springs are to be removed, mark their exact locations before doing so.
37 Clean all components and renew worn items.
38 Commence reassembly by lightly oiling the counterweight assembly and then fit the springs.
39 Fit the distance collar to the shaft.
40 Smear the shaft with light multi-purpose grease and insert it into the distributor body.

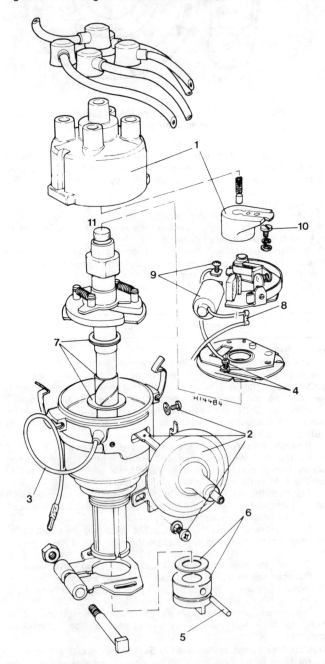

Fig. 4.11 Exploded view of Lucas 45 D4 distributor (Sec 8)

1 Cap and rotor
2 Vacuum unit
3 LT lead
4 Baseplate assembly
5 Pin
6 Drive dog and thrust washer
7 Shaft with distance collar and washer
8 Terminal connector
9 Capacitor
10 Contact set screw
11 Felt pad

Chapter 4 Ignition system

41 Fit the thrust washer and drivegear and drive in the retaining pin (offset dogs correctly aligned to rotor cut-out at the top of the shaft).
42 Apply grease to the pin on the moving plate then install the baseplate assembly by pushing it down into the body until its tang clicks into its cut-out.
43 Fit the earth lead and screw and tighten it to expand the baseplate tang.
44 Fit the vacuum capsule making sure that the link engages with the pin on the movable baseplate.
45 Pass the LT lead outwards through the hole in the body.
46 Refit the capacitor with earth lead from the movable plate.
47 Fit the contact breaker assembly.
48 Insert the felt pad to the top of the shaft and apply two drops of engine oil.
49 Set the breaker gap with a feeler gauge.
50 Refit the rotor and distributor cap.

9 Distributor (breakerless type) – overhaul

1 With the distributor removed from the car, take off the rotor and anti-flash cover.

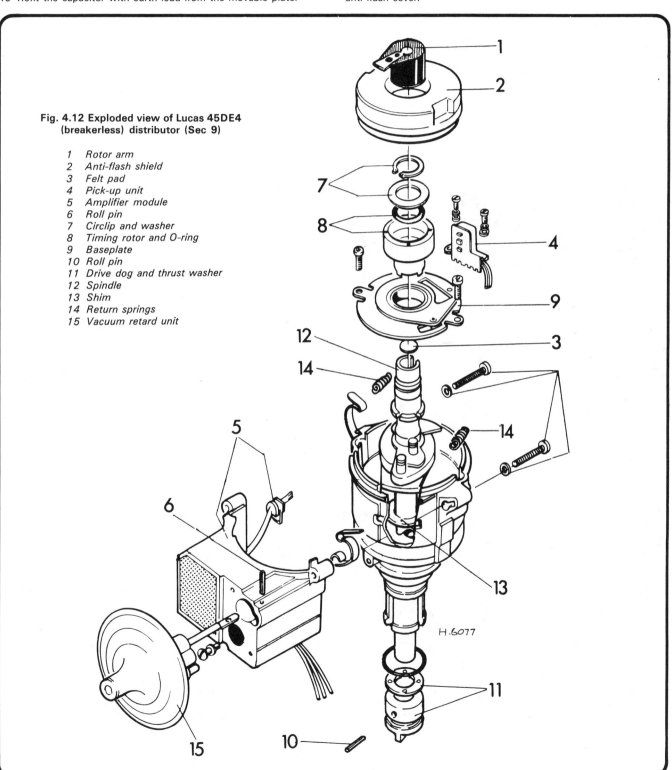

Fig. 4.12 Exploded view of Lucas 45DE4 (breakerless) distributor (Sec 9)

1 Rotor arm
2 Anti-flash shield
3 Felt pad
4 Pick-up unit
5 Amplifier module
6 Roll pin
7 Circlip and washer
8 Timing rotor and O-ring
9 Baseplate
10 Roll pin
11 Drive dog and thrust washer
12 Spindle
13 Shim
14 Return springs
15 Vacuum retard unit

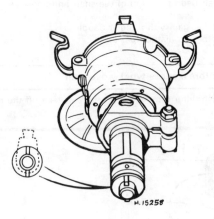

Fig. 4.13 Drive dog to rotor arm relationship (Lucas breakerless distributor) (Sec 9)

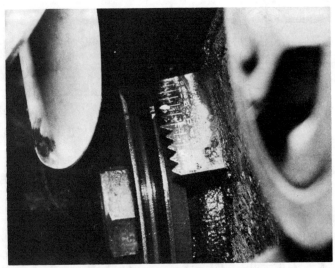

10.2a Timing at BTDC

10.2b Engine settings label (later models)

2 Pick out the felt pad from the top of the distributor shaft.
3 Release the two long screws which hold the upper part of the amplifier module.
4 Unscrew the remaining short amplifier fixing screw. Carefully withdraw the amplifier module and vacuum capsule whilst unhooking the link from the pivot pin on the movable plate.
6 Ease out the wiring grommet to fully remove the amplifier module with pick-up which are joined by wires.
7 Drive out the fixing pin and separate the vacuum unit from the amplifier module.
8 Extract the circlip from the top of the distributor shaft. Take off the washer and O-ring.
9 Withdraw the timing rotor.
10 Extract the two baseplate screws and remove the baseplate.
11 Drive out the drivegear retaining pin, remove the gear and thrust washer.
12 Withdraw the distributor shaft and take off the distance collar.
13 If necessary, remove the counterweight springs but do not dismantle further.
14 Reassembly is a reversal of dismantling, but observe the following points.
15 Lubricate with multi-purpose grease the distributor shaft and the pivot pin for the vacuum unit link.
16 When fitting the drivegear and its fixing pin to the shaft, make sure that the rotor arm to drive dog offset alignment is as shown in the illustration.
17 When inserting the timing rotor, make sure that the master projection enters the master groove correctly.
18 Adjust the pick-up air gap.
19 Apply two drops of engine oil to the felt pad at the top of the distributor shaft.

10 Ignition timing (mechanical breaker type distributor)

Static

1 Connect a test lamp between the negative terminal of the ignition coil and earth.
2 Remove the distributor cap and then turn the crankshaft until the notch on the pulley is opposite the appropriate timing mark on the timing scale (see Static settings in Specifications Section according to engine type or to timing label under bonnet) (photos).
3 Check that the rotor arm contact end is pointing to No. 1 contact in the distributor cap (as if fitted).
4 Release the distributor clamp plate bolt.
5 Switch on the ignition and rotate the distributor in an anti-clockwise direction until the test lamp lights up. Then turn it clockwise to the point where the lamp just goes out.
6 Tighten the distributor clamp plate bolt, switch off the ignition and remove the test lamp.

Dynamic

7 To set the ignition timing more accurately with the engine running, carry out the operations described in the following Section. With the mechanical breaker type ignition, the distributor vacuum pipe must be disconnected and plugged. The engine idle speed should be reduced to the lowest possible compatible with even tickover.
8 On very early models (Mk I and II) a vernier adjuster is incorporated for fine adjustment under road operating conditions.

11 Ignition timing (breakerless type distributor)

1 Timing on engines equipped with the Lucas breakerless electronic ignition system should only be carried out using the dynamic method described here. Static timing is not practical.
2 Connect a stroboscopic timing light in accordance with the manufacturer's instructions, making sure that timing is carried out on No. 1 cylinder.
3 Leave the distributor vacuum pipe connected.
4 Start the engine and hold the engine speed at 800 rpm as indicated on the car tachometer.
5 Point the timing light at the timing marks when the pulley notch

Are your plugs trying to tell you something?

Normal.
Grey-brown deposits, lightly coated core nose. Plugs ideally suited to engine, and engine in good condition.

Heavy Deposits.
A build up of crusty deposits, light-grey sandy colour in appearance.
Fault: Often caused by worn valve guides, excessive use of upper cylinder lubricant, or idling for long periods.

Lead Glazing.
Plug insulator firing tip appears yellow or green/yellow and shiny in appearance.
Fault: Often caused by incorrect carburation, excessive idling followed by sharp acceleration. Also check ignition timing.

Carbon fouling.
Dry, black, sooty deposits.
Fault: over-rich fuel mixture.
Check: carburettor mixture settings, float level, choke operation, air filter.

Oil fouling.
Wet, oily deposits. Fault: worn bores/piston rings or valve guides; sometimes occurs (temporarily) during running-in period.

Overheating.
Electrodes have glazed appearance, core nose very white – few deposits. Fault: plug overheating. Check: plug value, ignition timing, fuel octane rating (too low) and fuel mixture (too weak).

Electrode damage.
Electrodes burned away; core nose has burned, glazed appearance. Fault: pre-ignition. Check: for correct heat range and as for 'overheating'.

Split core nose.
(May appear initially as a crack). Fault: detonation or wrong gap-setting technique. Check: ignition timing, cooling system, fuel mixture (too weak).

WHY DOUBLE COPPER IS BETTER FOR YOUR ENGINE.

Unique Trapezoidal Copper Cored Earth Electrode — 50% Larger Spark Area — Copper Cored Centre Electrode

Champion Double Copper plugs are the first in the world to have copper core in both centre <u>and</u> earth electrode. This innovative design means that they run cooler by up to 100°C – giving greater efficiency and longer life. These double copper cores transfer heat away from the tip of the plug faster and more efficiently. Therefore, Double Copper runs at cooler temperatures than conventional plugs giving improved acceleration response and high speed performance with no fear of pre-ignition.

Champion Double Copper plugs also feature a unique trapezoidal earth electrode giving a 50% increase in spark area. This, together with the double copper cores, offers greatly reduced electrode wear, so the spark stays stronger for longer.

 FASTER COLD STARTING

 FOR UNLEADED OR LEADED FUEL

 ELECTRODES UP TO 100°C COOLER

 BETTER ACCELERATION RESPONSE

 LOWER EMISSIONS

 50% BIGGER SPARK AREA

 THE LONGER LIFE PLUG

Plug Tips/Hot and Cold.
Spark plugs must operate within well-defined temperature limits to avoid cold fouling at one extreme and overheating at the other.
Champion and the car manufacturers work out the best plugs for an engine to give optimum performance under all conditions, from freezing cold starts to sustained high speed motorway cruising.
Plugs are often referred to as hot or cold. With Champion, the higher the number on its body, the hotter the plug, and the lower the number the cooler the plug. For the correct plug for your car refer to the specifications at the beginning of this chapter.

Plug Cleaning
Modern plug design and materials mean that Champion no longer recommends periodic plug cleaning. Certainly don't clean your plugs with a wire brush as this can cause metal conductive paths across the nose of the insulator so impairing its performance and resulting in loss of acceleration and reduced m.p.g.
However, if plugs are removed, always carefully clean the area where the plug seats in the cylinder head as grit and dirt can sometimes cause gas leakage.
Also wipe any traces of oil or grease from plug leads as this may lead to arcing.

should appear stationary opposite the appropriate mark on the scale. If necessary, the marks may be painted beforehand with white paint to make them more visible.
6 If the timing marks do not appear to be in alignment, release the distributor clamp plate bolt and rotate the distributor one way or the other as required to bring the marks into alignment.
7 Tighten the clamp plate bolt, remove the timing light and switch off the ignition.

12 Capacitor (mechanical breaker distributor) – testing and renewal

1 The purpose of the capacitor, (sometimes known as a condenser) is to ensure that when the contact breaker points open there is no sparking across them which would waste voltage and cause wear.
2 The capacitor is fitted in parallel with the contact breaker points. If it develops a short circuit, it will cause ignition failure as the points will be prevented from interrupting the low tension circuit.
3 If the engine becomes very difficult to start or begins to miss after several miles running and the breaker points show signs of excessive burning, then the condition of the capacitor must be suspect. A further test can be made by separating the points by hand with the ignition switched on. If this is accompanied by a flash it is indicative that the capacitor has failed.
4 Without special test equipment the only sure way to diagnose capacitor trouble is to replace a suspected unit with a new one and note if there is any improvement.
5 To remove the capacitor from the distributor, remove the distributor cap and the rotor arm.
6 The capacitor is retained by a small screw. Its lead can be disconnected either by prising its end terminal from the terminal post or by unscrewing the terminal nut according to distributor type.

13 Ignition coil

1 High tension current should be negative at the spark plug terminals. Check that the LT lead from the distributor connects with the negative (–) terminal on the coil.
2 Without special equipment, the best method of testing for a faulty coil is by substitution of a new unit. Before doing this however, check the security of the connecting leads and remove any corrosion which may have built up in the coil HT socket.
3 Later vehicles are fitted with a six volt coil coupled with a six volt ballast resistance in series. Initially, the ballast resistor took the form of a separate unit, whereas later vehicles have the ballast resistor as a resistance wire built into the wiring harness feed supply to the coil. When starting, the voltage available to the coil is much lower then the nominal 12 volts of the system. The starter circuit therefore includes a change over mechanism to cut out the ballast resistance, and feed the full voltage to the coil. This slight overload of the coil ensures a good spark, yet it is for so short a time it does not overheat (photos).

13.3a Location of ignition coil

13.3b Ignition coil terminals and coil marking

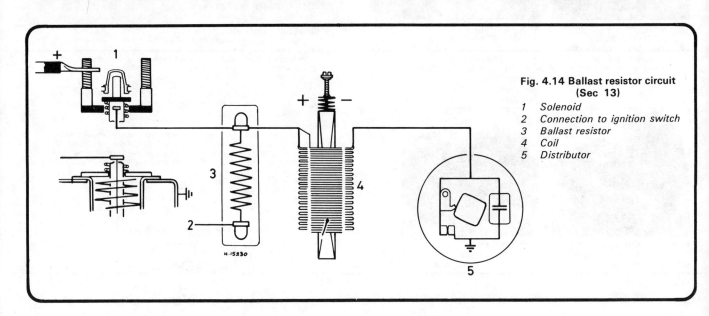

Fig. 4.14 Ballast resistor circuit (Sec 13)
1 Solenoid
2 Connection to ignition switch
3 Ballast resistor
4 Coil
5 Distributor

Chapter 4 Ignition system

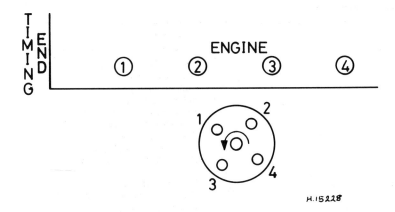

Fig. 4.15 HT lead connecting diagram (1500 cc) (Sec 14)

14 Sparking plugs and leads

1 The correct functioning of the sparking plugs is vital for the correct running and efficiency of the engine.
2 At the specified intervals the plugs should be removed, examined, cleaned, and if worn excessively, renewed. The condition of the sparking plug will also tell much about the overall condition of the engine.
3 If the insulator nose of the sparking plug is clean and white, with no deposits, this is indicative of a weak mixture, or too hot a plug. (A hot plug transfers heat away from the electrode slowly — a cold plug transfers it away quickly).
4 The plugs fitted as standard cannot be inproved upon. If the top and insulator nose is covered with hard black looking deposits, then this is indicative that the mixture is too rich. Should the plug be black and oily, then it is likely that the engine is fairly worn, as well as the mixture being too rich.
5 If the insulator nose is covered with light tan to greyish brown deposits, then the mixture is correct and it is likely that the engine is in good condition.
6 If there are any traces of long brown tapering stains on the outside of the white portion of the plug, then the plug will have to be renewed, as this shows that there is a faulty joint between the plug body and the insulator, and compression is being allowed to leak away.
7 Plugs should be cleaned by a sand blasting machine, which will free them from carbon more thoroughly than cleaning by hand. The machine will also test the condition of the plugs under compression. Any plug that fails to spark at the recommended pressure should be renewed.
8 The sparking plug gap is of considerable importance, as, if it is too large or too small, the size of the spark and its efficiency will be seriously impaired. The sparking plug gap should be set to that specified for the best results.
9 To set it, measure the gap with a feeler gauge, and then bend open, or close, the outer plug electrode until the correct gap is achieved. The centre electrode should never be bent as this may crack the insulation and cause plug failure if nothing worse.
10 When replacing the plugs, remember to use new plug washers, and replace the leads from the distributor in the correct firing order, which is 1, 3, 4, 2, No. 1 cylinder being the one nearest the radiator.
11 The plug leads require no routine attention other than being kept clean and wiped over regularly. At specified intervals, however, pull each lead off the plug in turn and remove them from the distributor by

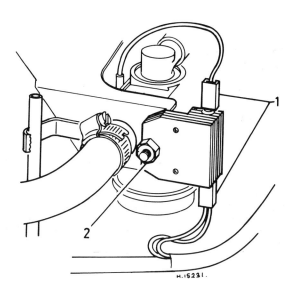

Fig. 4.16 Location of drive resistor

1 Spade connectors   2 Securing nut

unscrewing the knurled moulded terminal retaining rings or pulling them out (socket type). Water can seep down into these joints giving rise to a white corrosive deposit which must be carefully removed from the brass fitting at the end of each cable, through which the ignition wires pass.

15 Drive resistor (breakerless electronic ignition)

1 This is a remotely sited component of the ignition amplifier unit, its purpose is to function as an essential requirement of the amplifier transistors.
2 The drive resistor is mounted at the front end of the cylinder lead on the relief valve or air pump bracket and becomes hot with the ignition on.

16 Fault diagnosis – mechanical breaker ignition

Symptom	Reason(s)
Engine sluggish, hard to start	Contact breaker points not set correctly Plug gaps incorrect Ignition timing incorrect Wrong fuel used Loose battery connections Discharged battery Disconnected wires Damp sparking plug leads Damp distributor cap interior
Engine starts and runs but misfires	Incorrect points gap Worn distributor shaft or bearings Faulty sparking plugs Cracked distributor cap Cracked rotor Incorrect plug gap Faulty coil Worn advance mechanism Incorrect timing Poor earth connections
Engine overheats, lacks power	Seized centrifugal weights Poorly fitting vacuum pipe to carburettor or distributor Incorrect ignition timing (retarded)
Engine 'pinks' (pre-ignition)	Timing over-advanced Too low octane fuel Advance mechanism stuck open Broken centrifugal weight spring

17 Fault diagnosis – breakerless (electronic) ignition

1 Special equipment is required to test the major components of the system and to diagnose faults. It is therefore recommended that such work is left to your Lucas or British Leyland dealer.
2 This does not preclude your checking all the more obvious possibilities which could be responsible for a fault; these include:
    Air gap
    Loose wiring connections
    Cracked distributor cap
    Cracked rotor
    Loosely connected vacuum hose
    Wear in centrifugal weights or stretched or broken springs

Chapter 5 Clutch

Contents

Clutch – inspection and renovation	8
Clutch – refitting	9
Clutch – removal	7
Clutch hydraulic system – bleeding	5
Clutch pedal – removal and refitting	6
Clutch release bearing – renewal	10
Fault diagnosis	11
General description	1
Master cylinder – renewal, overhaul and refitting	3
Routine maintenance	2
Slave cylinder – removal, overhaul and refitting	4

Specifications

Up to engine No FC17136E (Mk I and Mk II)
Make .... Borg and Beck
Type .... Coil springs – single dry plate
Diameter .... $6\frac{1}{4}$ in (15.87 cm)
Clutch fluid .... Lockheed or Girling hydraulic fluid

From engine No FC17136E (Mk II and Mk III)
Make .... Borg and Beck
Type .... Diaphragm spring
Diameter .... $6\frac{1}{2}$ in (16.51 cm)
Clutch fluid .... Hydraulic fluid to SAE J1703 (Duckhams Universal Brake and Clutch Fluid)

Mk IV and 1500
Make/type .... Borg and Beck, single dry plate, diaphragm spring type
Release mechanism .... Hydraulically operated
Plate diameter .... $7\frac{1}{4}$ in (184 mm)
Facing material .... HK Porter 11046

Torque wrench settings

	lbf ft	Nm
Clutch to flywheel bolts	22	30
Bellhousing to engine bolts	14	19

1 General description

1 Early models are fitted with a Borg & Beck single dry plate clutch of $6\frac{1}{4}$ in diameter while later models are fitted with a $6\frac{1}{2}$ in diameter diaphragm spring unit.

2 The $6\frac{1}{4}$ in clutch comprises a steel cover cover which is bolted and dowelled to the rear face of the flywheel and contains the pressure plate, pressure plate springs, release levers, and clutch disc or driven plate.

3 The pressure plate, pressure springs, and release levers are all attached to the clutch assembly cover. The clutch disc is free to slide along the splined first motion shaft and is held in position between the flywheel and the pressure plate by the pressure of the pressure plate springs.

4 Friction lining material is riveted to the clutch disc and it has a spring cushioned hub to absorb transmission shocks and to help ensure a smooth take-off.

5 The $6\frac{1}{2}$ in diaphragm spring clutch is very similar to the $6\frac{1}{4}$ in unit but there are two main alterations. In place of the coil pressure springs there is just one diaphragm spring and this dispenses with the need for release levers.

6 The clutch is actuated hydraulically. The pendant clutch pedal, is connected to the clutch master cylinder and hydraulic fluid reservoir by a short push rod. The master cylinder and hydraulic reservoir are mounted on the engine side of the bulkhead in front of the driver.

7 Depressing the clutch pedal moves the piston in the master cylinder forwards, so forcing hydraulic fluid through the clutch hydraulic pipe to the slave cylinder.

8 The piston in the slave cylinder moves forward on the entry of the

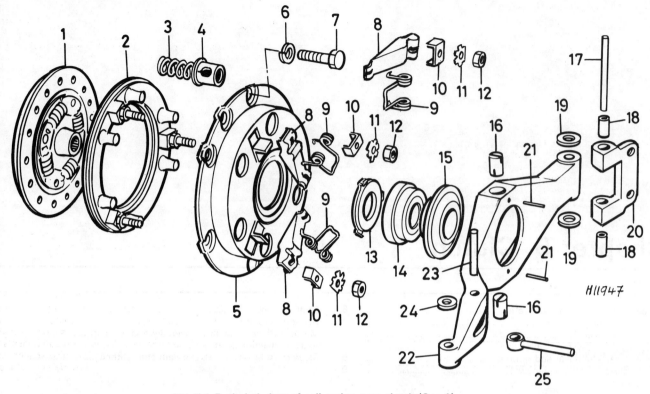

Fig. 5.1 Exploded view of coil spring type clutch (Sec 1)

1 Driven plate	8 Release levers	15 Release bearing mounting hub	20 Hinge plate
2 Pressure plate	9 Anti-rattle spring	16 Retaining plugs	21 Pin
3 Pressure spring	10 Bridge pieces	17 Hinge pin	22 Operating lever
4 Spring cup	11 Locking plates	18 Bushes	23 Pin
5 Pressure plate cover	12 Adjusting nuts	19 Spacer	24 Washer
6 Spring washer	13 Release lever plate		25 Pushrod
7 Setscrew	14 Release bearing		

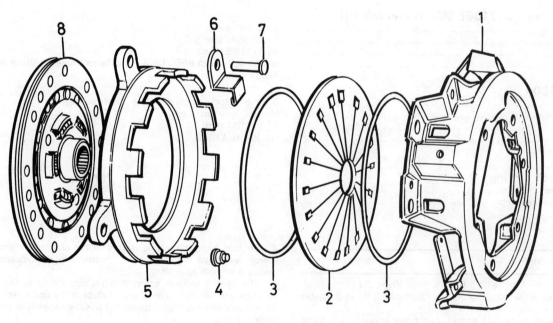

Fig. 5.2 Exploded view of diaphragm spring type clutch (Sec 1)

1 Pressure plate cover	3 Fulcrum ring	5 Pressure plate	7 Rivet
2 Diaphragm spring	4 Rivet	6 Clip	8 Driven plate

fluid and actuates the clutch release arm by means of a short pushrod.

9 On 6¼ in clutches the release bearing is pushed forwards to bear against the release bearing thrust plate and three clutch release levers. These levers are pivoted so as to move the pressure plate backwards against the pressure of the pressure plate springs, in this way disengaging the pressure plate from the clutch disc.

10 When the clutch pedal is released, the pressure plate springs force the pressure plate into contact with the high friction linings on the clutch disc, at the same time forcing the clutch disc against the flywheel and so taking up the drive.

11 On models fitted with the diaphragm spring clutch the release arm pushes the release bearing forwards to bear against the release plate, so moving the centre of the diaphragm spring inwards. The spring is sandwiched between two annular rings which act as fulcrum points. As the centre of the spring is pushed in the outside of the spring is pushed out, so moving the pressure plate backwards and disengaging the pressure plate from the clutch disc.

12 When the clutch pedal is released the diaphragm spring forces the pressure plate into contact with the high friction linings on the clutch disc and at the same time pushes the clutch disc a fraction of an inch forwards on its splines so engaging the clutch disc with the flywheel. The clutch disc is firmly sandwiched between the pressure plate and the flywheel so the drive is taken up.

13 As the friction linings on the clutch disc wear the pressure plate automatically moves closer to the disc to compensate. There is therefore no need to periodically adjust either type of clutch.

2 Routine maintenance

1 Routine maintenance consists of checking the level of the hydraulic fluid in the master cylinder reservoir every 1000 miles (1600 mm) and topping-up with hydraulic fluid if the level falls.

2 If it is noted that the level of the liquid has fallen then an immediate check should be made to determine the source of the leak.

3 Before checking the level of the fluid in the master cylinder reservoir, carefully clean the cap and body of the reservoir unit with a clean rag so as to ensure that no dirt enters the system when the cap is removed. On no account should paraffin or any other cleaning solvent be used in case the hydraulic fluid becomes contaminated.

4 Check that the level of the hydraulic fluid is up to within ¼ in (6.4 mm) of the filler neck and that the vent hole in the cap is clear. Do not overfill.

3 Master cylinder – removal, overhaul and refitting

1 Working within the engine compartment, at the rear bulkhead, pull back the rubber dust excluder and remove the split-pin from the clevis which holds the pushrod to the top of the clutch pedal. Pull out the clevis pin.

2 Place a rag under the master cylinder to catch any hydraulic fluid which may be spilt. Unscrew the union nut from the end of the hydraulic pipe where it enters the clutch master cylinder and gently pull the pipe clear.

3 Unscrew the two bolts and spring washers holding the clutch cylinder mounting flange to the mounting bracket.

4 Remove the master cylinder and reservoir, unscrew the filler cap, and drain the hydraulic fluid into a clean container.

5 Referring to Fig. 5.3 pull off the rubber boot to expose the circlip which must be removed so the pushrod complete with metal retaining washer can be pulled out of the master cylinder.

6 With a small electrical screwdriver lift the tag on the spring retainer which engages against the shoulder on the front of the piston shank and separate the piston from the retainer.

7 To dismantle the valve assembly manoeuvre the flange on the valve shank stem through the eccentrically positioned hole in the end face of the spring retainer. The spring, distance piece and valve spring seal washer can now be pulled off the valve shank stem.

8 Carefully ease the rubber seals from the valve stem and the piston respectively.

9 Clean and carefully examine all the parts, especially the piston cup and rubber washers, for signs of distortion, swelling, splitting, or other wear and check the piston and cylinder for wear and scoring. Replace any parts that are faulty and obtain a repair kit which will contain all the new seals and other renewable components. Use only clean hydraulic fluid or methylated spirit for cleaning internal components.

10 Rebuild the piston and valve assembly in the following sequence.

11 Fit the piston seal to the piston so the larger circumference of the rubber lip will enter the cylinder bore first. Then fit the seal in the same manner.

12 Fit the valve seal to the valve in the same way.

13 Place the valve spring seal washer so its convex face abuts the valve stem flange, and then fit the seat spacer and spring.

14 Fit the spring retainer to the spring which must then be compressed so the valve stem can be reinserted in the retainer.

15 Replace the front of the piston in the retainer and then press down the retaining leg so it locates under the shoulder at the front of the piston shank.

16 Generously lubricate the assembly with hydraulic fluid and carefully replace it in the master cylinder taking great care not to damage the rubber seals as they are inserted into the cylinder bore.

17 Fit the pushrod and washer in place and secure with the circlip. Replace the rubber boot.

18 Replacement of the unit in the car is a straightforward reversal of the removal sequence. Finally, bleed the system as described in Section 5.

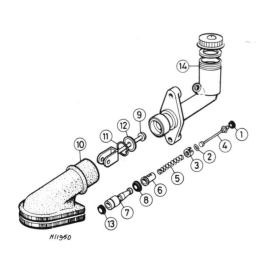

Fig. 5.3 Exploded view of clutch master cylinder (Sec 3)

1   Seal
2   Wave washer
3   Spacer
4   Valve
5   Spring
6   Spring retainer
7   Piston
8   Piston seal
9   Push-rod
10 Boot
11 Circlip
12 Washer
13 Seal
14 Body

4.3a Releasing slave cylinder clamp pinch bolt

4.3b Slave cylinder released from clamp

4 Slave cylinder – removal, overhaul and refitting

1   The clutch slave cylinder is positioned on the left-hand side of the bellhousing.
2   Before removing the cylinder take off the clutch reservoir cap and place a piece of thin polythene over the top of the reservoir. Screw down the cap tightly to create a vacuum which will reduce fluid loss when the pipeline is disconnected.
3   Release the clamp pinch-bolt and pull the slave cylinder from the clamp housing then disconnect the fluid line from the slave cylinder.
4   Referring to Fig. 5.4 pull off the rubber cover and take out the circlip. The piston, and the spring can then be shaken from the slave cylinder bore. Clean all the components thoroughly with hydraulic fluid or methylated spirit and then dry them off.
5   Carefully examine the rubber components for signs of swelling, distortion, splitting or other wear, and check the piston and cylinder wall for wear and score marks. Replace any parts that are found faulty and obtain a repair kit which will contain all the new seals and other renewable components.
6   Reassembly is a straightforward reversal of the dismantling procedure, but note the following points:-

   (a) As the compound parts are refitted to the slave cylinder barrel, smear them with hydraulic fluid

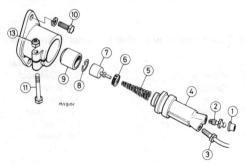

Fig. 5.4 Exploded view of slave cylinder (Sec 4)

1	Dust cap	8	Circlip
2	Bleed nipple	9	Sealing cup
3	Pipeline union	10	Bolt
4	Body	11	Pinch-bolt
5	Spring	12	Lockwasher
6	Seal	13	Self-locking nut
7	Piston		

   (b) When reassembling the operating piston, locate the piston seal at the end of the piston so that the sealing lip is towards the closed end of the slave cylinder bore
   (c) On completion of reassembly, top up the reservoir tank with the correct grade of hydraulic fluid and bleed the system, as described in Section 5

5 Clutch hydraulic system – bleeding

1   The method of bleeding is very similar to that described for the braking system in Chapter 9.
2   The bleed tube is connected to the bleed nipple at the rear of the clutch slave cylinder.

6 Clutch pedal – removal and refitting

1   If it is wished to renew the pivot bush or to remove the pedal first pull back the rubber dust excluder and then pull out the clevis pin.
2   Pull off the pedal return spring and with a pair of circlip pliers remove the circlip from the end of the pivot pin.
3   Push out the pivot pin from the bracket and pedal and take the pedal out of the bracket.
4   The old pivot bush if worn is simply pushed out of the pedal and a new one slid into place. Reassembly is a straightforward reversal of the removal sequence.

7 Clutch – removal

1   Remove the gearbox as described in Chapter 6, Section 3.
2   Remove the clutch assembly by unscrewing the six bolts holding the cover to the rear face of the flywheel. Unscrew the bolts diagonally half a turn at a time to prevent distortion to the cover flange. On later models, socket headed retaining screws are used.
3   With all the bolts and spring washers removed lift the clutch assembly off the locating dowels. The driven plate or clutch disc will fall out at this stage as it is not attached to either the clutch cover assembly or the flywheel (photo).

8 Clutch – inspection and renovation

1   Examine the clutch disc friction linings for wear and loose rivets and the disc for rim distortion, cracks, broken hub springs, and worn splines.

Chapter 5 Clutch

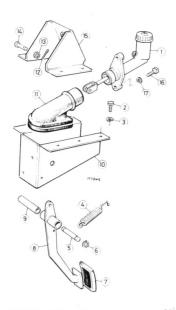

Fig. 5.5 Clutch pedal components (Sec 6)

1 Master cylinder	10 Bracket
2 Screw	11 Dust excluder
3 Spring washer	12 Split-pin
4 Return spring	13 Washer
5 Pivot pin	14 Clevis pin
6 Circlip	15 Bracket
7 Pedal rubber	16 Bolt
8 Pedal	17 Spring washer
9 Pivot bush	

7.3 Clutch removed from flywheel

2 It is always best to renew the clutch driven plate as an assembly to preclude further trouble. The manufacturers do not advise that only the linings are renewed and personal experience dictates that it is far more satisfactory to renew the driven plate complete than to try and economise by only fitting new friction linings.
3 Check the machined faces of the flywheel and the pressure plate. If either are badly grooved they should be machined until smooth. If the pressure plate is cracked or split it must be renewed, also if the portion on the other side of the plate in contact with the three release lever tips is grooved.
4 Check the release bearing thrust plate for cracks and renew it if any are found.
5 Examine the tips of the release levers which bear against the thrust plate, and renew the levers if more than a small flat has been worn on them.
6 On early models, renew any clutch pressure springs that are broken or shorter than standard.
7 Examine the depressions in the release levers which fit over the knife edge fulcrums and renew the levers if the metal appears badly worn.
8 Examine the clutch release bearing in the gearbox bellhousing and if it turns roughly or if it is cracked or pitted, it must be removed and replaced.
9 Also check the clutch withdrawal lever for slackness. If this is evident, withdraw the lever and renew the bush.

9 Clutch – refitting

1 It is important that no oil or grease gets on the clutch disc friction linings, or the pressure plate and flywheel faces. It is advisable to have clean hands when replacing the clutch and to wipe down the pressure plate and flywheel faces with a clean dry rag before assembly begins.
2 Place the clutch disc against the flywheel with the longer end of the hub facing towards the gearbox. On no account should the clutch disc be replaced with the longer end of the centre hub facing in to the flywheel as on reassembly it will be found quite impossible to operate the clutch with the friction disc in this position.
3 Replace the clutch cover assembly loosely on the dowels. Replace the six bolts and spring washers and tighten them finger tight so that the clutch disc is gripped but can still be moved.
4 The clutch disc must now be centralised so that when the engine and gearbox are mated, the gearbox input shaft splines will pass through the splines in the centre of the driven plate hub.
5 Centralisation can be carried out quite easily by inserting a round bar or long screwdriver through the hole in the centre of the clutch, so that the end of the bar rests in the small hole in the end of the crankshaft containing the input shaft bearing bush. Ideally an old Triumph input shaft or proprietary centring tool should be used.

9.2a Refitting clutch using old input shaft as alignment tool

9.2b Alignment of driven plate

9.2c Tightening pressure plate cover bolts

6 Using the input shaft bearing bush as a fulcrum, moving the bar sideways or up and down will move the clutch disc in whichever direction is necessary to achieve centralisation.
7 Centralisation is easily judged by removing the bar and viewing the driven plate hub in relation to the hole in the release bearing. When the hub appears exactly in the centre of the release bearing hole all is correct. Alternatively the input shaft, will fit the bush and centre of the clutch hub exactly, obviating the need for visual alignment.
8 Tighten the clutch bolts firmly in a diagonal sequence to ensure that the cover plate is pulled down evenly, and without distortion of the flange. The flywheel is prevented from turning by jamming the teeth of the starter ring gear.
9 Mate the engine and gearbox, bleed the slave cylinder if the pipe was disconnected and check the clutch for correct operation.

10 Clutch release bearing – renewal

1 Whenever the clutch is being renewed, the release bearing should be renewed at the same time even though it may appear to be in good order. This will save dismantling at a later date to replace a worn release bearing when the other clutch components are still serviceable.
2 With a thin metal drift drive the operating lever hinge pin out of the bellhousing.
3 Remove the operating lever complete with the release bearing and sleeve.
4 Place the operating lever on top of a vice and with the aid of a sawn off nail drive out the pins.
5 The returning plugs can then be partially levered out so as to release the release bearing mounting hub and bearing.
6 Place the old bearing in a vice and carefully lever off the bearing sleeve.
7 Note that the raised edge of the bearing is away from the sleeve and with the aid of a block of wood and a vice press the new bearing onto the old sleeve.
8 Refit the sleeve to the operating arm, tap back the retaining plugs, and refit the pins. Replace the arm in the bellhousing, and finally drift in the pin and lightly stake it in place.

11 Fault diagnosis – clutch

Symptom	Reason(s)
Judder when taking up drive	Loose engine mountings Worn or oil-contaminated driven plate friction linings Worn splines on driven plate hub or input shaft Worn crankshaft spigot bush
Clutch slip	Damaged or distorted pressure plate assembly Driven plate linings worn or oil-contaminated
Noise on depressing clutch pedal	Dry, worn or damaged clutch release bearing Excessive play in input shaft splines
Noise as clutch pedal is released	Distorted driven plate Broken or weak driven plate hub cushion coil springs Distorted or worn input shaft Release bearing loose on hub
Difficulty in disengaging clutch for gearchange	Fault in master cylinder or slave cylinder Air in hydraulic system Driven plate hub splines rusted on shaft

Chapter 6 Gearbox and overdrive

Contents

Fault diagnosis	38
Filters (Type J overdrive) – cleaning	33
Gear selectors (to 1970 part - synchro gearbox) – overhaul	9
Gearbox – dismantling general	4
Gearbox (to 1970 - part synchro) – examination and renovation	6
Gearbox components – examination and renovation	14
Gearbox (to 1970 - part synchro) – reassembly	11
Gearbox (1970 to 74 - all synchro) – reassembly	18
Gearbox/overdrive – removal and refitting	3
Gearbox (to 1970 - part synchro) – dismantling into major assemblies	5
Gearbox (1970 to 74 - all synchro) – description	12
Gearbox (1970 to 74 - all synchro) – dismantling into major assemblies	13
Gearbox (December 1974 on - single rail) – description	19
Gearbox (December 1974 on - single rail) – dismantling into major assemblies	21
Gearbox (December 1974 on - single rail) – reassembly	26
Gearbox (December 1974 on - single rail) – removal and refitting	20
Gearchange rod mechanism (December 1974 on - single rail gearbox) – overhaul	24
General description	1
Input shaft (to 1970 - part synchro gearbox) – overhaul	7
Input shaft (1970 to 74 - all synchro gearbox) – overhaul	15
Input shaft and layshaft/gear (December 1974 on - single rail gearbox) – overhaul	23
Inspection and preparation for reassembly – (December 1974 on - single rail gearbox)	25
Lubrication	2
Mainshaft (to 1970 - part synchro gearbox) – overhaul	8
Mainshaft (1970 to 74 - all synchro gearbox) – overhaul	16
Mainshaft (December 1974 on - single rail gearbox) – overhaul	22
Overdrive (Type D) – description	27
Overdrive (Type D) – operating lever adjustment	30
Overdrive (Type D) – overhaul	29
Overdrive (Type D) – removal and refitting	28
Overdrive valves (Type D) – removal, inspection and refitting	31
Overdrive (Type J) – description and maintenance	32
Overdrive (Type J) – overhaul	37
Overdrive (Type J) – removal and refitting	36
Pump non-return valve (Type J overdrive) – removal and refitting	35
Relief valve/dashpot (Type J overdrive) – removal and refitting	34
Remote control assembly (to 1970 - part synchro gearbox) – overhaul	10
Selector and remote control mechanism (1970 to 74 - all synchro gearbox) – description and overhaul	17

Specifications

Gearbox (Mk I, II, III and IV to October 1970)

Type .... Four forward speeds and reverse, no synchromesh on 1st gear. Three rail selector

Ratios
1st .... 3.75 : 1
2nd .... 2.16 : 1
3rd .... 1.39 : 1
4th .... 1.00 : 1
Reverse .... 3.75 : 1

Endfloat
Laygear .... 0.006 in (0.15 mm)
Mainshaft (2nd and 3rd gear endfloat on bushes) ...... 0.002 to 0.006 in (0.05 to 0.15 mm)
Mainshaft (overall endfloat of bushes) .... 0.004 to 0.010 in (0.10 to 0.25 mm)

Lubrication
Oil type/specification .... Hypoid gear oil, viscosity SAE 90EP (Duckhams Hypoid 90S)
Capacity:
  Without overdrive .... 1.5 Imp pts (1.8 US pts, 0.85 l)
  With overdrive .... 2.38 Imp pts (2.85 US pts, 1.35 l)

Overdrive (Laycock, Type D) .... Operational on 3rd and 4th gears

Overall ratios (overdrive engaged)
4th gear .... 3.30 : 1
3rd gear .... 4.60 : 1

Gearbox Mk IV, Oct 1970 to Dec 1974)
Type .... Four forward speeds and reverse. Synchromesh on all forward speeds. Three rail selector

Ratios
1st	3.50 : 1
2nd	2.16 : 1
3rd	1.39 : 1
4th	1.00 : 1
Reverse	3.99 : 1

Endfloat
Laygear	0.006 in (0.15 mm)
Mainshaft (1st gear endfloat on bushes	0.002 in (0.05 mm)
Mainshaft (2nd and 3rd gear endfloat on bushes)	0.006 in (0.15 mm)
Mainshaft overall endfloat	0.004 to 0.010 in (0.10 to 0.25 mm)

Lubrication
Oil type/specification .... Hypoid gear oil, viscosity SAE 90EP (Duckhams Hypoid 90S)

Capacity:
- Without overdrive .... 1.5 Imp pts (1.8 US pts, 0.85 l)
- With overdrive .... 2.7 Imp pts (3.25 US pts, 1.5 l)

Overdrive (Laycock, Type D)
Operational on 3rd and 4th gears

Overall ratios (overdrive engaged)
4th gear	2.89 : 1 (N. America 3.10 : 1)
3rd gear	4.03 : 1 (N. America 4.32 : 1)
Output shaft endfloat	0.005 to 0.010 in (0.13 to 0.25 mm)

Gearbox (1500 models Dec 1974 on)

Type .... Single rail selector, four forward speeds and reverse, synchromesh on all forward speeds

Ratios
1st	3.50 : 1
2nd	2.16 : 1
3rd	1.39 : 1
4th	1.00 : 1
Reverse	3.99 : 1

Endfloat
Laygear	0.007 to 0.015 in (0.18 to 0.38 mm)
Mainshaft (1st gear between split collars and thrust washer	0.004 to 0.013 in (0.10 to 0.33 mm)
Mainshaft (2nd gear endfloat on bush)	0.002 in (0.05 mm)
Mainshaft (3rd gear endfloat on bush)	0.002 to 0.006 in (0.05 to 0.15 mm)
Mainshaft overall endfloat (2nd and 3rd gear bushes)	0 to 0.006 in (0 to 0.15 mm)

Lubrication
Oil type/specification .... Hypoid gear oil, viscosity SAE 90EP (Duckhams Hypoid 90S)

Capacity:
- Without overdrive .... 1.5 Imp pts (1.8 US pts, 0.85 l)
- With overdrive .... 2.7 Imp pts (3.25 US pts, 1.5 l)

Overdrive (Laycock Type J)
Operational on 3rd and 4th gears

Overall ratios
4th gear	2.89 : 1 (N. America 3.10 : 1)
3rd gear	4.03 : 1 (N. America 4.32 : 1)

Torque wrench settings

	lbf ft	Nm

Gearbox (to Oct 1970)
	lbf ft	Nm
Bellhousing to engine bolts	14	19
Layshaft location bolt	25	34
Gearbox extension sset screws	15	20
Output flange nut	75	102
Operating shaft to gear lever	7	10
Reverse idler shaft	15	20
Speedometer sleeve attachment	15	20
Top cover attachment	7	10

Gearbox (Oct 1970 to Dec 1974)
	lbf ft	Nm
Drain plug	20	27
Extension housing bolts	20	27
Filler/level plug	20	27
Output flange nut	95	129
Bellhousing to gearcase bolts	30	41
Selector fork lock bolts	10	14

Chapter 6 Gearbox and overdrive

Top cover bolts	10	14
Bellhousing to engine bolts	14	19

Overdrive (Type D)
Top cover bolts	9	12
Support bracket bolts	20	27
Coupling flange castellated nut	100 to 130	136 to 173

Overdrive (Type J)
Adaptor to gearbox	9	12
Overdrive to adaptor	7	10
Overdrive to rear mounting	25	34
Restraining strap to overdrive	38	52
Sump plate bolts	6	8
Pressure filter base plug	16	22
Relief valve base plug	16	22
Pump non-return valve plug	16	22
Bridge piece nuts	8	11
Casing section nuts	15	20

Gearbox (single rail type Dec 74 on)
Clutch bellhousing to gear case bolts	32	44
Top cover bolts	9	12
Output flange nut	120	163
Drain plug	25	34
Filler/level plug	25	34
Reverse lamp switch	7	10
Seat belt switch	7	10
Reverse idle shaft locating screw	14	19
Selector shaft lock bolts	10	14
Bellhousing to engine bolts	14	19

1 General description

1   One of three types of gearbox may be fitted dependent upon the production date of the car.
2   On Mk I, II, III and IV cars built up until October 1970, the gearbox was of three rail type and did not have synchromesh on 1st gear.
3   Mk IV cars built between October 1970 and December 1974 were fitted with a similar gearbox but had the addition of synchromesh on 1st gear.
4   From December 1974, the 1500 model was equipped with a new single rail, all synchromesh gearbox.
5   An overdrive unit of Laycock manufacture has been available as a factory option on all models, a Type D unit up until 1974 and a Type J after that date.

2 Lubrication

1   At the intervals specified in 'Routine Maintenance', clean dirt from around the gearbox filler/level plug and unscrew the plug.
2   If necessary, top-up the oil level with the car on level ground, until it just begins to run out of the plug hole. Refit the plug.
3   Where an overdrive unit is fitted, it is lubricated by the oil supply common to the gearbox so no separate attention is required.
4   Also at the specified service intervals, renew the gearbox lubricant. Do this after a long run when the oil is hot.
5   Remove the drain plug and filler/level plug and allow the oil to drain into a suitable container.
6   When the oil has drained completely, refit the drain plug and fill the unit with oil of the specified type. Refit the filler/level plug.
7   The car manufacturers no longer specify routine oil changes.
8   It is still advisable however to change the lubricant periodically, say, every 25 000 miles (40 000 km) to ensure that any metal particles held in suspension in the oil are drained away. In addition, the additives contained in the lubricant to combat corrosion and to maintain other lubricating characteristics gradually deteriorate in time.

3 Gearbox/overdrive – removal and refitting

1   The gearbox may be removed together with the engine as described in Chapter 1 or on its own as described in the following paragraphs.
2   If an overdrive is fitted, this can be removed in conjunction with the gearbox and separated later, refer to Section 28 (Type D) or Section 36 (Type J).
3   To remove a gearbox fitted with a Type J overdrive if the intention is to separate the gearbox and overdrive after removal, raise the rear roadwheels clear of the ground. Start the engine and engage the transmission, working through the gears until the overdrive can be engaged at the recommended speed. Now disengage the overdrive while the clutch pedal is held depressed. This action releases the overdrive roller clutch spline loading to facilitate separation of the overdrive later on.
4   Disconnect the battery.
5   Drain the oil from the gearbox.
6   Working inside the car, release the locknut or ring. Unscrew and remove the knob from the gearchange lever. On later models with overdrive which have the control switch built into the gear lever knob, prise off the switch cap from the knob and disconnect the switch leads (photos)

3.6 Overdrive switch (later models)

Chapter 6 Gearbox and overdrive

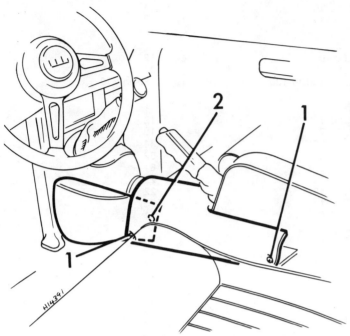

Fig. 6.1 Gearbox cover panel (Sec 3)

1 Trim pad screws 2 Trim pad screws

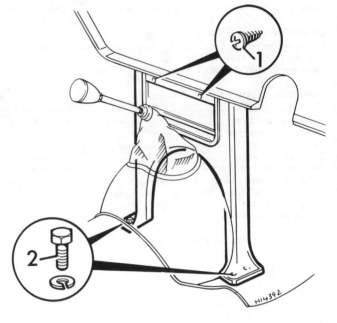

Fig. 6.2 Fascia support bracket (Sec 3)

1 Self-tapping screws 2 Lower mounting bolts

7 To make the following operations easier, remove the seats from inside the car as described in Chapter 12. The gearbox cover must now be removed. To do this, first extract the screws from the tunnel trim pads then unbolt the fascia support bracket. Remove the carpet. On three rail type gearboxes the gear lever can be removed either by unbolting the gearbox top cover complete or by disconnecting the bolt at its lower end and separating it from the remote control rod. Release the bayonet type cap. On single rail type gearboxes, remove the gearchange lever by releasing the gaiter, its retainer and the bayonet type securing cap. As the lever is withdrawn, take care not to lose the spring and nylon plunger. If overdrive is fitted, feed the previously disconnected switch leads through the lever. Cover the opening in the gearbox to prevent entry of dirt (photos).
8 Now extract the gearbox tunnel cover screws and the small reinforcement plates.
9 Withdraw the cover to expose the gearbox and front end of the propeller shaft.
10 Disconnect the front end of the propeller shaft from the gearbox output flange.
11 Disconnect the speedometer drive cable from the gearbox (photo).
12 Remove the pinch-bolt which retains the clutch slave cylinder, withdraw the cylinder and carefully tie it away from the gearbox. There is no need to disconnect the hydraulic pipeline, but do not bend or kink the pipe (photo).

13 Support the engine by placing a jack and a protective block of wood under the sump.
14 Disconnect the exhaust pipe from the gearbox support bracket (photo).
15 Remove the gearbox rear mounting bolts (photo) from the crossmember.
16 Disconnect the engine restraining cable from the clutch bellhousing.
17 Working under the car, unscrew and remove the bolts which connect the bellhousing to the engine.
18 Open the bonnet and unscrew the starter motor mounting bolts and pull the starter motor away from its mounting flange. Working inside the car, disconnect the electrical leads from the reverse lamp switch, the overdrive and seat belt warning system where fitted (photo).
19 Remove the mounting assembly from the gearbox.
20 Remove the remaining upper bellhousing bolts, noting the location of the engine earth bonding strap (photo).
21 With the help of an assistant, carefully withdraw the gearbox towards the rear of the car. The gearbox may be lifted carefully onto the floor by reaching through the aperture left by removal of the tunnel cover. Do not allow the weight of the gearbox to hang upon the input shaft while it is still engaged with the clutch driven plate. Protect the carpets from soiling by the gearbox.

3.7a Centre tunnel trim pads

3.7b Fascia support lower bolts

3.7c Fascia support upper screws

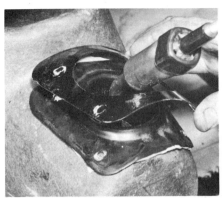
3.7d Gear lever retainer and gaiter (single rail gearbox)

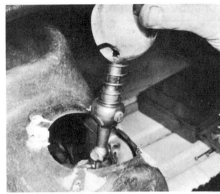
3.7e Gearchange lever assembly being removed (single rail gearbox)

3.7f Gearchange lever plunger and spring (single rail gearbox)

3.7g Gearchange lever assembly (single rail gearbox)

3.8 Transmission tunnel screws and plates

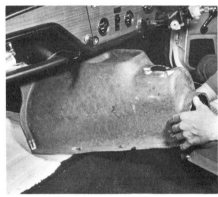
3.9 Removing transmission tunnel cover

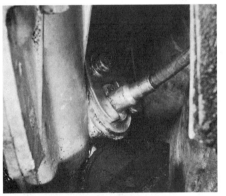
3.11 Speedometer cable connection

3.12 Removing clutch slave cylinder pinch bolt

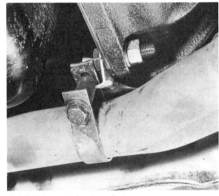
3.14 Exhaust pipe bracket

3.15 Gearbox rear mounting bolts

3.18 Reverse lamp switch

3.20 Engine earth bonding strap

Chapter 6 Gearbox and overdrive

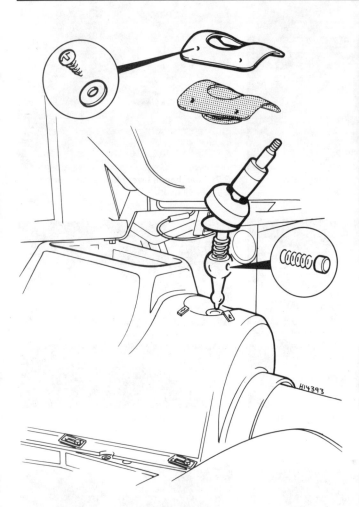

Fig. 6.3 Gearchange lever components (later type shown) (Sec 3)

5.1 Releasing remote control extension nuts

5.2 Removing gearbox top cover

5.3 Unscrewing speedometer pinion peg bolt

22 Refitting is a reversal of removal, but if the clutch has been disturbed, make sure that it has been centralised (see Chapter 5) or it will be quite impossible to fit the gearbox to the engine.
23 Refill the gearbox with the specified oil.

4 Gearbox – dismantling general

1 With the gearbox removed from the car, clean away all external dirt and grease using a water soluble solvent or paraffin and a stiff brush.
2 If an overdrive unit is fitted, remove this as described in Section 28.
3 From inside the bellhousing, remove the clutch release components as described in Chapter 5.
4 Finally, support the gearbox securely on a strong bench.

5 Gearbox (to 1970 - part synchro) – dismantling into major assemblies

1 Undo the four nuts and spring washers which hold the remote control extension in place and lift off the extension (photo).
2 Undo the eight bolts which hold the gearbox cover to the top of the gearbox and lift off the cover (photo).
3 The peg bolt which retains the speedometer drive housing is then unscrewed (photo).
4 Pull the housing complete with the speedometer drive pinion out of the gearbox extension (photo).
5 If you have a vice with wide enough opening jaws, grip the output flange in its jaws. Alternatively bolt a long lever to two of the holes in

Chapter 6 Gearbox and overdrive

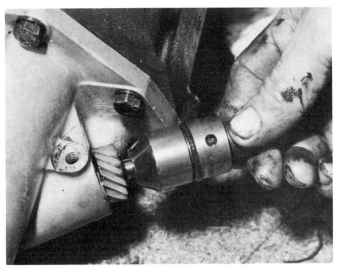

5.4 Withdrawing speedometer drive pinion

5.5 Unscrewing output flange nut

the flange to prevent it from turning when the nut is unscrewed. This should now be done and the washers removed. Withdraw the flange (photo).
6 Pull the extension away from the mainshaft flange (photo).
7 Undo the bolts which retain the aluminium alloy extension to the rear of the gearbox (photo). Note the position of the longer bolt.
8 Remove the extension by tapping the underside of the mounting lug with a rawhide hammer. Lift the extension off the gearbox (photo).
9 Then lift the reverse idler gear from the reverse gear shaft (photo).
10 From inside the bellhousing undo the bolts which hold the bellhousing to the front of the gearbox (photo).

11 Separate the bellhousing from the gearbox and place the former on one side (photo).
12 Then undo the bolt and washer which secure the layshaft in place (photo).
13 Pull the layshaft out of the gearbox so the laygear drops out of mesh with the mainshaft gear.
14 With a soft metal drift carefully tap the input shaft complete with bearing forwards from inside the gearbox.
15 As soon as the bearing is clear of the gearbox casing lift the input shaft out (photo).

5.6 Separating output flange from shaft

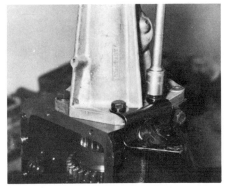

5.7 Unscrewing rear extension bolts

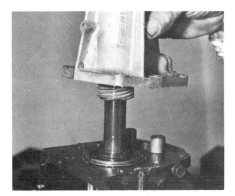

5.8 Removing rear extension from gearbox

5.9 Removing reverse idler gear

5.10 Unscrewing bellhousing bolts

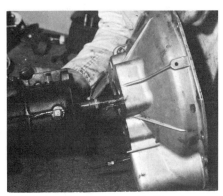

5.11 Separating gearbox and bellhousing

Chapter 6 Gearbox and overdrive

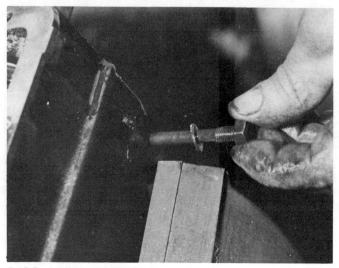
5.12 Removing layshaft lock bolt

5.15 Removing input shaft

16 Turning to the mainshaft, tap the end inside the gearbox with a rawhide or plastic headed hammer until the rear end ball bearing is clear of the casing.
17 The mainshaft can now be tilted and the synchronising unit slid off (photo).
18 Then remove the synchroniser ring (photo), which will probably have been left in place.
19 With a pair of fine nosed circlip pliers expand the circlip out of its retaining groove (photo).
20 With the aid of a couple of screwdrivers carefully ease the circlip off the nose of the mainshaft (photo).

21 The mainshaft can now be removed from the rear of the gearbox by driving it out with the aid of a plastic headed hammer (photo).
22 During the final stage hold the mainshaft gear cluster together while the shaft is pulled out (photo).
23 Then lift the mainshaft gear cluster out of the gearbox and place on one side (photo).
24 Raise the laygear so it is in its normal position and measure the endfloat with a feeler gauge. The laygear is then free t r be lifted out (photo).
25 Remove the reverse gear shaft by undoing the peg bolt and spring washer which holds it in place (photo).

5.17 Removing 3rd/4th synchro unit

5.18 Removing 3rd gear baulk ring

5.19 Extracting circlip from mainshaft groove

5.20 Easing circlip from mainshaft

5.21 Driving out mainshaft

5.22 Withdrawing mainshaft from gear cluster

Chapter 6 Gearbox and overdrive

5.23 Removing mainshaft gear cluster

5.24 Removing laygear

5.25 Removing reverse idler gear shaft and peg bolt

5.26 Removing reverse operating lever and nut

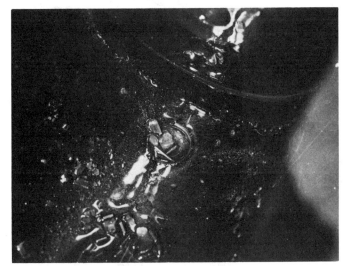

5.27 Condition of gearcase interior after stripping

26 Finally undo the nut and bolt and remove together with the operating lever (photo).

27 The gearbox is now stripped right out and must be thoroughly cleaned. Depending upon the quantity of metal chips and fragments in the bottom of the gearbox some items may be found to be badly worn. The component parts of the gearbox should now be examined for wear, and the laygear, input shaft and mainshaft assemblies broken down further as described in the following sections (photo).

6 Gearbox (to 1970 - part synchro) – examination and renovation

1 Carefully clean and then examine all the component parts for general wear, distortion, slackness of fit, and damage to machined faces and threads.

2 Examine the gearwheels for excessive wear and chipping of the teeth. Renew them as necessary. If the laygear endfloat is outside the stated tolerance range (see Specifications) the thrust washers must be renewed. New thrust washers will almost certainly be required on any car that has completed more than 50 000 miles (80 000 km).

3 Examine the layshaft for signs of wear where the laygear bushes bear, and check the laygear on a new shaft for worn bushes. These are simply drifted out if new bushes are to be fitted.

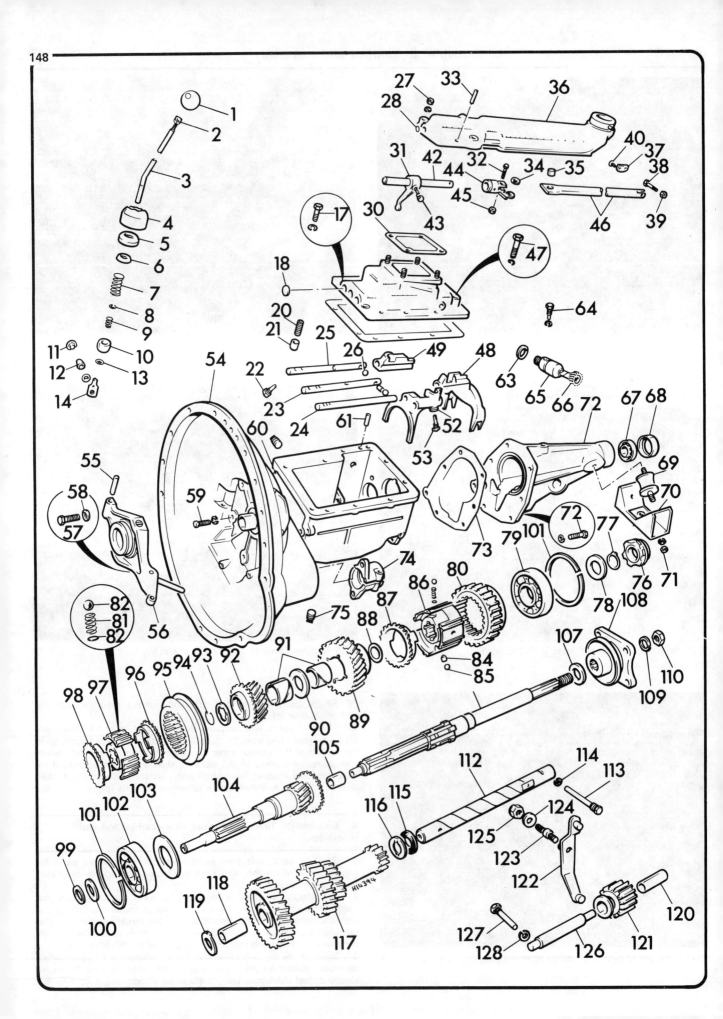

Fig. 6.4 Exploded view of part synchro gearbox used until 1970 (Sec 5)

1 Gear knob
2 Locknut
3 Gear lever
4 Cover
5 Shield
6 Plate
7 Spring
8 Circlip
9 Spring
10 Nylon ball
11 Stepped nylon washer
12 Bush
13 Washer
14 Gear lever end
15 Reverse stop pin
16 Locknut
17 Bolt
18 Plug
19 Gasket
20 Spring
21 Plunger
22 Taper locking pin
23 1st & 2nd gear selector shaft
24 3rd & 4th gear selector shaft
25 Reverse gear selector shaft
26 Interlock ball
27 Nut
28 Rubber O-ring
29 Gearbox top cover
30 Gasket
31 Selector ball-end
32 Bolt
33 Dowel
34 Washer
35 Bonded rubber bush
36 Gearchange extension
37 Reverse stop
38 Bolt
39 Nyloc nut
40 Screw
41 Pin
42 Front remote control rod
43 Taper locking pin
44 Fork
45 Nut
46 Rear remote control rod
47 Bolt
48 1st & 2nd gear selector fork
49 Reverse selector
50 Interlock ball
51 Interlock plunger
52 3rd & 4th gear selector fork
53 Taper locking pin
54 Clutch bellhousing
55 Pin
56 Clutch release mechanism
57 Wedgelock bolt
58 Plain washer
59 Bolt
60 Gasket
61 Dowel
62 Gearbox rear extension
63 Rubber O-ring
64 Peg bolt
65 Speedometer drive gear housing
66 Speedometer drive gear
67 Gearbox extension ball race
68 Oil seal
69 Gearbox mounting rubber
70 Mounting bracket
71 Nut
72 Bolt
73 Gasket
74 Clutch slave cylinder bracket
75 Gearbox drain plug
76 Speedometer driving gear
77 Circlip
78 Distance washer
79 Ball race
80 1st speed gear
81 Spring
82 Shim
83 Synchromesh ball
84 Plunger
85 Ball
86 2nd speed synchro hub
87 2nd speed synchro baulk ring
88 Thrust washer
89 2nd speed mainshaft gear
90 Thrust washer
91 Bushes
92 3rd speed mainshaft gear
93 Thrust washer
94 Circlip
95 3rd & 4th speed synchro sleeve
96 3rd speed synchro baulk ring
97 3rd & 4th speed inner synchro hub
98 4th speed synchro baulk ring
99 Circlip
100 Distance washer
101 Circlip
102 Ball race
103 Oil deflector
104 Input shaft
105 Needle roller bearing
106 Mainshaft
107 Distance washer
108 Driving flange
109 Spring washer
110 Nut
112 Layshaft
113 Peg bolt
114 Spring washer
115 Rear fixed thrust washer
116 Rear rotating thrust washer
117 Layshaft gear cluster
118 Layshaft bush
119 Front fixed thrust washer
120 Reverse gear bush
121 Reverse gear
122 Reverse gear actuator
123 Actuator pivot
124 Plain washer
125 Nyloc nut
126 Reverse gear shaft
127 Reverse shaft retaining bolt
128 Spring washer

Chapter 6 Gearbox and overdrive

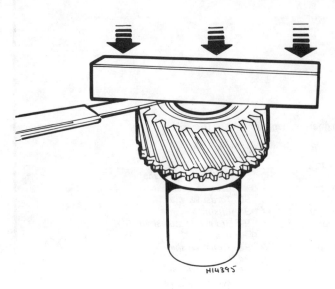

Fig. 6.5 Checking mainshaft 2nd and 3rd gear to bush endfloat (Sec 6)

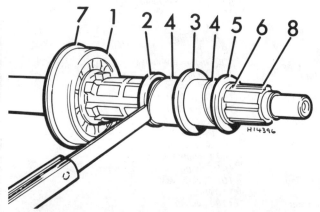

Fig. 6.6 Checking mainshaft bush overall endfloat (Sec 6)

1 Ball bearing
2 Thrust washer
3 Washer
4 Bush
5 Thrust washer
6 Circlip
7 Bearing outer circlip
8 Mainshaft

4 The three synchroniser baulk rings are bound to be badly worn and it is a false economy not to renew them. New rings will improve the smoothness and speed of the gearchange considerably.
5 The needle roller bearing and cage located between the nose of the mainshaft and the annulus in the rear of the input shaft is also liable to wear, and should be renewed as a matter of course.
6 Examine the condition of the three ball bearing assemblies, one on the input shaft, one on the mainshaft and the other in the tail of the gearbox extension. Check them for noisy operation, looseness between the inner and outer races, and for general wear. Normally they should be renewed on a gearbox that is being rebuilt.
7 Fit the mainshaft 2nd and 3rd gears to their bushes and using feeler blades and a straight edge, measure the endfloat which should be as shown in the Specifications. If the endfloat is too small, fit a new bush. If the endfloat is too large, reduce the length of the bush by carefully rubbing it square on an oilstone. Once this has been done, the mainshaft overall endfloat must be checked. To do this, assemble the thrust washer, bush, washer, bush and thrust washer in that order to the mainshaft. Secure the assembly with a circlip. Measure the total endfloat of the bushes and thrust washers on the shaft. Substitute selective thrust washers if necessary to bring the endfloat within the specified tolerance (see Specifications).
8 To dismantle the synchromesh units, first wrap a length of clean rag completely round a unit and then pull off the outer synchro sleeve. The cloth will catch the spring loaded balls and springs which are bound to fly out. Compare the length of the old springs with new and replace any that are worn. Note that an interlock plunger and ball is fitted to the second speed synchromesh hub.
9 The remote control gearchange is bound to be worn but this is dealt with in Section 10.

7 Input shaft (to 1970 - part synchro gearbox) – overhaul

1 Place the input shaft in a vice splined end uppermost and, with a pair of circlip pliers, remove the circlip which retains the ball bearing in place (photo), and then take off the distance washer.
2 Slightly close the vice with the bearing resting on top of the jaws. Tap the shaft through the bearing with a soft headed hammer and remove the oil deflector plate.
3 Prise out the old bearing from the annulus and fit a new roller bearing assembly in place (photo).
4 With the aid of a socket spanner of slightly smaller diameter than the annulus tap the roller bearing into place (photo).
5 Fit the circlip over a new ball bearing so the concave side of the clip faces the narrowest portion of the bearing rim (photo).
6 With the aid of a block of wood and the vice tap the bearing into place on the shaft as shown (photo).
7 Finally refit the distance washer and the circlip.

7.1 Input shaft circlip

7.3 Input shaft needle bearing

Chapter 6 Gearbox and overdrive

7.4 Fitting a new input shaft needle bearing

7.5 Fitting input shaft outer circlip

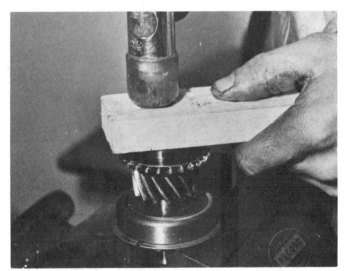
7.6 Fitting bearing to input shaft

8.2 Removing speedometer gear from mainshaft

8 Mainshaft (to 1970 - part synchro gearbox) – overhaul

1 The mainshaft has to be partially dismantled before it is possible to remove it from the gearbox (Section 5, paragraphs 16 to 23). Final mainshaft dismantling consists of removing the nylon speedometer drivegear and the ball bearing.
2 To remove the nylon speedometer drivegear, select an open ended spanner which just fits over the mainshaft and then lay the spanner across the jaws of the vice under the drivegear. Carefully tap the mainshaft downwards with a soft headed hammer, so driving off the drivegear (photo).
3 Remove the bearing retaining circlip and distance washer and tap the shaft out of the bearing as described previously.
4 Fit a new bearing onto the mainshaft and with the aid of the vice tap the new bearing into place.
5 Then fit the distance washer and the retaining circlip. Finally tap the nylon speedometer drivegear into place using the side of the jaws of the vice as a press.

9 Gear selectors (to 1970 - part synchro gearbox) – overhaul

1 Position the selector shafts so that they are as far forward as possible and then drive out the welch plugs with a $\frac{1}{8}$ in punch

9.1 Detent plug removal holes

Chapter 6 Gearbox and overdrive

9.3 Selector shaft detent holes

10.2 Removing gear lever fixing cap

positioned in turn through the small holes just inside the end of the cover (photo).
2 Undo the threaded tapered locking bolts from the selector shafts and forks.
3 Push the reverse gear selector shaft out, followed by the other two. The two interlock balls, plunger, three selector plungers and springs can then be removed. One spring and one plunger will emerge from each of the three holes indicated in the photo.
4 Examine the selector forks for wear. It is usually the ends which engage in the synchro sleeve groove which wear rather than the sleeve groove. If undecided on whether to renew a fork, compare it with a new one at your dealer's parts department.
5 Check the selector rods for scoring or wear in the detent ball grooves.
6 Reassembly commences by fitting the springs and plungers in place and then sliding the third and top selector shaft through the third and top selector fork in the top cover. Press down the selector plunger to allow the shaft to pass over it, and continue pushing the shaft home until it is in the neutral position, ie, the plunger is resting in the centre one of the three cut-outs on the shaft.
7 Replace the reverse gear selector shaft and selector fork in the same way ensuring it too is in neutral.
8 Then fit the interlock plunger to the first and second gear selector shaft, and slide the shaft into place in the selector fork, noting that the shaft also passes through the third and top selector fork. Before the shaft is fully home, drop the two interlock balls in through the centre selector shaft hole so that one ball seats each side of the transverse bore which connects the selector shaft bores. The centre selector shaft can then be pushed further in until the plunger is resting in the centre of the three cut-outs, and the interlock balls and plunger are held by the shafts.
9 Refit the threaded tapered lock bolts and refit the welch plugs using sealing compound to give a leakproof joint.

10 Remote control assembly (to 1970 - part synchro gearbox) – overhaul

1 Certain items in the remote control gearchange are prone to wear and these should always be renewed when the gearbox is being overhauled. The items concerned are small and relatively inexpensive and comprise the plate, reverse gear spring, a smaller spring, the nylon ball, the bonded rubber bush and washers, and the bush and washers.
2 To renew the items on the gear lever press down and twist off the cover and remove the cap and larger spring (photo).
3 Undo and remove the nut and bolt which holds the gear lever to the rear remote control rod (photo).
4 Lift out the gear lever and place it in a vice so the circlip which retains the small spring in place can be removed, and the spring and nylon ball removed (photo).

10.3 Disconnecting gearchange lever from remote control rod

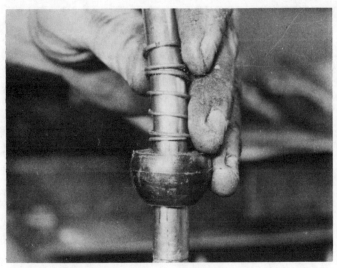

10.4 Removing gearchange lever spring and ball

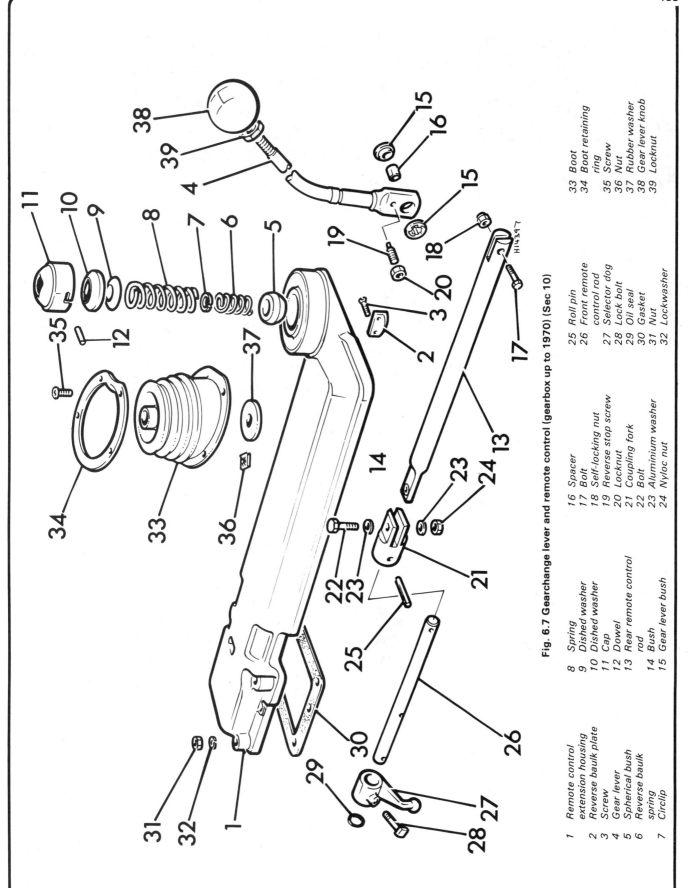

Fig. 6.7 Gearchange lever and remote control (gearbox up to 1970) (Sec 10)

1 Remote control extension housing
2 Reverse baulk plate
3 Screw
4 Gear lever
5 Spherical bush
6 Reverse baulk spring
7 Circlip
8 Spring
9 Dished washer
10 Dished washer
11 Cap
12 Dowel
13 Rear remote control rod
14 Bush
15 Gear lever bush
16 Spacer
17 Bolt
18 Self-locking nut
19 Reverse stop screw
20 Locknut
21 Coupling fork
22 Bolt
23 Aluminium washer
24 Nyloc nut
25 Roll pin
26 Front remote control rod
27 Selector dog
28 Lock bolt
29 Oil seal
30 Gasket
31 Nut
32 Lockwasher
33 Boot
34 Boot retaining ring
35 Screw
36 Nut
37 Rubber washer
38 Gear lever knob
39 Locknut

10.5 Fitting gearchange lever ball circlip

10.6 Fitting bush and washer

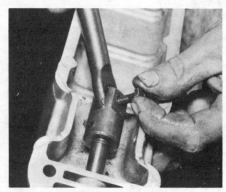
10.7 Remote control rod to fork bolt

10.8 Remote control rod bush broken up

10.9 Fitting new remote control rod bush

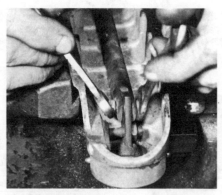
10.10 Connecting gearchange lever to remote control rod

10.11 Fitting gearchange lever components

5 Fit a new nylon ball and spring and with the aid of a spanner which just fits over the gear lever tap the circlip hole until it rests in its groove (photo).
6 Then fit a new bush and washer to the end of the gear lever (photo).
7 Undo the nut and bolt which holds the rear remote control rod to the fork (photo).
8 As can be seen in the photograph the old bush (bottom) had completely broken up. Press out the remains of the old bush (photo).
9 This is most easily done in a vice using two sockets, one considerably larger, and one fractionally smaller than the bush. Place the sockets either side of the rod and use the small socket to push the bush into the larger socket (photo).
10 Carefully press the new bush into place and reconnect the rod to the lever using new washers. Fit the gear lever to the extension and refit the nut and bolt which secures the gearchange lever to the remote control rod (photo).
11 Then refit the larger spring and the remaining components to complete the assembly (photo).

11 Gearbox (to 1970 - part synchro) – reassembly

1 Refit the gear selector lever to the side of the gearbox and replace the securing nut and plain washer.
2 Smear the laygear endface adjacent to the small straightcut gear with thick grease and fit the special thrust washer.
3 Fit a new laygear thrust washer to each end of the gearbox with the aid of thick grease so the bronze faces are adjacent to the laygear and the thrust washer tags rest in the recesses in the casing.

Chapter 6 Gearbox and overdrive

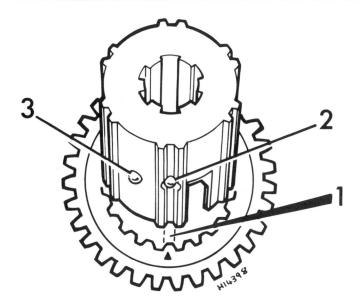

Fig. 6.8 2nd speed synchro unit (Sec 11)

1 Master spline
2 Plunger
3 Synchro ball

4 Carefully lower the laygear into place taking care not to disturb any of the washers. The laygear endfloat will already have been checked (see Section 6, paragraph 2).
5 Reassemble a new synchroniser ring to the second gear synchroniser hub making sure that the three lugs on the ring locate in the cutouts in the hub. Then fit the rear thrust washer with its scrolled face upwards and fit second gear and bush inside the synchroniser ring.
6 Next fit the centre thrust washer and third gear and bush and finally the front thrust washer with its scrolled face down.
7 Pass the end of the mainshaft into the gearbox and fit the assembled gear cluster over the end of the shaft.
8 With the aid of a screwdriver carefully work the circlip into place. During this operation press each side of the circlip in turn to keep it square on the mainshaft.
9 Then fit a new synchroniser ring with the three lugs facing forwards.
10 Fit the third and top synchroniser hub to the mainshaft with the longer boss of the inner synchro member facing forwards.
11 Screw a nut over the end of the mainshaft to protect the threads and carefully drive the mainshaft assembly complete with the rear ball bearing race into place.
12 Fit a new synchroniser ring to the front of the third and top synchro hub so the tabs in the former engage with the slots in the latter and then fit the input shaft in place.
13 Tap the shaft and bearing fully home with the aid of a soft faced hammer (photo).
14 To align the thrust washers and the laygear push a tapering rod through the gearbox and laygear (photo).
15 With the thrust washers and laygear correctly positioned feed in the layshaft, oiling it as it enters the gearbox (photo). Ensure the lockpin hole enters the casing last.
16 Align the lockpin hole in the shaft with the hole in the casing and fit the lockpin and washer (photo).
17 Then fit the reverse gear shaft into the casing and align its locating hole with the hole in the casing and insert the lockpin and washer.
18 Carefully engage the pin on the lower arm of the operating lever, with the groove cut on one side of the reverse gear.
19 Finally push the reverse gear fully home (photo). Take great care that reverse gear does not move forwards and disengage with the operating lever when the rear extension is being fitted.
20 If the gearbox is being rebuilt it is a false economy not to renew the rear oil seal and ball race. Mount the rear extension in a vice and with the aid of a drift drive out the oil seal and ball race from inside the extension (photo).

11.13 Tapping mainshaft and bearing into gearcase

11.14 Aligning thrust washers and laygear

11.15 Inserting layshaft

11.16 Fitting layshaft lockpin

11.19 Fitting reverse gear

11.20 Removing rear extension bearing and oil seal

11.21 Fitting rear extension bearing

21 Carefully tap a new bearing into place ensuring it is square in the bore (photo).
22 Then tap the oil seal into place (photo) with the sealing lip facing forward.
23 Fit the special distance washer in place on the end of the mainshaft (photo).
24 Then fit a new gasket to the rear face of the gearbox casing using gasket cement if wished (photo).
25 Carefully lower the rear extension into position (photo) tapping it home with a soft headed hammer if need be.
26 Replace the bolts and washers securing the extension in place not omitting the special bracket (photo).
27 Then refit the mainshaft drive flange, grip it as described in Section 5, tighten the nut and washer which holds it in place (photo).
28 Refit the speedometer drivegear assembly, making sure the rubber washer is in place in the groove, and that the hole for the lock bolt and washer is in line with the hole in the casing (photo).
29 Fit a new gasket to the bellhousing/gearbox flange. If a new gasket is not to hand cut a new gasket from stiff brown paper. Lay the paper over the bellhousing endface, and holding the paper taut make a series of rapid gentle taps with a ball-headed hammer. This will soon cut the paper to the desired shape (photo). Proper gasket paper is even better than brown paper.
30 Offer up the bellhousing to the gearbox (photo), and replace the securing bolts and washers.

11.22 Fitting rear extension oil seal

11.23 Fitting distance washer to mainshaft

11.24 Fitting rear extension joint gasket

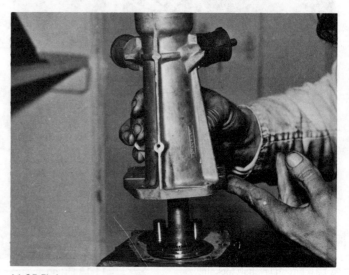

11.25 Fitting rear extension housing

11.26 Fitting rear extension housing special bracket

11.27 Tightening output flange nut

11.28 Refitting speedometer drive pinion

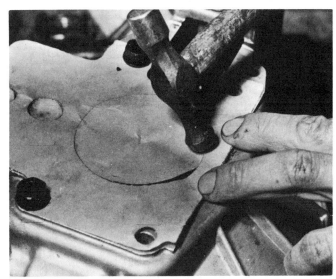
11.29 Cutting a new bellhousing gasket

11.30a Connecting bellhousing to gearcase

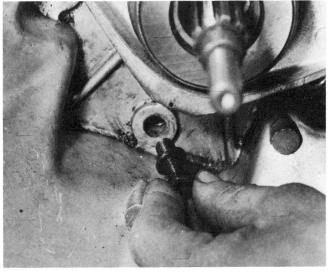
11.30b Fitting bellhousing bolts

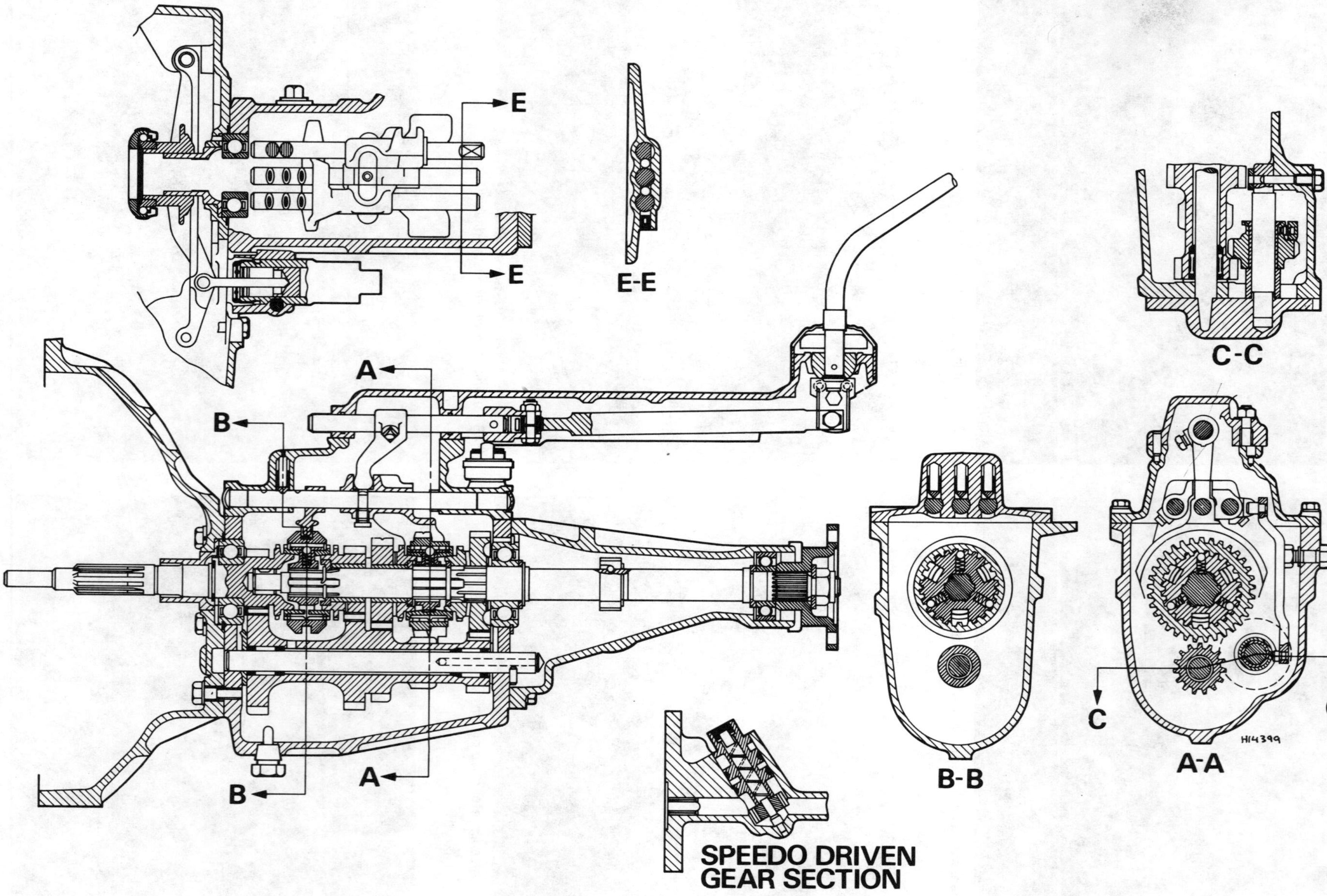

SPEEDO DRIVEN GEAR SECTION

Fig. 6.9 Section views of all-synchro gearbox (1970-74) (Sec 12)

Chapter 6 Gearbox and overdrive

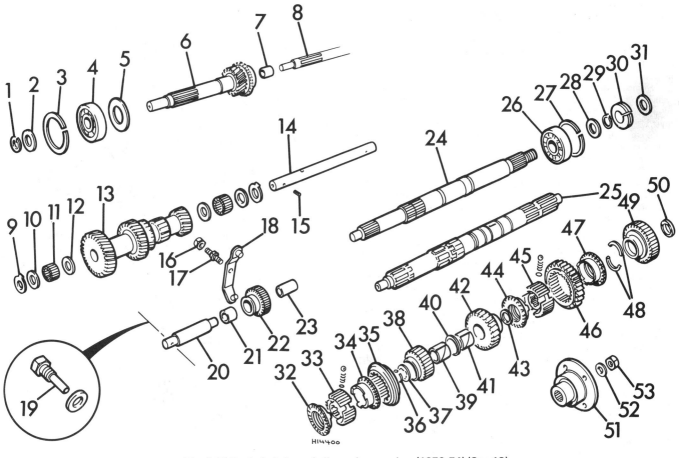

Fig. 6.10 Exploded view of all-synchro gearbox (1970-74) (Sec 12)

1 Circlip	15 Roll pin	27 Bearing outer track circlip	42 2nd gear
2 Thrust washer	16 Nut	28 Thrust washer	43 Thrust washer
3 Bearing outer track circlip	17 Pivot pin	29 Circlip	44 2nd gear baulk ring
4 Bearing	18 Reverse gear actuator	30 Speedometer drivegear	45 1st/2nd gear synchro hub
5 Washer	19 Reverse idler shaft locating pin	31 Washer	46 1st/2nd speed synchro sleeve with reverse
6 Input shaft	20 Reverse idler shaft	32 4th speed baulk ring	47 1st gear baulk ring
7 Needle roller bearing	21 Distance collar	33 3rd/4th synchro hub	48 Semi-circular thrust washers
8 Front end of mainshaft	22 Reverse idler gear	34 3rd speed baulk ring	49 1st gear
9 Thrust washer	23 Sleeve	35 3rd/4th synchro sleeve	50 Thrust washer
10 Needle retainer	24 Mainshaft (overdrive not fitted)	36 Circlip	51 Output flange
11 Needle bearings	25 Mainshaft (overdrive fitted)	37 Thrust washer	52 Nut
12 Needle retainer	26 Bearing	38 3rd gear	53 Washer
13 Layshaft cluster		39 3rd gear bush	
14 Layshaft		40 Thrust washer	
		41 2nd gear bush	

31 Finally, ensuring that all the selectors and gears are in neutral, fit the top cover and remote control extension in place using new gaskets.
32 Refit the clutch release lever and bearing.

12 Gearbox (1970 to 74 – all synchro) – description

1 Apart from the addition of a synchro facility for 1st gear, this gearbox is very similar to the earlier design.
2 The gearbox contains four forward gears and one reverse gear. Synchromesh is fitted to all forward gears. The gear lever is mounted on the extension housing and operates the selector mechanism in the gearbox by a long shaft. When the gear lever is moved sideways the shaft is rotated so that the pins in the gearbox end of the shaft locate in the appropriate selector fork. Forward or rearward movement of the gear lever moves the selector fork, which in turn moves the synchromesh unit outer sleeve until the gear is firmly engaged. When reverse gear is selected, a pin on the selector shaft engages with a lever, and this in turn moves the reverse idler gear into mesh with the laygear reverse gear and mainshaft. The direction of rotation of the mainshaft is thereby reversed.
3 The gearbox input shaft is splined and it is onto these splines that the clutch driven plate is located. The gearbox end of the input shaft is in constant mesh with the laygear cluster, and the gears formed on the laygear are in constant mesh with the gears on the mainshaft, with the exception of the reverse gear. The gears on the mainshaft are able to rotate freely, which means that when the neutral position is selected the mainshaft does not rotate.
4 When the gear lever moves the synchromesh unit outer sleeve via the selector fork, the synchromesh baulk ring first moves and friction caused by the conical surfaces meeting takes up initial rotational movement until the mainshaft and gear are both rotating at the same speed. This condition achieved, the sleeve is able to slide over the dog teeth of the selected gear, thereby giving a firm drive. The synchromesh unit hub is splined to the mainshaft and, because the outer sleeve is splined to the inner hub, engine torque is passed to the mainshaft and propeller shaft.

13 Gearbox (1970 to 74 - all synchro) – dismantling into major assemblies

1 The operations are similar to those described in Section 5, but note the addition of the 1st gear synchro unit and the fact that the laygear runs on needle roller bearings.
2 Removal of this type of laygear requires the following procedure. Make up a dummy rod slightly less in diameter than the layshaft, but of similar length to the laygear plus thrust washers. Remove the layshaft securing bolt and then tap the dummy shaft into the gearcase to displace the layshaft proper. The dummy shaft will retain the needle rollers in position during the laygear removal operations.
3 When reassembling, insert a retaining ring into the laygear at one end. Then using thick grease, stick the needle rollers in position followed by the second retaining ring. Insert an equal number of needle rollers at the other end. Then insert the dummy shaft and stick the thrust washers to the inside of the gearcase. Tap in the layshaft and displace the dummy shaft. Fit the shaft bolt.

14 Gearbox components – examination and renovation

1 The gearbox has been stripped, presumably because of wear or malfunction, or possibly excessive noise, ineffective synchromesh, or failure to stay in a selected gear. The cause of most gearbox ailments is failure of the ball bearings on the input or mainshaft and wear on the synchro-rings, both the bore surfaces and dogs. The nose of the mainshaft which runs in the needle roller bearing in the input shaft is also subject to wear. This can prove very expensive as the mainshaft would need renewal and this represents about 20% of the total cost of a new gearbox.
2 Examine the teeth of all gears for signs of uneven or excessive wear and, of course, chipping. If a gear on the mainshaft requires renewal check that the corresponding laygear is not equally damaged,. If it is the whole laygear may need renewing as well.
3 All gears should be a good running fit on the shaft with no signs of rocking. The hubs should not be a sloppy fit on the splines.
4 Selector forks should be examined for signs of wear or ridging on the faces which are in contact with the operating sleeve.
5 Check for wear on the selector rod and interlock spool.
6 The ball bearings may not be obviously worn, but if one has gone to the trouble of dismantling the gearbox it would be shortsighted not to renew them. The same applies to the four synchroniser rings, although for these the mainshaft has to be completely dismantled for the new ones to be fitted.
7 The input shaft bearing retainer is fitted with an oil seal and this should be renewed if there are any signs that oil has leaked past it into the clutch housing or, of course, if it is obviously damaged. The rear extension has an oil seal at the rear as well as a ball bearing race. If either have worn or oil has leaked past the seal, the parts should be renewed.
8 Before finally deciding to dismantle the mainshaft and renew parts, it is advisable to make enquiries regarding the availability of parts and their costs. It may still be worth considering an exchange gearbox even at this stage. You should reassemble it before exchange.

15 Input shaft (1970 to 74 - all synchro gearbox) – overhaul

Refer to Section 7.

16 Mainshaft (1970 to 74 - all synchro gearbox) – overhaul

1 Refer to Section 8, but note the two semi-circular collars which are located at the 1st gear synchro baulk ring.
2 The mainshaft circlip which is fitted between 3rd gear and 3rd/4th synchro unit should have its offset visible when fitted.
3 Note that the speedometer drivegear is retained to the shaft by a locking ball and circlips.
4 Check the laygear and mainshaft gear endfloat as described in Section 6, using the values given for this type of gearbox in the Specifications Section.

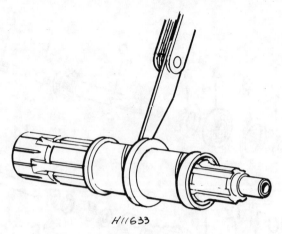

Fig. 6.11 Checking mainshaft endfloat on all-synchro gearbox with 2nd and 3rd gear bushes temporarily assembled (Sec 16)

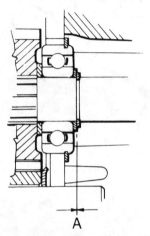

Fig. 6.12 Measuring centre bearing endfloat (Sec 16)

A = Insertion point for feeler gauge

17 Selector and remote control mechanism (1970 to 74 - all synchro gearbox) – description and overhaul

The components and overhaul operations are virtually identical to those for the earlier part-synchromesh gearbox and described in Sections 9 and 10 of this Chapter.

18 Gearbox (1970 to 74 - all synchro) – reassembly

1 This is very similar to the operations described in Section 11 to which reference should be made.
2 Fill the gearbox after it has been installed in the car.

19 Gearbox (December 1974 on - single rail) – description

1 This gearbox fitted to 1500 models from December 1974 is of single rail selector type with synchromesh on all forward speeds.
2 The selector mechanism provides very smooth positive selection and engagement of all gears. If an overdrive unit is installed, a higher 3rd and 4th speed ratio is provided to give what is virtually a six speed unit.

20 Gearbox (December 1974 on - single rail) – removal and refitting

This is as described in Section 3.

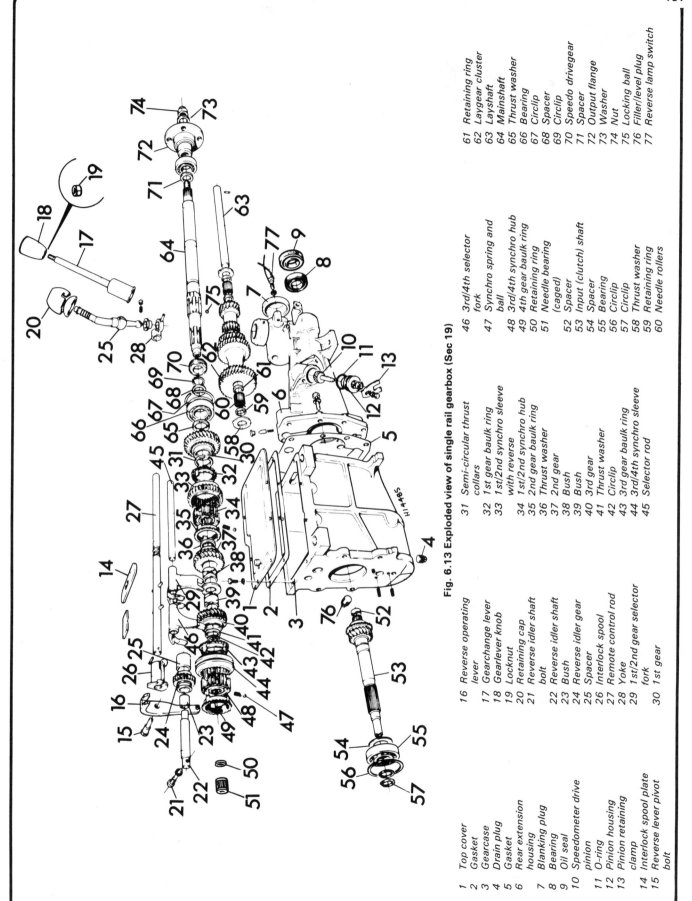

Fig. 6.13 Exploded view of single rail gearbox (Sec 19)

1 Top cover
2 Gasket
3 Gearcase
4 Drain plug
5 Gasket
6 Rear extension housing
7 Blanking plug
8 Bearing
9 Oil seal
10 Speedometer drive pinion
11 O-ring
12 Pinion housing
13 Pinion retaining clamp
14 Interlock spool plate
15 Reverse lever pivot bolt
16 Reverse operating lever
17 Gearchange lever
18 Gearlever knob
19 Locknut
20 Retaining cap
21 Reverse idler shaft bolt
22 Reverse idler shaft
23 Bush
24 Reverse idler gear
25 Spacer
26 Interlock spool
27 Remote control rod
28 Yoke
29 1st/2nd gear selector fork
30 1st gear
31 Semi-circular thrust collars
32 1st gear baulk ring
33 1st/2nd synchro sleeve with reverse
34 1st/2nd synchro hub
35 2nd gear baulk ring
36 Thrust washer
37 2nd gear
38 Bush
39 Bush
40 3rd gear
41 Thrust washer
42 Circlip
43 3rd gear baulk ring
44 3rd/4th synchro sleeve
45 Selector rod
46 3rd/4th selector fork
47 Synchro spring and ball
48 3rd/4th synchro hub
49 4th gear baulk ring
50 Retaining ring
51 Needle bearing (caged)
52 Spacer
53 Input (clutch) shaft
54 Spacer
55 Bearing
56 Circlip
57 Circlip
58 Thrust washer
59 Retaining ring
60 Needle rollers
61 Retaining ring
62 Laygear cluster
63 Layshaft
64 Mainshaft
65 Thrust washer
66 Bearing
67 Circlip
68 Spacer
69 Circlip
70 Speedo drivegear
71 Spacer
72 Output flange
73 Washer
74 Nut
75 Locking ball
76 Filler/level plug
77 Reverse lamp switch

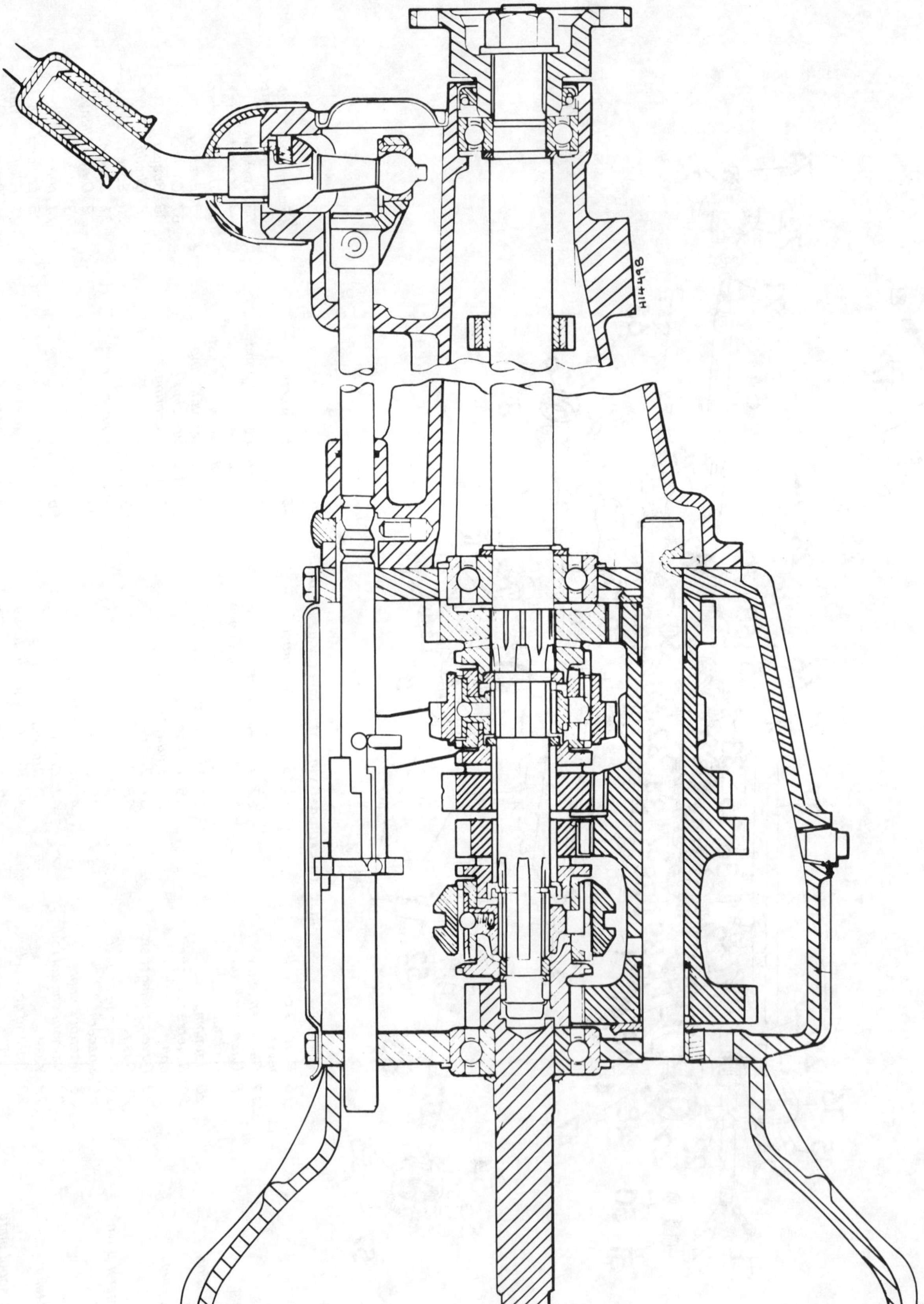

Fig. 6.14 Cutaway view of single rail gearbox (Sec 19)

Chapter 6 Gearbox and overdrive

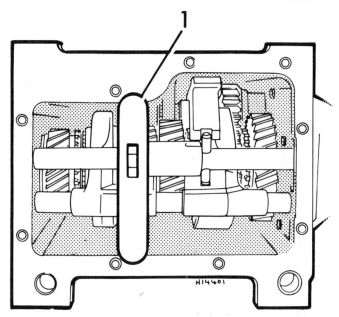

Fig. 6.15 Location of spool interlock plate (1) (Sec 21)

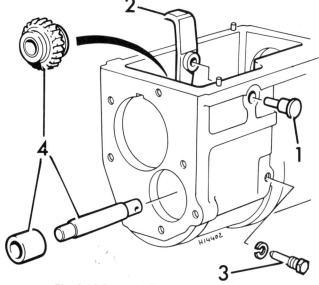

Fig. 6.16 Reverse idler components (Sec 21)

1 Pivot pin
2 Reverse lever
3 Idler shaft locating pin
4 Idler shaft, spacer, reverse idler gear

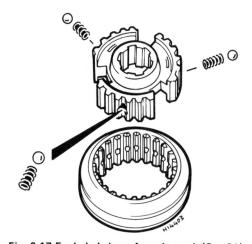

Fig. 6.17 Exploded view of synchro unit (Sec 21)

21 Gearbox (December 1974 on - single rail) – dismantling into major assemblies

1 With the unit removed from the car and cleaned externally, remove the clutch release components from inside the bellhousing as described in Chapter 5.
2 Unbolt and remove the bellhousing from the gearcase.
3 Unbolt and remove the top cover with spool interlock plate.
4 Unscrew and remove the reverse lamp switch.
5 Drive out the selector rod roll pin.
6 Unbolt and withdraw the speedometer drive pinion.
7 If you have a vice with wide enough jaws, grip the output flange in the vice, alternatively bolt a long lever to two of the holes in the output flange, to prevent the flange from turning when the nut is unscrewed. This should now be done and the washer removed.
8 Withdraw the output flange.
9 Looking into the open top of the gearcase, move the selector to the reverse gear position.
10 Check to see that the selector shaft pins clear the interlock spool and selector forks.
11 Remove the rear extension housing fixing bolts and exhaust pipe support bracket.
12 Withdraw the rear extension housing taking care not to displace the layshaft or to allow the selector shaft pins to foul the interlock spool or selector forks (photo).
13 Remove the interlock spool as it becomes free.
14 Remove the joint gasket.
15 Take the distance washer from the end of the mainshaft and then draw off the speedometer drivegear.
16 Remove the selector shaft and forks.
17 Using a dummy shaft of lightly smaller diameter than the layshaft, push out the layshaft from the gearcase. If the dummy shaft is cut to length to correspond with the internal length of the gearcase, then the laygear assembly can be lowered to the bottom of the gearcase while the dummy shaft retains the needle rollers in place.
18 Remove the input shaft. This can be done by driving out the bearing by applying a brass drift to its outer track. Alternatively, fit two hose clips to the shaft and use them as fulcrum points for leverage or impact.
19 Remove the spigot bearing and spacers from the front end of the mainshaft.
20 Remove 4th gear synchro baulk ring.
21 Remove the bolt and spring washer which secures the reverse idler gear spindle.
22 Withdraw the reverse idler spindle, spacer and take off reverse gear.
23 Release the circlip which secures the mainshaft rear bearing to the shaft.

21.12 Selector shaft pins

Chapter 6 Gearbox and overdrive

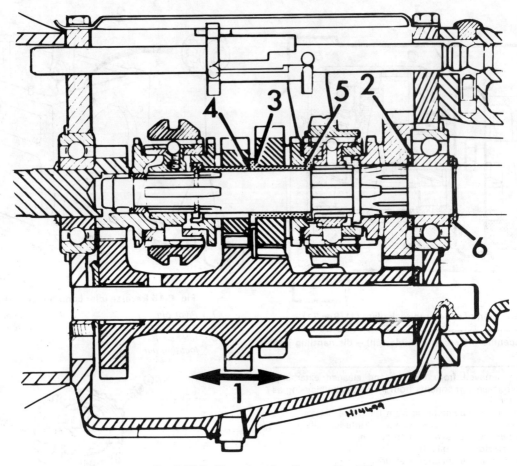

Fig. 6.18 Endfloat checking diagram (Sec 22)

1. Laygear cluster endfloat
2. 1st gear endfloat
3. 2nd gear endfloat
4. 3rd gear endfloat
5. Mainshaft overall endfloat
6. Mainshaft bearing circlip

24 A special extractor will now be required to withdraw the mainshaft bearing, but success can usually be achieved using two screwdrivers inserted under the outer circlip of the bearing while the rear end of the mainshaft is gently tapped with a plastic or copper faced hammer. By tapping and levering the bearing can be gradually worked out of the casing and up the mainshaft.
25 Tilt the mainshaft, remove the complete assembly through the top of the gearcase.
26 Remove the laygear and thrust washers from the gearcase.
27 Remove reverse gear operating lever.
28 Remove the gearcase rear mountings if necessary.
29 Examine the geartrains for obvious signs of wear or damage and if evident, dismantle further as described in the following Sections. A history of noisy gearchanging will indicate the need for attention to the synchro units.

22 Mainshaft (December 1974 on - single rail gearbox) – overhaul

1 From the rear end of the mainshaft, remove the thrust washer and 1st gear.
2 Remove 1st gear synchro baulk ring and the two split collars.
3 Take off 1st/2nd synchro unit with reverse.
4 From the front end of the mainshaft, take off the 3rd/4th synchro unit.
5 Remove the 3rd speed synchro baulk ring.
6 Extract the retaining circlip and remove 3rd gear, bush and thrust washer.
7 Remove 2nd gear and bush followed by 2nd gear synchro baulk ring.

8 Remove the circlip and extract the shaft ball using a magnet. Remove 1st/2nd synchro unit (photo).
9 If the synchro units are to be dismantled, mark the relationship of hub to sleeve before pushing the hub from the sleeve. Catch the balls and springs as they are ejected (photo).
10 With the mainshaft dismantled, renew any worn components such as gears with chipped teeth, slack bearings, worn synchros.
11 Temporarily assemble the necessary components to the mainshaft so that the following endfloats can be checked.

1st gear (between split collars and thrust washer)
2nd gear (gear on bush)
3rd gear (gear on bush)
Mainshaft (2nd and 3rd gears) overall endfloat

12 Work to the endfloat tolerances given in Specifications for this particular gearbox.
13 To reassemble the mainshaft, first locate the locking ball in the hole in the shaft (photo).
14 Now slide 1st/2nd synchro (which incorporates reverse gear) onto the front end of the mainshaft so that the greater synchro hub projection is towards the front end of the mainshaft.
15 Fit the thrust washer so that its cut-out engages with the ball (photo).
16 Fit 2nd gear synchro baulk ring (photo).
17 Fit 2nd gear with bush. Bush collar to be towards front end of mainshaft (photo).
18 Fit 3rd gear with its bush (photo).
19 Fit the thrust washer, rim towards front end of mainshaft (photo).
20 Fit the shaft circlip (photos).
21 Fit 3rd gear baulk ring (photo).

22.8 Mainshaft

22.9 Fitting synchro ball and spring

22.13 Fitting mainshaft locking ball

22.14 Fitting 1st/2nd synchro with reverse to mainshaft

22.15 Fitting thrust washer to engage with locking ball

22.16 Fitting 2nd gear synchro baulk ring

22.17 Fitting 2nd gear and bush

22.18 Fitting 3rd gear and bush

22.19 Fitting thrust washer

22.20a Pushing shaft circlip into position

22.20b Shaft circlip engaged in groove

22.21 Fitting 3rd gear baulk ring

Chapter 6 Gearbox and overdrive

22.22 Fitting 3rd/4th synchro

22.23 Semi-circular thrust collars in position

22.24 1st speed gear baulk ring

22.25 Fitting 1st gear

22.26 Fitting thrust washer next to 1st gear

22.27 Mainshaft geartrain secured with wire

23.1 Input shaft bearing and circlips

23.2 Laygear needle rollers and retaining rings

24.2 Extension housing blanking plug

22 Fit 3rd/4th synchro so that the greater hub projection is towards the front end of the shaft (photo).
23 To the rear end of the mainshaft fit the two semi-circular thrust collars (photo).
24 Fit 1st gear baulk ring (photo).
25 Fit 1st gear (photo).
26 Fit the thrust washer (photo).
27 Secure the geartrain with a length of wire to prevent it falling apart during installation to the gearcase (photo).

23 Input shaft and layshaft/gear (December 1974 on - single rail gearbox) – overhaul

1 The input shaft bearing can be renewed by extracting the circlip and then tapping or drawing the bearing off and pressing a new bearing into position (photo).

2 If the laygear needle rollers are worn, prise out the retaining rings and renew both sets of 25 rollers (photo).
3 If the gears on either the input or layshaft are worn then the complete assembly will have to be renewed.

24 Gearchange rod mechanism (December 1974 on - single rail gearbox) – overhaul

1 Support the gearbox extension housing and then slide the gearchange rod towards the rear until it is felt to make contact with the blanking plug.
2 Tap the end of the gearchange rod with a plastic-faced hammer to eject the plug (photo).
3 Slide the gearchange rod out of the blanking plate hole until the yoke is exposed.
4 Tap out the roll pin and remove the yoke from the rod.

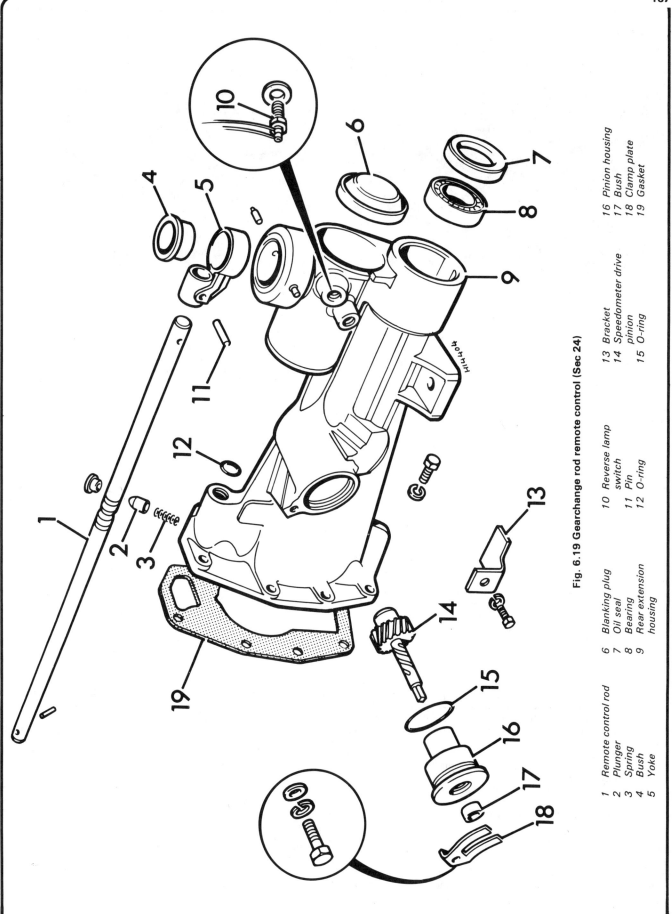

Fig. 6.19 Gearchange rod remote control (Sec 24)

1 Remote control rod
2 Plunger
3 Spring
4 Bush
5 Yoke
6 Blanking plug
7 Oil seal
8 Bearing
9 Rear extension housing
10 Reverse lamp switch
11 Pin
12 O-ring
13 Bracket
14 Speedometer drive pinion
15 O-ring
16 Pinion housing
17 Bush
18 Clamp plate
19 Gasket

24.7 Extension housing oil seal removal tool

25.4 Laygear needle rollers retained with grease

5 Withdraw the gearchange rod from the front end of the extension housing, but keeping the roll pin hole horizontal.
6 Remove the nylon plug, plunger, spring and O-ring from the extension housing.
7 Remove the bearing and oil seal from the extension housing using a long rod as a drift or by using an extractor (photo).
8 Reassembly is a reversal of dismantling, but always use a new oil seal for the gearchange rod and the output flange. The flange oil seal should be fitted flush with the surface of the gearcase and the seal lips inwards and filled with grease.

25 Inspection and preparation for reassembly – (December 1974 on - single rail gearbox)

1 Clean all the components and inspect for wear, (chipped gear teeth, scored shafts, worn forks).
2 Renew the gaskets and oil seals as a matter of routine.
3 If new laygear needle roller retaining rings are to be fitted, they must be installed in accordance with the diagram.
4 The needle rollers can be retained using thick grease until the dummy layshaft is pressed through them to retain them in position (photo).
5 Obtain new laygear thrust washers which should provide a laygear endfloat as specified.

Fig. 6.20 Laygear needle roller retaining ring positioning diagram (Sec 25)

26 Gearbox (December 1974 on - single rail) – reassembly

1 Commence reassembly by sticking the laygear thrust washers to the inside of the gearcase with thick grease. Make sure that the tabs engage in the casing cut-outs (photo).
2 Lower the laygear assembly complete with dummy shaft into the bottom of the gearcase. Make sure that the large gear is towards the front of the casing (photo).
3 Tilt the rear end of the mainshaft downward and pass the geartrain into the gearcase through the top cover aperture (photo).
4 Place reverse idler gear in the bottom of the gearcase and then fit reverse operating lever (photos).
5 Fit the outer circlip to the mainshaft bearing and slide the bearing onto the mainshaft so that the circlip is towards the rear end of the shaft (photo).
6 Using a suitable drift, drive the mainshaft bearing into position both in the gearcase and down the shaft simultaneously. Fit the thrust washer (photos).
7 Select and fit the thickest circlip which will fit into the shaft groove, to retain the bearing to the mainshaft. Four different thicknesses of circlip are available. The mainshaft endfloat should not exceed that specified if the correct circlip has been fitted (photo).
8 Fit the speedometer drivegear to the mainshaft by driving it hard up against the shaft shoulder using a length of tubing (photo).
9 To the front end of the mainshaft fit the 4th gear synchro baulk ring (photo).
10 To the recess in the end of the input shaft fit the needle roller bearing and thrust ring (photos).
11 Fit the input shaft (photo).
12 Raise the laygear assembly into alignment with the holes in the gearcase and insert the layshaft to displace the dummy shaft. Take care not to displace the needle rollers and the thrust washers retained temporarily by grease. Fit the layshaft roll pin (photos).
13 Raise reverse idle gear from the bottom of the gearcase and check that its teeth are towards the rear of the gearbox, then fit reverse idler shaft and the spacer. Screw in reverse shaft lock bolt (photos).
14 Fit the selector forks and shaft (photo).
15 Stick a new flange gasket, to the rear face of the gearcase.
16 Fit the distance washer to the rear end of the mainshaft against the rear face of the speedo gear.
17 Offer up the rear extension housing to the gearcase. Guide the

26.1 Laygear thrust washer

26.2 Laygear resting in bottom of gearcase

26.3 Installing mainshaft assembly

26.4a Reverse idler gear resting in bottom of gearcase

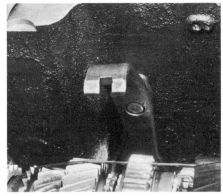
26.4b Reverse operating lever

26.5 Fitting mainshaft rear bearing

26.6a Driving mainshaft bearing into position

26.6b Fitting rear mainshaft thrust washer

26.7 Fitting mainshaft circlip

26.8 Driving speedo gear onto mainshaft using tubing

26.9 Fitting 4th gear synchro baulk ring

26.10a Input shaft needle roller bearing

26.10b Input shaft needle roller bearing thrust ring

26.11 Fitting input shaft

26.12a Layshaft roll pin being located

26.12b Layshaft roll pin in position

26.13a Reverse idler shaft

26.13b Reverse idler shaft lock bolt

26.14 Passing selector shaft through forks

26.17 Fitting selector spool

26.20a Fitting output flange

26.20b Output flange nut

26.21 Selector shaft roll pin

26.22 Fitting speedometer drive pinion

Chapter 6 Gearbox and overdrive

26.26a Interlock spool plate

26.26b Fitting gearbox top cover

26.26c Gearbox rear mounting

selector rail into position, remembering to fit the selector spool (photo).
18 Insert and tighten the connecting bolts, fit the exhaust bracket.
19 The rear oil seal will already have been renewed (see Section 24).
20 Fit the output flange, tightening the nut to the specified torque while the flange is prevented from rotating using one of the methods described in Section 5 (photos).
21 Drive the roll pin into the front end of the selector rail. Position the pin so that it extends equally from both sides of the rail (photo).
22 Fit the speedometer drive pinion into the rear extension housing and secure it temporarily with its clamp plate and bolt (photo).
23 Screw the reverse lamp switch into the extension housing.
24 Using a new gasket, bolt the clutch bellhousing to the gearcase. Tape the input shaft splines before this operation to prevent damage to the lips of the oil seal.
25 Fit the clutch release lever and bearing (see Chapter 5).
26 Fit the interlock spool plate, a new gasket and the top cover. Refit the rear mountings if they were removed (photos).
27 Filling the gearbox with oil is best left until the gearbox is refitted to the car.

27 Overdrive (Type D) – description

1 The overdrive unit is attached to the extension on the rear of the gearbox by eight studs and nuts, and takes the form of a hydraulically operated epicyclic gear. Overdrive operates on third and fourth speeds to provide fast cruising at lower engine revolutions. The overdrive 'IN-OUT' switch on the right of the steering wheel actuates a solenoid attached to the side of the overdrive unit. In turn the solenoid operates a valve which opens the hydraulic circuit which pushes the cone clutch into contact with the annulus when overdrive is engaged.
2 Attached to the end of the extended gearbox mainshaft are the inner components of a unidirectional clutch. The hydraulic pressure which enables the overdrive clutch to be engaged is provided by a hydraulic pump operated by an eccentric on the front of the mainshaft.
3 Behind the cam is a steady bearing with a plain phosphor-bronze bush carried in the main housing. Next to this is the sun wheel of the epicyclic gear which is carried on a Clevite bush. The planet carrier and unidirectional clutch come next and are mounted on splines cut in the mainshaft. The smaller diameter portion of the end of the mainshaft turns in a needle roller bearing fitted inside the larger diameter output shaft.
4 Two roller bearings mounted in the rear of the overdrive casing support the output shaft. A ball beraing housed in a flanged ring is held to the cone clutch member. A bolt at each of the four corners of the flange pass through one of the four clutch return springs by which the ring, together with the clutch cone is held against the annulus.
5 The pressure of the springs prevents freewheeling on over-run and they are also strong enough to handle reverse torque. Also attached to the bolts are two bridge pieces which rest against two hydraulic operating pistons working in cylinders cast in the main casing.
6 The sun wheel and pinions are case hardened and the annulus heat treated. The pinions have needle roller bearings and run on case

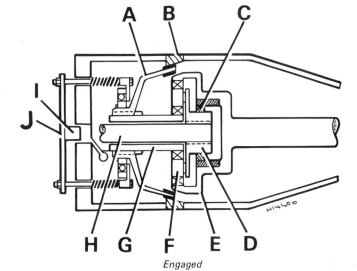

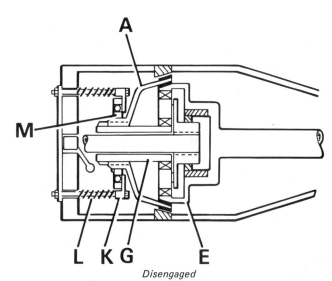

Fig. 6.21 Overdrive conditions (Type D) (Sec 27)

A Cone clutch
B Brake ring
C Unidirectional clutch
D Planet carrier
E Annulus
F Planet gears
G Sungear
H Input shaft
J Bridge pieces
K Thrust ring
L Clutch spring
M Bearing

Chapter 6 Gearbox and overdrive

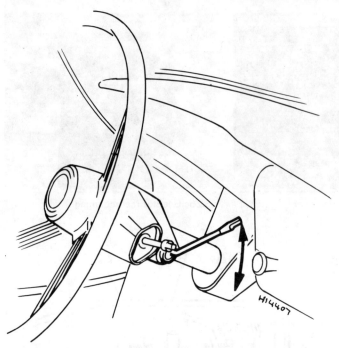

Fig. 6.22 Overdrive control switch (Sec 27)

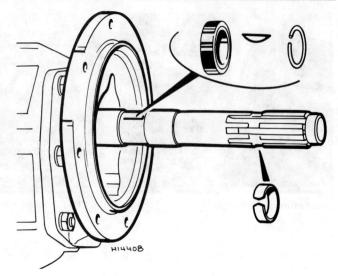

Fig. 6.23 Adaptor plate and mainshaft (Sec 27)

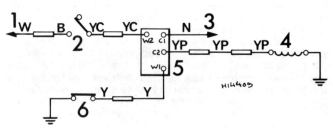

Fig. 6.24 Overdrive (Type D) circuit (Sec 27)

1 To coil SW terminal
2 Overdrive switch
3 To No 1 terminal on ignition switch
4 Solenoid
5 Relay
6 Gearbox isolator switch

hardened pins. The gear teeth are helical. The outer ring of the unidirectional clutch is pressed into the annulus. The clutch is of the caged roller type and is loaded by a round wire lock type spring.
7 The overdrive unit works in the following way. Under normal running conditions with the overdrive switched out the cone clutch is held against the centre annulus by the four clutch return springs, so locking the sun wheel to the annulus. In this way the complete gear train rotates together as a solid unit giving direct drive.
8 On switching in the electrically operated solenoid the centre rod moves inwards operating the linkage mechanism which lifts a tube so raising the ball valve off its seating against the pressure of the spring which normally holds the ball in place.
9 With the ball valve lifted, oil under pressure is free to travel along the drillings in the casing to the two operating cylinders. The pistons are pushed forward against the two bridge pieces which move the clutch cone into contact with the cast iron brake ring sandwiched between the main and tail casings. This brings the sun wheel to rest and allows the annulus to over-run the unidirectional clutch and so give an increased speed to the output shaft.
10 When the overdrive is switched out the rod in the centre of the solenoid moves outwards allowing the valve tube to drop and the ball valve to return to its seat. Oil from the two cylinders is then free to return through the centre of the tube to the bottom of the overdrive casing. To ensure direct drive is re-engaged smoothly the cylinders are emptied slowly because of a small restrictor jet in the base of the valve tube. The oil is then free to flow round in open circuit.
11 The oil for the hydraulic system is supplied by a pump which is pressed into the main housing and held by a grub screw. The pump supplies oil through a non-return valve to a relief valve, in which a piston moves back against a compression spring, till the correct pressure is obtained when a hole in the relief valve is uncovered. Excess oil from the relief valve is then led through drilled passages to an annular groove in the mainshaft steady bush. Radial holes in the shaft feed oil through axial drillings to the needle roller bearing, thrust washer, and unidirectional clutch.
12 The overdrive is normally a very reliable unit and trouble is usually due to either the solenoid sticking; a fault in the hydraulic system due to dirt or insufficient oil; or incorrect solenoid operating lever adjustment.

28 Overdrive (Type D) – removal and refitting

1 It is not necessary to remove the overdrive from the car in order to attend to the following: the hydraulic lever setting; the relief valve;

the non-return valve; the solenoid; and the operating valve.
2 If the unit as a whole requires overhaul it must be removed from the car together with the gearbox as described in Section 3.
3 To separate the overdrive from the gearbox undo the eight nuts from the $\frac{1}{4}$ in (6.35 mm) diameter studs (noting the extra length of one of the studs) to separate the main overdrive casing from the gearbox rear extension. Carefully pull the overdrive off the end of the mainshaft.
4 To mate the overdrive and gearbox start by placing the overdrive in an upright position and then line up the splines of the clutch and planet carrier by eye, turning them anti-clockwise only, with the aid of a long thin screwdriver. Make certain that the locating ring is correctly positioned in its groove in the mainshaft and that it does not protrude above the mainshaft splines.
5 Under normal circumstances if everything is in line the gearbox mainshaft should enter the overdrive easily. If trouble is experienced do not try and force the components together but separate them and re-align the components. Place the gearbox in top gear while refitting.
6 As the mainshaft is fed into the overdrive gently rotate the input shaft to and fro to help in mating the mainshaft into the splines. At the same time make certain that the lowest portion of the cam on the mainshaft will rest against the pump and make certain as the gearbox extension and overdrive come together that the end of the mainshaft enters into the needle roller bearing in the tailshaft.
7 The remainder of the replacement procedure is a straightforward reversal of the removal sequence.

29 Overdrive (Type D) – overhaul

1 Unscrew the operating valve plug and take out the spring, plunger and ball.
2 Bend back the tabs on the four lock washers and undo the nuts

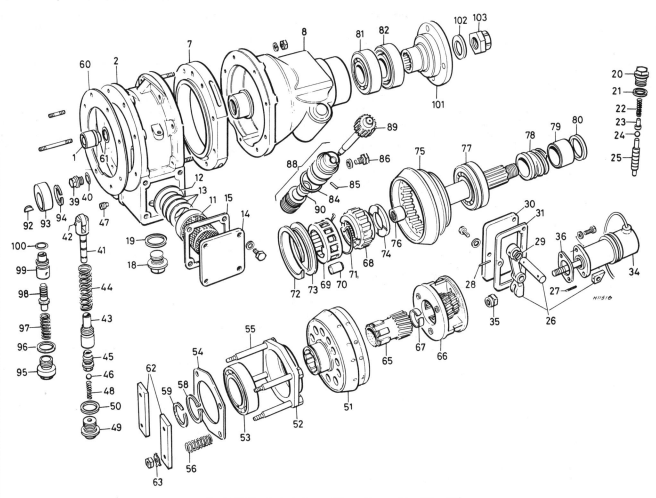

Fig. 6.25 Exploded view of Type D overdrive (Sec 28)

1 Operating piston
2 Main casing assembly
7 Brake ring
8 Intermediate casing
11 Filter
12 Sealing plate
13 Magnetic rings
14 Side cover plate
15 Gasket
18 Drain plug
19 Washer
20 Plug
21 Washer
22 Spring
23 Plunger
24 Steel ball
25 Operating valve
26 Operating valve lever assembly
27 Mills pin
28 Mills pin
29 O-ring
30 Cover - solenoid
31 Gasket
34 Solenoid
35 Self-locking nut
36 Gasket
39 Plug
40 Washer
41 Plump plunger
42 Pin

43 Pump body
44 Pump plunger spring
45 Non-return valve body
46 Steel ball
47 Screw
48 Spring
49 Plug
50 Washer
51 Clutch assembly
52 Bearing housing
53 Thrust bearing
54 Plate
55 Bolt
56 Spring
58 Circlip
59 Circlip
60 Gasket
61 O-ring
62 Bridge piece
63 Lockwasher
65 Sun wheel assembly
66 Planet carrier
67 Locating ring - 3rd motion shaft
68 Inner member - unidirectional clutch
69 Cage - unidirectional clutch
70 Roller - unidirectional clutch
71 Spring for clutch
72 Circlip
73 Oil thrower

74 Thrust bearing
75 Annulus assembly
76 Bearing
77 Inner bearing
78 Speedometer driving gear
79 Bush
80 Thrust washer
81 Bearing
82 Oil seal
83 Steady bush - 3rd motion shaft
84 O-ring
85 Pin
86 Locking screw
88 Speedometer drive bearing assembly
89 Speedometer driven gear
90 Oil seal - speedometer bearing
92 Key
93 Cam
94 Circlip
95 Plug
96 Washer
97 Spring
98 Plunger
99 Body
100 O-ring
101 Flange
102 Washer
103 Nut

from the bolts. Then remove the two bridge pieces. If wished the two operating pistons can now be pulled out of their cylinders in the main casing assembly.
3 Cut the locking wire on the non-return relief valve plug; undo the plug and remove the spring and ball. The non-return valve body can then be unscrewed from the pump body. Undo the grub screw and pull the pump body from the casing.
4 Undo the eight nuts and spring washers a turn at a time from the studs which hold the main casing assembly to the rear casing. As the nuts are undone the pressure of the springs will be gradually released. Take off the main casing together with the brake ring, pull the four clutch springs off their guide bolts and remove the clutch together with the sun wheel assembly.
5 It is likely that the brake ring will stick to the casing. To separate the ring gently tap it on its flange with a soft faced hammer.
6 To free the sun wheel from the centre of the clutch assembly release the circlip on the splined end of the sun wheel and push out the sun wheel.
7 Remove the large circlip with a pair of circlip pliers and pull the bearing housing complete with thrust bearing off the clutch assembly.
8 Lift out the planet carrier from the annulus and if it is wished to remove the unidirectional roller clutch (this is not recommended unless the rollers are thought to be chipped or worn), take off the circlip and brass retaining washer in front of the clutch.
9 As the inner member is removed the roller bearings will fall out. Gather them together carefully. On no account should the outer bearing ring be removed as it is expanded into the annulus.
10 If it is wished to renew the needle roller bearing in the centre of the annulus, carefully lever it out or use an extractor if available.
11 To remove the combined output shaft and annulus from the rear casing first undo the locking screw to free the speedometer pinion and bush, undo and remove the nut from the rear of the output shaft, and pull off the coupling flange; with the aid of a drift tap out the output shaft and annulus from the rear of the casing. Note that the larger inner bearing will come away with the output shaft, but the smaller outer bearing will remain in position. To remove it, it must be extracted from the rear of the casing.
12 Thoroughly clean all the component parts and then examine them carefully. Check that the oil pump plunger and pin are not worn and that the spring has not contracted. (Free length should not be less than 2 in (50.8 mm). Examine the O-rings from the operating pistons and renew them if worn or if they are becoming hard, and check that the cylinder bores are free from score marks and wear. Check the ball bearings for roughness when turned and for looseness between the inner and outer races. Examine the splines for burrs and wear, and the rollers of the clutch for chips and flat spots.
13 Renew the clutch linings if they are burnt or worn and carefully examine the main and rear casings for cracks and other damage. Renew the steady bush if it is worn and examine the gear teeth for cracks, chips, and general wear. Examine the sealing balls for ridges which will prevent them seating properly and check the free length of the springs, the measurement being given in the specification.
14 Assembly of the unit can commence after any damaged or worn parts have been exchanged and new gaskets and seals obtained. Start by replacing the annulus and output shaft in the rear casing. Do not omit to fit the speedometer drivegear, distance piece, and thrust washer.
15 If it is necessary to fit a new bearing note that four different thicknesses of shim are available. The correct endfloat of the output shaft should be as specified. Shims are available in the following sizes:

0.090 in (2.28 mm) 0.095 in (2.41 mm)
0.100 in (2.54 mm) 0.105 in (2.67 mm)

16 Refit the coupling flange, the flange washer, and the castellated nut, tightening it to the specified torque and locking it in place with a split-pin.
17 Replace the speedometer pinion and the pinion bush securing the latter with the washer and lock screw.
18 Reassemble the components of the unidirectional clutch holding the rollers in place in the cage with grease prior to fitting the inner member. Ensure that the spring is fitted in such a way that it pushes the rollers up the ramp on the inner member. Do not omit to replace the thrust bearing. Finally fit the oiler thrower and circlip in place.
19 Turn each of the planet gearwheels so the line etched on one tooth of each of the gearwheels lines up with one of the three corresponding lines on the periphery of the planet carrier. Insert the sun wheel into the carrier to keep the planet gearwheels in the correct positions and carefully fit the complete sun wheel and carrier assembly to the annulus. When fitted, the sun wheel can be withdrawn. Note that the sun wheel can now be inserted or removed as frequently as required, but if the planet carrier is removed from the annulus, the carrier gear wheels will have to be reset as described at the beginning of the paragraph.
20 Slide the splined end of the sun wheel into the centre of the clutch assembly and secure the sun wheel with the circlip. If the thrust bearing was removed for renewal press the new bearing in its housing; insert the four bolts into the housing threaded ends facing forwards, and fit the bearing and housing assembly over the centre of the clutch assembly. Lock the bearing and housing in place on the clutch with the bearing retaining circlip which fits in a groove on the clutch.
21 Carefully fit the clutch and sun wheel assembly to the planet carrier in the annulus. Refit the retainer plate and clutch springs over the bolts.
22 Coat the mating faces of the main casing and the brake ring with jointing compound noting that the smaller diameter of the brake ring abuts the main casing, and fit the brake ring and main casing together.
23 Coat the mating faces of the brake ring and rear casing with jointing compound and offer up the rear casing to the brake ring and front casing. Slide the four bolts through their holes in the main casing and replace and tighten down a turn at a time the nuts and washers which hold the casings together.
24 Fit the rubber O-rings to the two operating pistons and with the pistons generously lubricated with oil, slide them into their cylinders in the housing, so the spigoted ends face the front of the overdrive assembly. Slip the two bridge pieces over the ends of the four bolts and refit the washers and nuts not omitting to turn up the lock washer tabs when the nuts are correctly tightened.
25 Replace the oil pump body smaller end first into the centre hole at the bottom of the casing making sure the oil inlet faces the rear. Gently tap it into position until the groove lines up with the grub screw hole. Fit and tighten the grub screw.
26 Refit the component parts of the relief valve, operating valve, and non-return valve and do up the three plugs. Lock together the heads of the relief valve and non-return valve with wire. Assembly is now complete.

30 Overdrive (Type D) – operating lever adjustment

1 If the overdrive does not engage, or will not release when it is switched out, providing the solenoid is not at fault, the trouble is likely to be that the operating lever is out of adjustment. Adjustment can be made without removing the overdrive.
2 Undo the three bolts and washers holding the solenoid cover plate

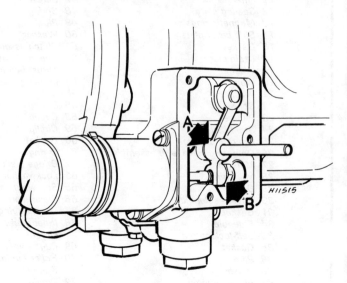

Fig. 6.26 Overdrive operating plunger adjustment (Sec 30)

A Gauge rod *B Self-locking nut*

Chapter 6 Gearbox and overdrive

in position, to give access to the operating lever and solenoid plunger.
3 Procure a short length of mild steel rod of $\frac{3}{16}$ in (4.76 mm) diameter. Switch the ignition on, put the car in top gear, and flick the actuating switch to the overdrive position.
4 If the $\frac{3}{16}$ in rod can now be passed through the hole A in the operating arm into the hole in the casing, adjustment is correct.
5 If the solenoid does not move the arm far enough for the rod to be inserted into the hole in the casing, or if it moves the arm too far, hold the solenoid plunger from turning by means of the two flats machined on its shank, and pressing the plunger tightly into the solenoid, screw the self locking nut B in Fig. 6.26 in or out until the $\frac{3}{16}$ in test rod can be pushed fully home into the hole in the casing.
6 Operate the switch several times, checking with the test rod to ensure adjustment is correct. Measure the current consumed by the solenoid which, with the operating arm correctly set, should be 2 amps. If a reading of about 20 amps is obtained this shows that the solenoid plunger is not moving sufficiently to switch to the holding coil from the operating coil.
7 Continuous high current will cause premature solenoid failure.
8 With the solenoid again de-energised, re-align the setting holes and insert the test rod. Hold the solenoid plunger against the blanking plug and check that dimension A is between 0.150 and 0.155 in (3.81 and 3.937 mm).
9 To obtain this dimension, alter the thickness of the blanking plug washer (on early models) or on later units turn the adjuster screw (32A in Fig. 6.25).

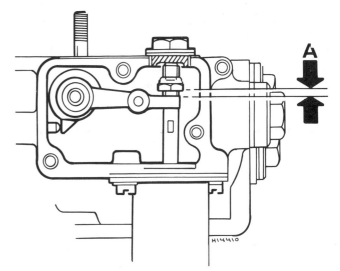

Fig. 6.27 Overdrive solenoid plunger setting diagram

A = 0.150 to 0.155 in (3.81 to 3.94 mm) (Sec 30)

31 Overdrive valves (Type D) – removal, inspection and refitting

1 Access to the relief and non-return valves which are located in the bottom of the overdrive is obtained in the following way. First drain the oil from the gearbox and overdrive.
2 Cut through the locking wire, unscrew the plugs and remove and clean the components. Note that the valve cap and non-return valve body are unscrewed from the pump, and that the relief valve body is removed with circlip pliers.
3 Examine the seatings for pits or chips, and the balls for wear, and ridges. The steel ball in the non-return valve is very hard and if the ball is undamaged and the seating is suspect tap the ball firmly into its seat with a soft metal drift.
4 Reassembly is a straightforward reversal of the removal sequence. Do not omit to fit the copper washer on the relief valve between the cap and main casing, and hold the non-return valve ball to its spring with petroleum jelly during refitment.
5 Access to the operating valve can only be gained after removing the remote control assembly from inside the car. Undo the plug and check that the ball is lifted $\frac{1}{32}$ in when the solenoid is actuated. Failure to move, points to a fault in the solenoid or operating arm.
6 The ball can be removed with a magnet and the valve with a piece of $\frac{1}{8}$ in (3.2 mm) wire. Check the ball and seat and clean out the small hole in the side of the valve tube. Check if the oil pump is working by jacking the rear of the car off the ground, placing the car in top gear, engage overdrive and with the engine running watch if oil is being pumped into the valve chamber. Replacement is a reversal of the removal procedure.

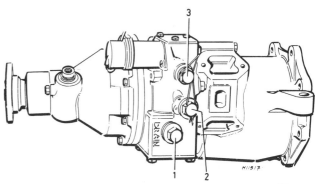

Fig. 6.28 View of overdrive main casing from underneath (Sec 31)

1 Filter cover plate   3 Non-return valve plug
2 Drain plug   4 Relief valve plug

32 Overdrive (Type J) – description and maintenance

1 This type of overdrive is optionally available on later models.
2 Its purpose is as described earlier for the Type D overdrive, but the control switch is incorporated in the knob of the gearchange lever.
3 The overdrive gears are epicyclic and comprise a central sun wheel which meshes with three planet gears. These in turn mesh with an internally-toothed annulus. All gears are in constant mesh.
4 The planet carrier is attached to the input shaft while the annulus is integral with the output shaft.
5 An extended gearbox mainshaft provides the overdrive input shaft.
6 Forward direct drive power is transmitted to a clutch which operates by means of rollers which wedge between the inner and outer clutch members.
7 A spring-loaded cone clutch is mounted on the splined extension of the sun wheel and causes locking of the gear train during over run

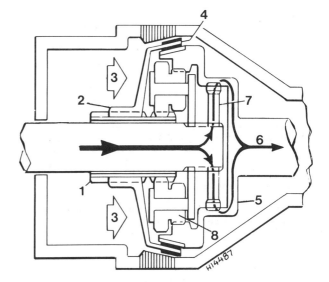

Fig. 6.29 Overdrive (Type J) in direct drive (Sec 32)

1 Sun wheel   5 Annulus
2 Sliding cone clutch   6 To propeller shaft
3 Spring pressure   7 Uni-directional roller clutch
4 Annulus and sun wheel locked   8 Planet wheels

and reverse torque conditions when otherwise the unidirectional clutch would give a free-wheel condition.
8 The cone clutch is moved in a forward direction by hydraulic pressure acting on two pistons applied by opening a valve controlled by the gear lever knob switch.
9 This pressure overcomes the force of the clutch springs and engages the clutch with the brake ring so holding the sun wheel at rest. The pressure valve is operated by a solenoid.
10 Reference to the diagram will indicate the position of the cone clutch when in contact with the brake ring, which itself is part of the overdrive casing. With the sun wheel held stationary, the output shaft and the annulus continue to rotate at the same speed, but the planet wheels continue to rotate on their axes around the stationary sun wheel so reducing the speed of the planet carrier and input shaft. The unidirectional clutch allows the outer member to overrun the inner member. These conditions provide a lower engine speed for a given road speed.
11 Hydraulic pressure for the overdrive is generated by a plunger type pump, operated by a cam on the input shaft.
12 Maintenance consists of keeping the oil level (common oil supply) in the gearbox up to its maximum. The most likely cause of a fault occurring is due to dirt or grit. Observe absolute cleanliness during oil replenishment and changing. If the gearbox oil is drained, then the residue of oil in the overdrive can only be removed if the overdrive unit sump is removed.
13 Although routine renewal of the gearbox lubricant is not specified by the manufacturer, your decision to do so may be influenced by the comments contained in Section 2, paragraph 7.

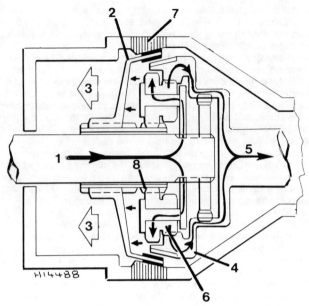

Fig. 6.30 Type J overdrive engaged (Sec 32)

1 From gearbox
2 Sliding cone clutch
3 Hydraulic pressure
4 Annulus driven by planet gears
5 To propeller shaft
6 Planet wheels
7 Locked cone clutch holding sun wheel
8 Sun wheel

33 Filters (Type J overdrive) – cleaning

1 The following 'in situ' operations will normally only be required in the event of a malfunction or during overhaul of the overdrive unit.

Sump filter
2 Unscrew and remove the six bolts and their washers which secure the sump.
3 Remove the sump and the gasket.
4 Prise out the filter.
5 Cleaning of the filter mesh should be done with either petrol or paraffin.

Pressure filter
6 With the sump filter removed as just described, remove the pressure filter base plug. This is the largest plug and should be unscrewed using a suitable pin wrench. The filter element will be withdrawn with the plug.
7 Extract the aluminium washer from the shoulder in the filter bore.
8 Wash the filter element with petrol or paraffin.
9 Refitting is a reversal of removal, but tighten components only to the specified torque.

34 Relief valve/dashpot (Type J overdrive) – removal and refitting

1 Remove the overdrive sump and plug as previously described and refer to Fig. 6.31.
2 Unscrew and remove the relief valve plug using a suitable pin wrench.
3 Withdraw the dashpot piston complete with springs and cup. According to date of production, the assembly may be one of two types.
4 Remove the residual pressure spring.
5 Withdraw the relief valve piston by drawing it out of the overdrive with a pair of pliers.
6 Withdraw the relief valve and dashpot sleeve using either the special tool shown or a suitable substitute.
7 Do not dismantle the removed components any further, but clean and examine them and if they are worn or damaged, renew the assemblies complete.
8 Refitting is a reversal of removal, but use new O-rings and tighten the base plug to the specified torque.

35 Pump non-return valve (Type J overdrive) – removal and refitting

1 Remove the sump and filter as previously described.
2 Unscrew and remove the centre plug now exposed, using a suitable pin wrench.
3 Remove the valve components taking care not to lose the spring and ball.
4 Remove the valve seat.
5 Refitting is a reversal of removal; tighten the plug to the specified torque.

36 Overdrive (Type J) – removal and refitting

1 Before commencing removal operations, raise the rear wheels of the car and run the engine with transmission engaged. Engage the overdrive, depress the clutch pedal and holding it depressed, disengage the overdrive. Switch off the engine. This action releases the spline loading between the planet carrier and the unidirectional roller clutch which sometimes makes separation of the overdrive from the gearbox difficult.
2 To make the removal operations easier, take out the seats (Chapter 12). Disconnect the battery.
3 Slacken all radiator mounting nuts and bolts.
4 Remove the gearbox tunnel cover as described in Section 3 of this Chapter.
5 Disconnect the electrical leads from the reverse lamp switch, the overdrive and the seat belt warning system (where fitted).
6 Clean any dirt around the gearbox top cover plate, remove the bolts and lift the cover plate away together with the gasket.
7 Remove the selector rail locking bolt.
8 Remove the bolts securing the remote control rear support bracket to the overdrive unit.
9 Withdraw the remote control selector rail.
10 Remove the four bolts from the propeller shaft front coupling and disconnect the speedometer cable.
11 Raise the front of the car and disconnect the exhaust system front flange and the gearbox mounting point.

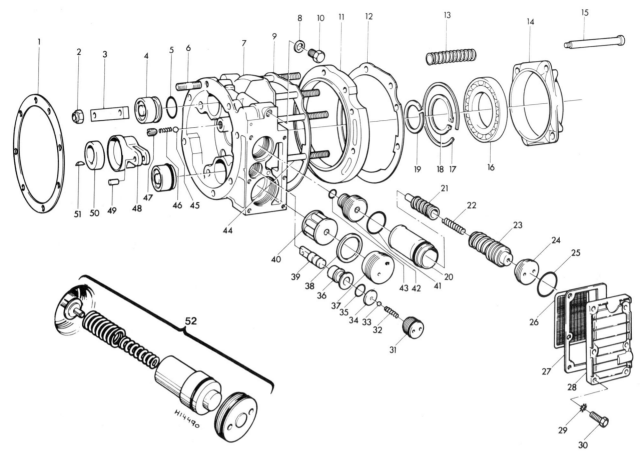

Fig. 6.31 Main casing components of Type J overdrive (Sec 33)

1 Gasket
2 Locknut
3 Bridge piece
4 Operating piston
5 O-ring
6 Stud
7 Main case
8 Washer
9 Gasket
10 Pressure tapping plug
11 Brake ring
12 Gasket
13 Clutch return spring
14 Thrust ring
15 Thrust pin
16 Thrust ball race
17 Retaining circlip
18 Circlip for sliding member
19 Circlip for sun wheel
20 Dashpot sleeve
21 Relief valve assembly
22 Double dashpot spring
23 Dashpot piston assembly
24 Dashpot plug
25 O-ring
26 Sump filter
27 Sump gasket
28 Sump
29 Star washer
30 Bolt
31 Pump plug
32 Non-return valve spring
33 Steel ball
34 Non-return valve seat
35 O-ring
36 Pump body
37 Pressure filter plug
38 Pressure filter washer
39 Pump plunger
40 Pressure filter
41 O-ring
42 Relief valve body
43 O-ring
44 Stud
45 Steel ball
46 Lubrication relief valve spring
47 Lubrication relief valve plug
48 Pump strap
49 Pump pin
50 Cam
51 Woodruff key
52 Later type relief valve/dashpot

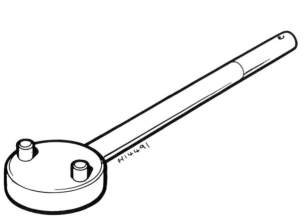

Fig. 6.32 Typical pin wrench (Sec 33)

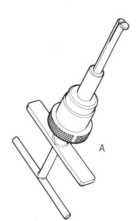

Fig. 6.33 Overdrive relief valve and dashpot sleeve removal tool (L.401A) (Sec 34)

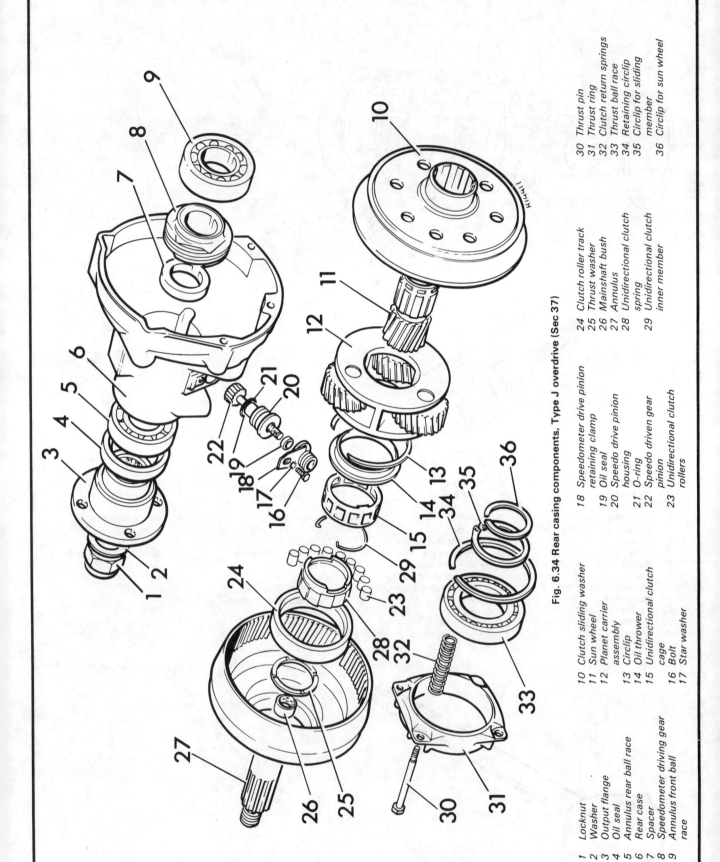

Fig. 6.34 Rear casing components, Type J overdrive (Sec 37)

1 Locknut
2 Washer
3 Output flange
4 Oil seal
5 Annulus rear ball race
6 Rear case
7 Spacer
8 Speedometer driving gear
9 Annulus front ball race
10 Clutch sliding washer
11 Sun wheel
12 Planet carrier assembly
13 Circlip
14 Oil thrower
15 Unidirectional clutch cage
16 Bolt
17 Star washer
18 Speedometer drive pinion retaining clamp
19 Oil seal
20 Speedo drive pinion housing
21 O-ring
22 Speedo driven gear pinion
23 Unidirectional clutch rollers
24 Clutch roller track
25 Thrust washer
26 Mainshaft bush
27 Annulus
28 Unidirectional clutch spring
29 Unidirectional clutch inner member
30 Thrust pin
31 Thrust ring
32 Clutch return springs
33 Thrust ball race
34 Retaining circlip
35 Circlip for sliding member
36 Circlip for sun wheel

Chapter 6 Gearbox and overdrive

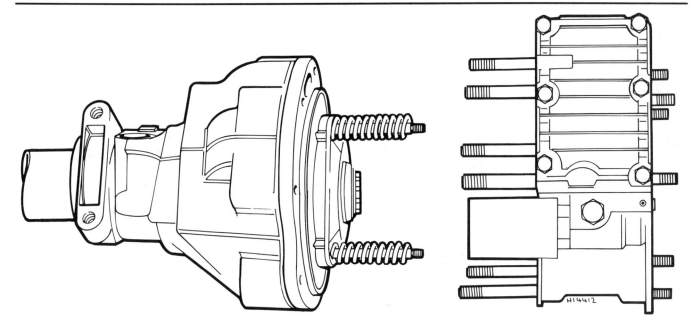

Fig. 6.35 Separating main and rear overdrive casings (Sec 37)

12 Disconnect the engine restraint cable and drain the gearbox oil.
13 Position a jack under the gearbox casing and raise the jack, checking that the fan blades do not damage the radiator as you do so.
14 Remove the two bolts which secure the overdrive to the gearbox mounting.
15 Remove the eight nuts and spring washers securing the overdrive unit to the gearbox adaptor plate.
16 Remove the overdrive unit from the mainshaft leaving the adaptor plate in position.
17 Refitting is a reversal of removal. Use new flange gaskets and top-up the combined gearbox/overdrive oil level after the overdrive unit has been refitted. Recheck the oil level after the car has been driven for the first few miles.

37 Overdrive (Type J) – overhaul

Dismantling
1 With the overdrive unit removed from the car, clean away external dirt using paraffin and a stiff brush.
2 Unscrew the four nuts which hold the bridge pieces. Remove the bridge pieces.
3 Unscrew the six nuts around the main casing, a half turn at a time to progressively release the clutch return spring pressure. Note the two copper washers located on the two uppermost studs, and then remove all the washers.
4 Separate the main and rear casings.
5 Lift the sliding member complete with sun wheel out of the main casing.
6 Lift the planet carrier out of the main casing, taking care not to damage the oil catcher which is attached to the underside of the carrier.
7 Tap the brake ring from its spigot in the main casing using a drift.
8 With a pair of pliers, withdraw the operating pistons.
9 Unbolt and remove the sump plate and filter screen. Allow any residual oil to drain.
10 Remove the relief valve assembly (Section 34).
11 Remove the pump non-return valve assembly (Section 35).
12 Remove the oil pump by twisting it out of the main casing.
13 Remove the pressure filter (Section 33).
14 Unscrew and remove the solenoid valve.
15 Extract the circlip from the sun wheel extension and remove the sun wheel.
16 Extract the circlip from its groove on the hub of the cone clutch.

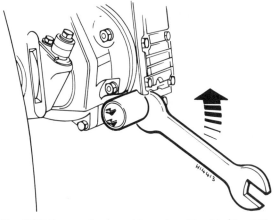

Fig. 6.36 Unscrewing overdrive solenoid valve (Sec 37)

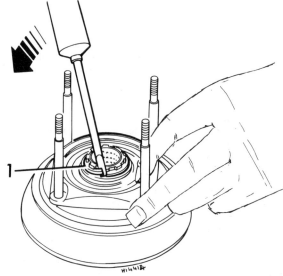

Fig. 6.37 Extracting overdrive sun wheel extension circlip (Sec 37)

Chapter 6 Gearbox and overdrive

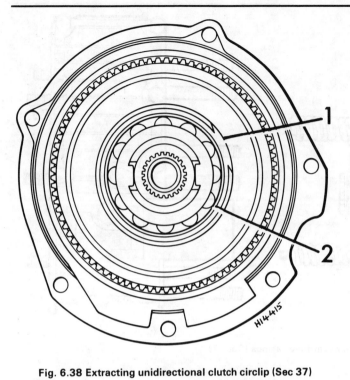

Fig. 6.38 Extracting unidirectional clutch circlip (Sec 37)

1 Circlip   2 Oil thrower

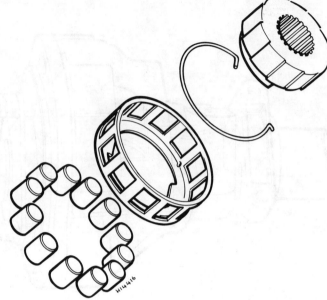

Fig. 6.39 Unidirectional clutch dismantled (Sec 37)

17 Using a plastic faced mallet tap the clutch from the thrust ring bearing.
18 Now extract the large circlip and press the bearing from its housing.
19 To dismantle the unidirectional clutch, extract the circlip and lift out the oil thrower.
20 Withdraw the inner member complete with rollers. Remove the bronze thrust washer.
21 Remove the speedometer gear and bush.
22 Hold the output flange either by bolting a length of flat steel to two of its bolt holes or if you have a vice with wide enough opening jaws, grip it and then unscrew the nut and remove it with its washer.
23 Remove the output flange. This may require the use of an extractor.
24 Drive the annulus from the rear casing by applying a plastic-faced hammer to the end of the output shaft. The front bearing, speedometer drivegear and spacer will be withdrawn with the annulus.
25 Remove the oil seal.
26 Drive out the rear bearing.
27 Clean and examine all components for wear or damage, particularly the cylinder bores for scoring. If the clutch linings on the sliding member are worn or burnt, then the complete member must be renewed. Renew all O-rings, oil seals and gaskets. These are usually supplied in the form of a repair kit.

Reassembly

28 Commence reassembly by locating the speedometer drivegear in the rear casing so that its plain boss faces the front bearing. Press the bearing into the rear casing until it is fully seated.
29 Now press the front bearing with the rear casing and speedometer drivegear onto the annulus until the bearing is fully seated.
30 Fit the spacer onto the annulus.
31 Using a piece of suitable tubing, press the rear bearing onto the annulus and into the rear casing simultaneously.
32 Drive in a new oil seal, lips filled with grease.
33 Press the output flange into position.
34 Fit the washer and then holding the flange from rotating, tighten the nut to the specified torque.
35 Locate the spring and inner member of the unidirectional clutch into its cage. Locate the spring so that the cage is spring-loaded in an anti-clockwise direction when viewed from the front. The clutch rollers

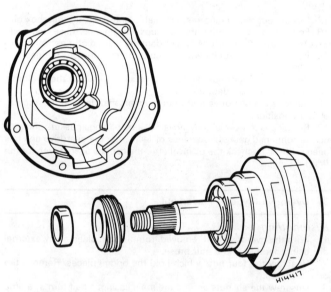

Fig. 6.40 Annulus removed from rear casing (Sec 37)

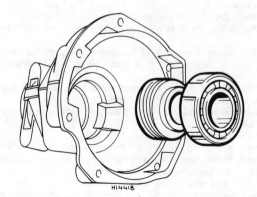

Fig. 6.41 Fitting speedo drivegear and bearing into overdrive rear housing (Sec 37)

Chapter 6 Gearbox and overdrive

are best retained in their cage for refitting by applying thick grease and a worm drive clip not too tightly fitted.
36 Fit the bronze thrust washer into the recess in the annulus.
37 Slide the clutch inner member into the outer member and as it moves into position the temporary clip will slide off.
38 Fit the oil thrower and secure with the circlip.
39 Check that the clutch rotates only in an anti-clockwise direction.
40 Fit the bearing and housing, with retaining circlips, onto the hub of the cone clutch.
41 Fit the sun wheel and the circlip to the sun wheel extension.
42 Apply oil to the operating pistons and fit them into their cylinder bores.
43 Fit the solenoid valve, applying torque only to the mounting nut.
44 Fit the pressure filter, oil pump, non-return valve and relief valve assemblies.
45 Fit the sump plate and filter screen.
46 Fit the planet carrier into the rear casing, meshing the gears in any position.
47 Place the sliding member assembly complete with clutch return springs onto the cone of the annulus. Engage the sun wheel with the planet gears.
48 Apply jointing compound to new gaskets located on both sides of the brake ring. Note that these gaskets are not interchangeable.
49 Fit the brake ring onto its spigot in the rear casing and align the stud holes.
50 Place the main casing over the thrust housing pins. At the same time enter the studs in the brake ring.
51 Screw on the main and rear casing connecting nuts. Tighten them progressively to the specified torque. Make sure that jointing compound is applied to the threads of the two uppermost studs and to their copper washers.
52 Connect the earth lead to the stud which is located above the solenoid aperture.
53 Fit the two bridge pieces and tighten their fixing nuts to the specified torque.

38 Fault diagnosis – gearbox and overdrive

Symptom	Reason(s)
Gearbox	
Weak or ineffective synchromesh	Baulk ring synchromesh dogs worn, or damaged
Jumps out of gear	Broken detent springs
	Gearbox coupling dogs badly worn
	Selector fork rod groove badly worn
	Selector fork loose on shaft
Excessive noise	Incorrect grade of oil in gearbox or oil level too low
	Bush or needle roller bearings worn or damaged
	Gearteeth excessively worn or damaged
	Laygear thrust washers worn allowing excessive endplay
Excessive difficulty in engaging gear	Clutch pedal adjustment incorrect
	Clutch not fully disengaging
Overdrive	
Does not engage	Low oil level
	Faulty solenoid or wires
	Low hydraulic pressure
Does not disengage	Sticking solenoid valve
	Cone clutch sticking
	Control orifice between solenoid valve and relief valve clogged (clean only with air pressure)
	Faulty circuit in electrical system
Cone clutch slipping in overdrive	Low oil level
	Low oil pressure
	Sticking solenoid valve
	Worn or burnt clutch linings
Slow disengagement, free wheeling on overrun, slip in reverse gear	Sticking relief valve
	Sticking control valve
	Control orifice blocked

Chapter 7 Propeller shaft, driveshaft, universal joints

Contents

Description and maintenance 1	Fault diagnosis 8
Driveshaft – removal and refitting 4	Propeller shaft – balancing 7
Driveshaft hub – overhaul 5	Propeller shaft – removal and refitting 2
Driveshaft universal joints – overhaul 6	Propeller shaft universal joints – overhaul 3

Specifications

Propeller shaft type .... Single piece, tubular with universal joints. Cars with overdrive have modified shaft incorporating strap or resident coupling

Driveshaft type .... Open with universal joint at inboard end and splined to hub at outboard end

Torque wrench settings

	lbf ft	Nm
Propeller shaft flange bolts	34	46
Driveshaft inboard flange screws	46	63
Driveshaft to hub nut	120	163
Transverse spring eye bolt	45	61
Radius arm to vertical link	32	44
Shock absorber lower mounting	38	52

1 Description and maintenance

1 Drive is transmitted from the gearbox to the differential unit by means of a finely balanced propeller shaft.
2 Fitted at each end of the shaft is a universal joint which allows for slight movement of the engine, gearbox and differential units in their mountings. A further universal joint is fitted to each of the differential unit drive flanges to accommodate the movements of the swing axle rear suspension.
3 Each universal joint comprises a four legged centre spider, four needle roller bearings and cups, two yokes, and the necessary seals, retainer, and circlips.
4 On vehicles equipped with overdrive, a slightly modified type of propeller shaft is fitted which incorporates a resilient coupling or strap.
5 The propeller shaft and universal joints are fairly simple components and to overhaul and repair them is not difficult. It should be noted that when the propeller shaft universal joints become worn and develop radial play, to obtain the best result always fit a complete exchange propeller shaft which will already be balanced to the fine degree necessary for smooth running. Overhauling the universal joints on the existing shaft may or may not throw the shaft out of balance. Refer to Section 7.
6 The universal joints are all of the sealed type and require no maintenance. On earlier models however there are threaded plugs in the spiders of the joint.
7 These plugs can be removed and grease nipples fitted for lubrication purposes if wished, but note that some of the plugs have left-hand threads and they are very difficult to get at with the shaft in situ. Do not run the car on the road with grease nipples fitted instead of plugs as there is a possibility of fouling.
8 On later models, grease plugs are not fitted and the joints are lubricated for life.
9 Periodically, check that the propeller shaft flange nuts and bolts are tight.

1.7 Propeller shaft rear universal joint showing grease nipple

Chapter 7 Propeller shaft, driveshaft, universal joints

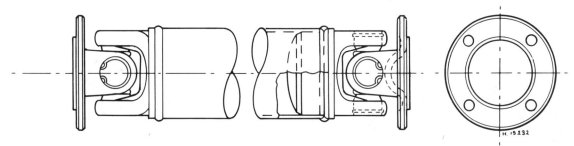

Fig. 7.1 Standard propeller shaft (Sec 1)

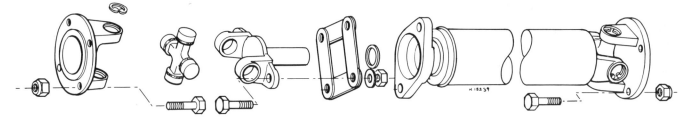

Fig. 7.2 Strap type coupling propeller shaft (Sec 1)

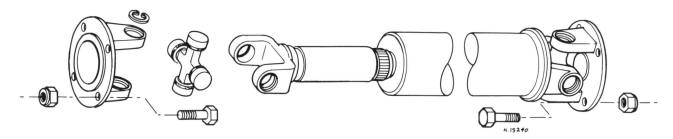

Fig. 7.3 Alternative type propeller shaft used on overdrive models (Sec 1)

2 Propeller shaft – removal and refitting

1 Jack-up the rear of the car, or position the rear of the car over a pit or on a ramp.
2 If the rear of the car is jacked-up supplement the jack with support blocks so that danger is minimised should the jack collapse.
3 If the rear wheels are off the ground place the car in gear or put the handbrake on to ensure that the propeller shaft does not turn when an attempt is made to release the nuts.
4 If the original shaft is being refitted, mark the relative position of the shaft flanges to the gearbox output and final drive coupling flanges by scribing a mark across the flange edges.
5 Remove the exhaust system excluding the front pipe.
6 Unscrew the flange nuts, remove the bolts and lift the propeller shaft from under the car.
7 Refitting is a reversal of removal, but align the marks made during removal.
8 If a new or overhauled shaft is fitted and any imbalance is detected when driving the car on the road, refer to Section 7.
9 Tighten flange nuts to the specified torque.

3 Propeller shaft universal joints – overhaul

1 Wear in the needle roller bearings is characterised by vibration in the transmission, 'clonks' on taking up the drive, and in extreme cases of lack of lubrication, metallic squeaking, and ultimately grating and shrieking sounds as the bearings break up.
2 Check the needle roller bearings for wear. This can be done with the propeller shaft in position. To check the rear universal joint turn the shaft with one hand and hold the rear axle flange with the other hand.

To check the front universal joint turn the shaft with one hand whilst holding the front gearbox coupling. Any movement detected in the universal joint indicates wear in the needle roller bearing.
3 If worn, the old bearings and spiders will have to be discarded and a repair kit, comprising new universal joint spiders, bearings, oil seals,

2.6 Propeller shaft front connection to gearbox output drive flange

Chapter 7 Propeller shaft, driveshaft, universal joints

3.5 Extracting circlip

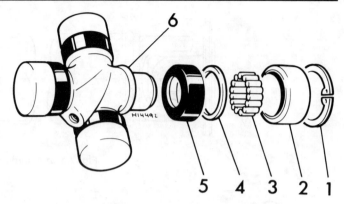

Fig. 7.4 Universal joint components (Sec 3)

1 Circlip
2 Bearing cup
3 Needle rollers
4 Washer
5 Seal
6 Spider

and retainers purchased. Check also by trying to lift the shaft and noticing any movement in the joints.

4 Remove the propeller shaft from the car.

5 Clean away all traces of dirt and grease from the circlips located on the ends of the bearing cups, and remove the clips by pressing their open ends together with a pair of pliers and lever them out with a screwdriver. If they are difficult to remove tap the bearing cup face resting on top of the spider with a mallet which will ease the pressure on the circlip (photo).

6 Strike the yoke, at the point indicated, to partially eject each bearing cup in turn.

7 Remove the cup with a pair of pliers. Remove the opposite cup, and then free the yoke from the propeller shaft.

8 To remove the remaining two cups repeat the instructions.

9 Smear jointing compound on the journal shoulders of the new spider if a retainer is used. With the aid of a tubular drift fit the oil seal retainers to the trunnions and then fit the oil seals to the retainers.

10 On some models it will be found that there is no retainer, merely a rubber seal. The spider must always be fitted so the lubricating plug holes are towards the propeller shaft.

11 On replacement great care must be taken to ensure the journals and associated parts are absolutely clean. Fill the grease holes in the spider journal with multi-purpose grease making sure all air bubbles are eliminated. Fill each bearing assembly to a depth of approximately $\frac{1}{8}$ in (3 mm).

12 Fit new rubber seals to the spiders and then replace the spiders and bearings in the yokes. Press the bearing cups into the yoke using a vice.

13 Fit new circlips.

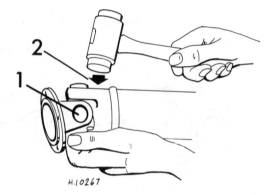

Fig. 7.5 Ejecting universal joint bearing cup (Sec 3)

1 Bearing cup 2 Striking point

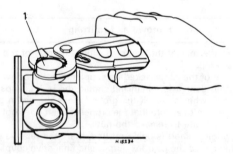

Fig. 7.6 Removing bearing cup (Sec 3)

3.13a Assembling spider to yoke

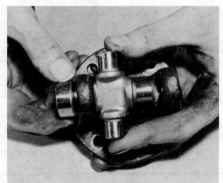

3.13b Positioning bearing cup

3.13c Pressing bearing cups into yoke

Chapter 7 Propeller shaft, driveshaft, universal joints

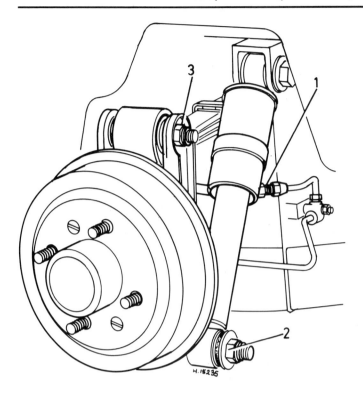

Fig. 7.7 Rear hub and link disconnection points (Sec 4)

1 Brake pipeline union
2 Shock absorber lower mounting nut
3 Spring eye bolt

4 Driveshaft – removal and refitting

1 The drive shafts are removed complete with the hubs and vertical links. It is necessary to remove the shafts if attention is required to either the universal joints or the hub bearings.
2 Working on one side of the car, remove the hub caps, and loosen the wheel nuts. Jack-up the rear of the car and support it on stands. Take off the road wheel.
3 Disconnect the hydraulic pipeline from the brake wheel cylinder and quickly cap the end of the pipe to prevent loss of fluid (a bleed nipple dust cap is useful for this).
4 Free the clevis pin from the lever on the backplate which is attached to the handbrake cable lever.
5 Place a jack under the vertical links and screw the jack up slightly to relieve the mounting pin and nut which secures the lower end of the damper, of all load. Remove the nut and pull the damper off the pin.
6 Undo the bolt which holds the radius arm to its bracket on the vertical link and undo the four bolts and nuts which hold the differential drive flange to the flange on the universal joint after having made a mating mark across the flanges to ensure identical positioning on reassembly.
7 Remove the jack, undo the nut which holds the spring eye bolt in place, and supporting the brake drum with one hand pull out the spring eye bolt as shown in Fig. 7.8.
8 The drive shaft can now be removed complete with the hub and vertical link.
9 Replacement commences by refitting the vertical link to the spring, tightening the nyloc nut on the spring eye bolt finger tight only at this stage.
10 Jack-up the vertical link so the damper can be refitted and reconnect the universal joint to the differential flange and the radius arm to the bracket on the front of the vertical link.
11 Reconnect and bleed the brakes, refit the road wheel, lower the car to the ground and with one person sitting in the car to give a 'static laden' condition tighten the nyloc nut in the spring eye bolt to the specified torque.

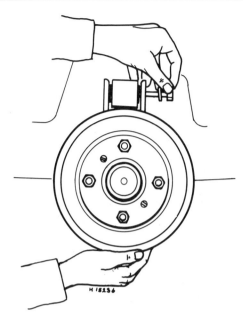

Fig. 7.8 Removing spring eye bolt (Sec 4)

5 Driveshaft hub – overhaul

1 Due to the exceptionally tight fit of the rear hub on the shaft, removal can normally only be carried out if the correct tool (S4221A) is available, or a hydraulic press used with suitable supporting adaptors. Failing the availability of these tools, the hub/shaft will have to be taken to a dealer for removal of the hub.
2 Place the driveshaft between the protected jaws of a vice, and take off the brake drum after undoing the countersunk screws and backing off the brake shoe adjustment.
3 Referring to Fig. 7.9 undo the hub nut and washer and then pull off the hub using a hub extractor. Note the key.
4 Bend back the lock tabs and undo the four bolts which hold the grease trap, brake backplate, outer seal housing, oil seal and gasket to the trunnion housing.
5 To free the vertical link from the trunnions, undo the nyloc nut and then wriggle out the bolt.
6 Now push the trunnion further onto the driveshaft so that the ball race protrudes enough for the fitting of an extractor. With the ball race bearing removed, the shaft can be pulled out of the trunnion.
7 Carefully remove the oil seal and then with the trunnion placed on a block of wood with its flanged face downwards drift out the needle roller bearing with the aid of a suitable drift.
8 Thoroughly clean all the parts and renew the bearings if they are worn, rough or chipped. Examine the hub for cracks, a worn taper and keyway, and worn outer oil seal contacting faces. Check the shaft for straightness and wear. If necessary get expert advice from your local Triumph distributor, as wear that appears to be very small can have a considerable effect on performance.
9 Reassembly commences by fitting the needle roller bearng radiused end first (i.e. pressing against the lettered end) into the trunnion with the aid of a suitable drift until the bearing is 0.5 in. (12.7 mm.) from the small end of the trunnion face.
10 Drift the inner oil seal into the trunnion housing with the sealing lips trailing and then drive the oil flinger onto the axle shaft, positioning it as shown in the diagram.
11 Thoroughly grease the needle rollers and slide the drive shaft through the trunnion taking great care that the lips of the seal are not turned back or damaged.
12 Fit the driveshaft between the protected jaws of a vice and then drift on the ball race which should be prepacked with grease. Fit a new oil seal ensuring that the sealing lip is trailing.
13 Fit a new gasket to the trunnion flange face and then fit the seal housing, brake backplate with the wheel cylinder uppermost, and the grease trap with the duct facing downwards.
14 Then fit the four securing bolts and tab washers. Check that the

Chapter 7 Propeller shaft, driveshaft, universal joints

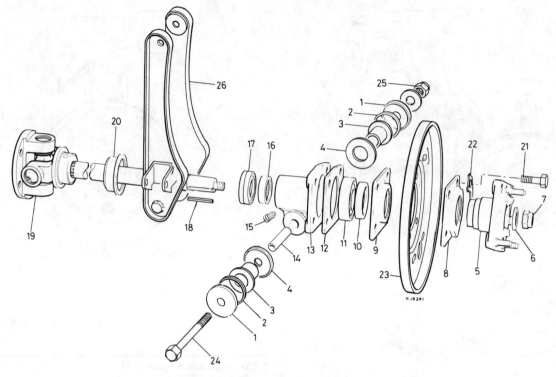

Fig. 7.9 Driveshaft and hub components (Sec 5)

1 Washer	8 Grease trap	15 Grease plug	21 Bolt
2 Seal	9 Seal housing	16 Needle roller bearing	22 Lockplate
3 Nylon bush	10 Oil seal	17 Oil seal	23 Brake backplate
4 Flanged washer	11 Bearing	18 Woodruff key	24 Bolt
5 Hub	12 Gasket	19 Universal joint	25 Self-locking nut
6 Washer	13 Trunnion	20 Oil seal flinger	26 Vertical link
7 Nut	14 Sleeve		

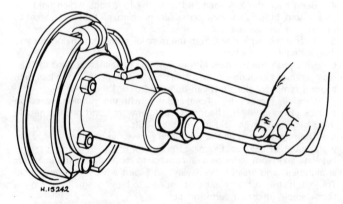

Fig. 7.10 Using official tool to remove hub (Sec 5)

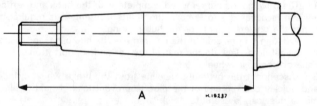

Fig. 7.11 Shaft oil flinger (20) setting diagram (Sec 5)

A 5.740 to 5.760 in (145.8 to 146.3 mm)

key fits tightly in the slot in the driveshaft and also the hub. Drive the hub on squarely, replace the washer and nut and tighten to the specified torque. Refit the brake drum and replace the countersunk screws.

15 Finally reassemble the vertical links to the trunnion housing. Ensure that the various bushes, rubber seals and steel sleeve are in the order shown in Fig. 7.9. Refit the bolt and tighten the nut to the specified torque.

6 Driveshaft universal joints – overhaul

1 The joints can be removed and overhauled as described in Section 3 for the propeller shaft joints.

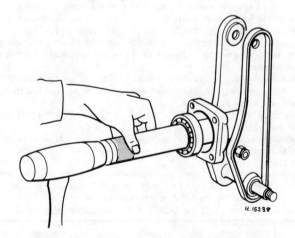

Fig. 7.12 Fitting driveshaft ball race (Sec 5)

Chapter 7 Propeller shaft, driveshaft, universal joints

7 Propeller shaft – balancing

1   Factory reconditioned or new propeller shafts are finely balanced, but where the shaft joints have been renewed by installing repair kits then it is possible that vibration may be detected once the car is back on the road and being driven through its speed range.

2   To eliminate this condition, first try releasing the flange couplings and turning the propeller shaft through 180°.

3   If this does not effect a cure, a worm drive hose clip can be fitted around the shaft and the position of the worm housing set at different positions around the shaft. Employing a trial and error method, the best setting may be found for minimum vibration. Moving the position of the clip by sliding it up and down the shaft may also have a beneficial effect.

8 Fault diagnosis – propeller shaft

Symptom	Reason(s)
Vibration when car running on road	Out-of-balance shaft Loose flange bolts Worn shaft joints

Chapter 8 Differential and final drive

Contents

Differential carrier – removal and refitting	4
Differential/final drive mounting plate rubber cushions – renewal	8
Differential/final drive unit – removal and refitting	3
Fault diagnosis – differential/final drive	9
General description	1
Maintenance	2
Output shaft oil seal – renewal	7
Pinion oil seal (bearing preload by collapsible spacer) – renewal	6
Pinion oil seal (bearing preload by shims) – renewal	5

Specifications

Type .... Hypoid, chassis-mounted drive transmitted by open universally-jointed shafts to rear roadwheels

Ratio:
Mk I, II, III and IV .... 4.11 : 1
1500 (except N. America) .... 3.63 : 1
1500 (N. America) .... 3.89 : 1

Lubrication
Oil type/specification .... Hypoid gear oil, viscosity SAE 90EP (Duckhams Hypoid 90S)
Oil capacity .... 1.0 Imp pt (1.2 US pts, 0.57 l)

Torque wrench settings

	lbf ft	Nm
Differential to mounting plate bolts	38	52
Differential carrier to final drive housing bolts	20	27
Drain and filler/level plugs	25	34
Roadspring centre plate nuts	34	46
Output and driveshaft flange bolts	46	63
Radius arm to anchor bracket	32	44
Final drive housing rear mounting bolt	34	46
Mounting plate to chassis	45	61
Pinion drive/flange nut	85	115

1 General description

The main rear axle component is the hypoid differential unit which is fixed to the chassis at four points. Swing axle driveshafts, pivoting at their inner ends on universal joints attached to the differential drive flanges, carry the drive to the hubs via needle roller and ball bearings carried in trunnions mounted in the centre of vertical links. The upper ends of the links are attached to a single transverse leaf spring and the lower ends to longitudinal tie rods. The tie rods make use of flexible bushes at each end and provide fore and aft location for the links.

The crown wheel and pinion each run on opposed taper roller bearings, the bearing preload and meshing of the crown wheel and pinion being controlled by shims on early models or a collapsible spacer on later 1500 models.

Spring loaded oil seals, of the type normally found at the front of the differential nose piece, prevent oil loss from the differential at the pinion and driveshaft holes.

2 Maintenance

1 At the specified intervals, wipe the dirt from around the filler/level plug and unscrew it.
2 The oil level should be level with the bottom of the filler plug hole with the lubricant cold, the car not having been run for at least one hour.
3 Where necessary, top up with the specified oil.
4 Frequent topping up will indicate an oil leak probably from a faulty oil seal, rectify immediately.

Chapter 8 Differential and final drive

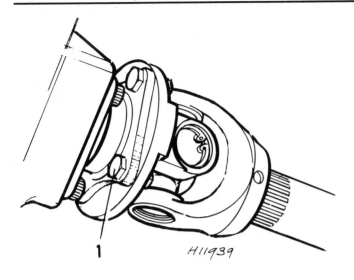

Fig. 8.1 Driveshaft to output flange connecting bolt (1) (Sec 3)

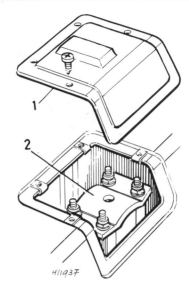

Fig. 8.2 Rear roadspring access cover (1) and spring plate (2) (Sec 3)

5 On very early models fitted with a drain plug, the oil should be drained (hot) from the differential unit at the intervals specified in *Routine Maintenance*. Refill with the specified oil.
6 On later models, the differential is filled for life and a drain plug is not fitted, topping up, if necessary, only being required after a level check at the specified mileage intervals.
7 On completion of topping up or refilling, always check that the filler/level and drain plug if fitted, are tight.

3 Differential/final drive unit – removal and refitting

1 Remove the rear wheel hub caps and loosen the wheel nuts one turn.
2 Jack-up the rear of the car and place stands or other supports under the chassis. For safety's sake disconnect the battery.
3 Drain the oil from the differential unit and then remove both rear wheels placing the wheel nuts in the hub caps for safe keeping.
4 Place jacks under the vertical link on each side of the car and raise the jack slightly to take the spring load. Take off the nyloc nuts and washers from the shock absorbers' lower attachment eyes and then pull the bottom of the shock absorbers clear of the mounting pins. Lower the jacks.
5 Remove the rear exhaust pipe and silencer by undoing the pipe clip at the rear of the expansion chamber, free the mounting from the hypoid front plate and take out the nut and bolt which holds the fabric strap in place.
6 Mark the adjacent differential to propeller shaft and driveshaft flanges and undo and remove the four bolts and nyloc nuts from these three sets of couplings.
7 From behind the centre of the front seats undo the four attachments which hold the spring access cover in place and lift the cover away.
8 Access can now be gained to the nyloc nuts which hold the spring securing plate in place. Undo the nuts and remove the plate. Unscrew the studs from the top of the differential unit, using two nuts locked together on the threads.
9 Take the weight of the hypoid unit on a jack, or better still get an assistant to hold it, undo the nyloc nuts, and remove the large plain washers and rubber bushes which hold each end of the front mounting plate to the studs on the chassis. Undo and remove the two nuts and bolts which hold the differential casing lugs to the chassis brackets and carefully manoeuvre the differential unit out from under the car. On later models a single bolt is used at the mounting lugs (photo).
10 Replacement commences by refitting the differential unit lugs to the rear mounting points on the chassis, doing up the nuts on the retaining bolts finger tight. Refit the rubber pads to the ends of the front mounting plate taking great care to fit the rubbers the right way

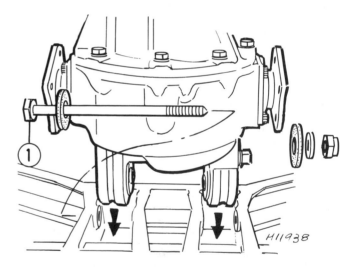

Fig. 8.3 Later type rear mounting lugs and bolts (1) (Sec 3)

3.9 Single bolt at differential mounting lugs on later models

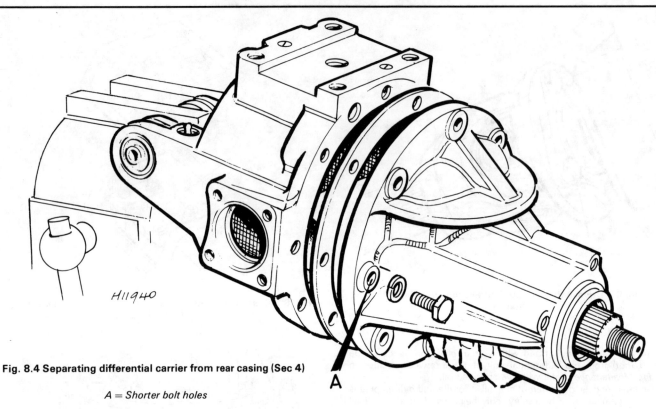

Fig. 8.4 Separating differential carrier from rear casing (Sec 4)

A = Shorter bolt holes

up and renew any of the rubbers that are worn or perished. Replace the plain washers and tighten down all the nyloc nuts securely.
11 Then refit the spring studs, and spring plate followed by the driveshaft and propeller shaft couplings, the dampers, exhaust tail pipe and silencer and wheels.
12 Fill the unit with oil.

4 Differential carrier – removal and refitting

1 Remove the complete differential/final drive unit as described in the preceding Section and clean away external dirt with paraffin or a water soluble solvent using a stiff brush.
2 Undo the ring of bolts and spring washers which hold the front casing and differential carrier assembly to the rear casing.
3 Turn the points so that the two chamfered parts on the edge of the differential carrier allow removal of the differential housing.
4 Further dismantling of the unit is not recommended as, for it to function satisfactorily, a large number of specialised tools are required, including a special hypoid housing spreader which has to be fitted before the differential unit can be removed from its housing.
5 Obtain a new or reconditioned assembly or if this proves too expensive it may be possible to obtain a serviceable unit from a vehicle dismantler.
6 On replacement thoroughly clean the differential unit joint faces and fit a new paper joint. Note specially that the two short flange bolts must be fitted next to the output shafts in holes (A) as shown in Figure 8.4.

5 Pinion oil seal (bearing preload by shims) – renewal

1 On Mk I, II, III and IV models, the pinion bearing preload is controlled by shims.
2 If oil is leaking from the front of the differential nose piece it will be necessary to renew the pinion oil seal. If a pit is not available, jack and chock up the rear of the car. It is much easier to do this job over a pit, or with the car on a ramp. First remove the exhaust tail pipe.
3 Mark the propeller shaft and pinion drive flanges to ensure their replacement in the same relative positions.
4 Unscrew the nuts from the four bolts holding the flanges together, remove the bolts and separate the flanges.
5 If the oil seal is being renewed with the differential in position, drain the oil and check that the handbrake is firmly on to prevent the pinion flange moving.
6 Pull out the split-pin and unscrew the nut in the centre of the pinion drive flange. Although it is tightened down to the specified torque, it can be moved fairly easily with a long extension arm fitted to the appropriate socket spanner. Remove the nut and spring washer. If necessary bolt a long lever to two of the flange bolt holes to prevent the flange from turning.
7 Pull off the splined drive flange, which may be a little stubborn, in which case it should be tapped with a hide mallet from the rear and prise out the oil seal with a screwdriver taking care not to damage the lip of its seating.
8 Replacement is a reversal of the above procedure. Note that the new seal must be pushed into the differential nose piece with the edge of the sealing ring facing inwards, and take great care not to damage the edge of the oil seal when replacing the end cover and drive flange. Smear the face of the flange which bears against the oil seal lightly with oil before driving the flange onto its splines. Tighten the nut to the specified torque.

6 Pinion oil seal (bearing preload by collapsible spacer) – renewal

1 On later 1500 models, the pinion bearing preload is controlled by a collapsible spacer.
2 Disconnect the propeller shaft at its rear flange.
3 Remove the nut cover from the centre of the pinion drive flange.
4 Using a centre punch mark the pinion shaft, and nut in relation to each other and to the drive flange. Make sure that these marks are precisely in alignment.
5 Having drilled two holes in a long piece of flat steel, bolt it to the pinion flange and use it as a lever to prevent the flange rotating when the nut is unscrewed.
6 As the nut is unscrewed, count the number of turns required to unscrew it.
7 Remove the pinion drive flange and prise out the oil seal. If the oil seal is stuck tight, try drilling two or three small holes into its metal

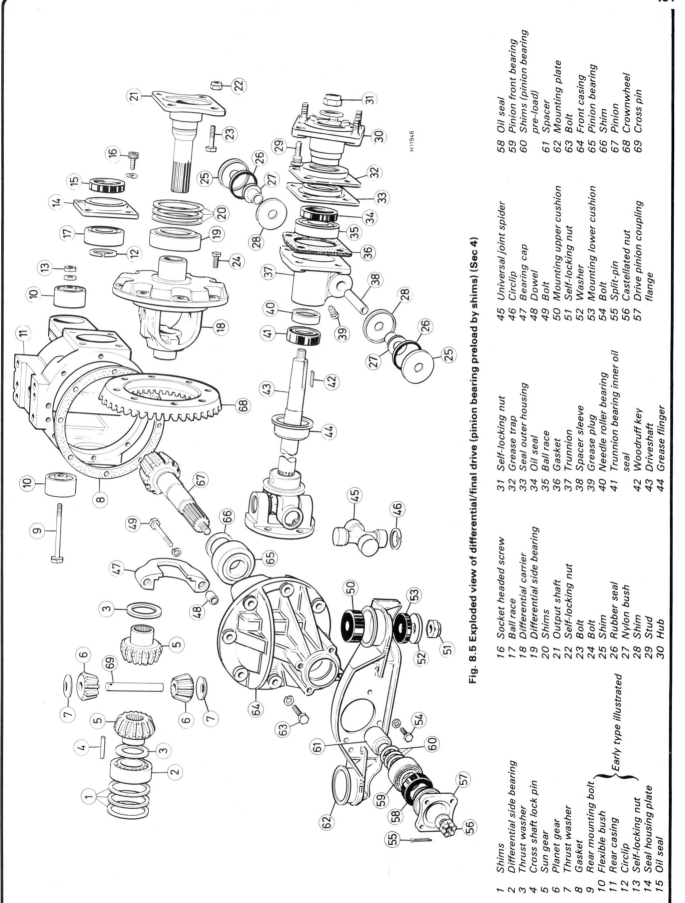

Fig. 8.5 Exploded view of differential/final drive (pinion bearing preload by shims) (Sec 4)

1 Shims
2 Differential side bearing
3 Thrust washer
4 Cross shaft lock pin
5 Sun gear
6 Planet gear
7 Thrust washer
8 Gasket
9 Rear mounting bolt
10 Flexible bush
11 Rear casing
12 Circlip
13 Self-locking nut
14 Seal housing plate
15 Oil seal
16 Socket headed screw
17 Ball race
18 Differential carrier
19 Differential side bearing
20 Shims
21 Output shaft
22 Self-locking nut
23 Bolt
24 Bolt
25 Shim
26 Rubber seal
27 Nylon bush
28 Shim
29 Stud
30 Hub
31 Self-locking nut
32 Grease trap
33 Seal outer housing
34 Oil seal
35 Ball race
36 Gasket
37 Trunnion
38 Spacer sleeve
39 Grease plug
40 Needle roller bearing
41 Trunnion bearing inner oil seal
42 Woodruff key
43 Driveshaft
44 Grease flinger
45 Universal joint spider
46 Circlip
47 Bearing cap
48 Dowel
49 Bolt
50 Mounting upper cushion
51 Self-locking nut
52 Washer
53 Mounting lower cushion
54 Bolt
55 Split-pin
56 Castellated nut
57 Drive pinion coupling flange
58 Oil seal
59 Pinion front bearing
60 Shims (pinion bearing pre-load)
61 Spacer
62 Mounting plate
63 Bolt
64 Front casing
65 Pinion bearing
66 Shim
67 Pinion
68 Crownwheel
69 Cross pin

Early type illustrated

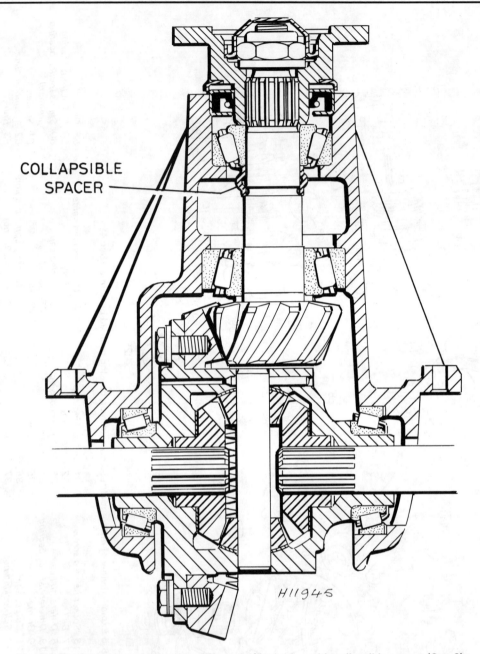

Fig. 8.6 Sectional view of later type differential/final drive with collapsible spacer (Sec 6)

face and then unscrewing in some self-tapping screws which can be used as leverage points.

8 Drive the new oil seal squarely into the pinion housing recess so that the seal lips are towards the pinion gear. It is recommended that before fitting, the oil seal should be soaked in engine oil for one hour.

9 Fit the pinion flange with its mark aligned with the one made on the end of the shaft.

10 Fit the washer and nut, screwing the nut on one half a turn less than the number of turns recorded at removal. Now very carefully tighten the nut until the alignment mark is in line with the ones made on the flange and shaft. On no account overtighten this nut or the collapsible spacer will be over compressed. This will mean a complete axle strip to renew it and then adjusting the pre-load with special equipment.

7 Output shaft oil seal – renewal

1 Around or underneath the differential output shafts is another likely place for oil leakage caused by a worn oil seal.

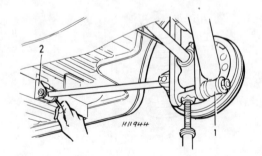

Fig. 8.7 Supporting rear suspension vertical link (Sec 7)

1 Shock absorber lower mounting
2 Radius rod anchor bracket

Chapter 8 Differential and final drive

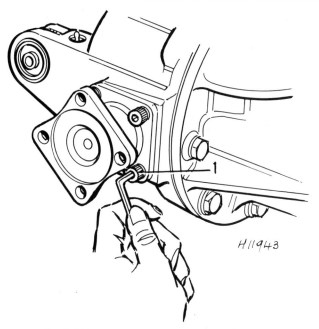

Fig. 8.8 Unscrewing seal housing plate (Sec 7)

1   Socket-headed screw

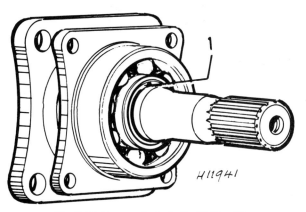

Fig. 8.10 Output shaft circlip (1) (Sec 7)

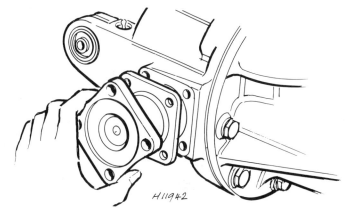

Fig. 8.9 Removing output shaft (1) (Sec 7)

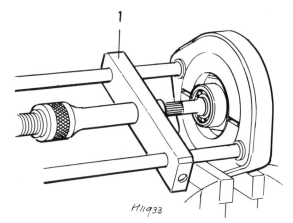

Fig. 8.11 Removing output shaft bearing (Sec 7)

1   Typical extractor

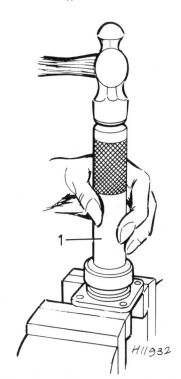

Fig. 8.12 Removing seal from seal housing plate (Sec 7)

1   Removal tool

2   Raise the rear of the car, support securely under the chassis members and remove the rear roadwheel.
3   Disconnect the inboard end of the driveshaft by removing the flange connecting bolts.
4   Place a jack under the vertical link and by raising the jack, relieve the tension in the shock absorber.
5   Disconenct the shock absorber mounting and push the shock absorber aside.
6   Unbolt the radius rod from the vertical link and them remove the jack.
7   Unscrew and remove the bolt which connects the spring eye to the vertical link.
8   Using an Allen key, unscrew and remove the socket-headed screws which holds the seal housing plate to the differential casing.
9   Pull the output shaft from the differential casing, but be prepared for some loss of oil as it is withdrawn. If both output shafts are being removed at the same time, do not interchange them.
10 Extract the shaft circlip.
11 Remove the bearing from the shaft either with an extractor or by pressing the shaft out of the bearing using a press.
12 Take off the oil seal plate and drive out the oil seal.
13 Refitting is a reversal of removal and dismantling, apply grease to the lips of the new oil seal and tighten all bolts and screws to the specified torque.

Chapter 8 Differential and final drive

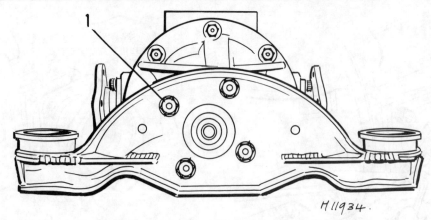

Fig. 8.13 Mounting plate to differential casing bolts (Sec 8)

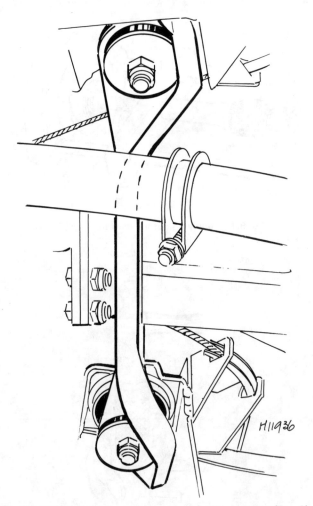

Fig. 8.14 Differential mounting plate viewed from below (Sec 8)

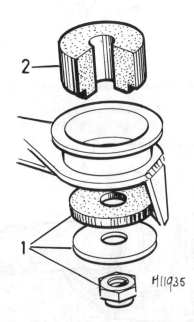

Fig. 8.15 Components of mounting plate end fixings (Sec 8)

1 Lower cushion, washer and nut
2 Upper cushion

8.8 Differential mounting plate attachment to chassis

8 Differential/final drive unit mounting plate rubber cushions – renewal

1 After a high mileage, the mounting plate rubbers may become compressed or hardened with resultant increase in transmission noise being transmitted to the vehicle interior.

2 To renew the cushions, first place the car over an inspection pit or raise the rear end on ramps.

3 Disconnect the silencer mountings and exhaust pipe rear mounting.

Chapter 8 Differential and final drive

4 Raise the rear roadwheels further, to clear the ramps on the ground.
5 Disconnect the propeller shaft rear flange as described in Chapter 7.
6 Remove the pinion nut and drive flange paying particular attention to Section 5 or 6 of this Chapter according to model.
7 Unscrew and remove the four bolts which hold the mounting plate to the differential casing. Support the front end of the casing as necessary with a jack.
8 Unscrew the self-locking nut from each end of the mounting plate which holds the plate to the chassis. Take off the washer and lower rubber cushion.
9 Using a lever, prise the mounting plate downwards until the upper rubber cushions can be removed from the studs.
10 Fitting the new cushions and reassembly is a reversal of the dismantling and removal operations.
11 Tighten nuts and bolts to the specified torque.
12 When correctly assembled and tightened, it should still be possible to rotate the lower rubber cushions with the fingers, with the weight of the car on the roadwheels.

9 Fault diagnosis – differential/final drive

Symptom	Reason(s)
Oil leakage	Faulty pinion oil seal Faulty output shaft oil seal Defective cover gasket Blocked axle casing breather
Noise	Lack of oil Worn bearings General wear
'Clonk' on taking up drive, and excessive backlash	Incorrectly tightened pinion nut Worn components Worn axleshaft splines Elongated roadwheel bolt holes

Chapter 9 Braking system

Contents

Brake disc – inspection and renewal	7
Brake drum – inspection and renewal	8
Brake hydraulic system – bleeding	16
Brake pedal – removal and refitting	21
Caliper – removal, overhaul and refitting	6
Disc pads – inspection and renewal	4
Fault diagnosis – braking system	22
Flexible hydraulic hoses – inspection and renewal	14
General description	1
Handbrake cables – renewal	20
Handbrake lever – removal and refitting	18
Handbrake pawl and ratchet – removal and refitting	19
Maintenance	2
Master cylinder (single type) – removal, overhaul and refitting	11
Master cylinder (tandem type) – removal, overhaul and refitting	12
Pressure differential warning actuator (PDWA)	13
Rear brakes – adjustment	3
Rear brake shoes – inspection and renewal	5
Rear wheel cylinder – overhaul without removal	9
Rear wheel cylinder – removal and refitting	10
Rigid brake pipes – inspection and renewal	15

Specifications

System type .... Four wheel hydraulic, discs front, drums rear. Handbrake mechanical to rear wheels. Dual circuit on N. American models

Front disc brakes
Diameter of disc .... 9.0 in (229.0 mm)
Lining area .... 14.8 in^2 (95.0 cm^2)
Minimum lining thickness .... 0.079 in (2.0 mm)
Front hub end float .... 0.003 to 0.005 in (0.08 to 0.13 mm)

Rear drum brakes
Diameter of drum (internal) .... 7.0 in (178.0 mm)
Width of drum (internal) .... 1.25 in (32.0 mm)
Lining area .... 34.0 in^2 (220.0 cm^2)
Minimum lining thickness .... 0.079 in (2.0 mm)

Brake fluid .... Hydraulic fluid to SAE J1703 (Duckhams Universal Brake and Clutch Fluid)

Torque wrench settings	lbf ft	Nm
Caliper mounting bolts	65	88
Disc to hub bolts	34	46
Tandem master cylinder (earlier models) – tipping valve nut	40	54

Chapter 9 Braking system

1 General description

1 Disc brakes are fitted to the front wheels of all models together with single leading shoe drum brakes at the rear. The mechanically operated handbrake works on the rear wheels only.
2 The brakes fitted to the front two wheels are of the rotating disc and static caliper type, with one caliper per disc, each caliper containing two piston operated friction pads, which on application of the footbrake pinch the disc rotating between them.
3 Application of the footbrake creates hydraulic pressure in the master cylinder and fluid from the cylinder travels via steel and flexible pipes to the cylinders in each half of the calipers, the fluid so pushing the pistons, to which are attached the friction pads, into contact with either side of each disc.
4 As the friction pad wears so the pistons move further out of the cylinders and the level of the fluid in the hydraulic reservoir drops, but disc pad wear is then taken up automatically and eliminates the need for periodic adjustments by the owner.
5 The rear brakes are of the single leading shoe type, with one brake cylinder per wheel for both shoes. The cylinder is free to float on the backplate. Attached to each of the rear wheel operating cylinders is a mechanical expander operated by the handbrake lever through a cable which runs from the brake lever to a compensator and thence to the wheel operating levers.
6 Drum brakes have to be adjusted periodically to compensate for wear in the linings. It is unusual to have to adjust the handbrake system as the efficiency of this system is largely dependent on the condition of the brake linings and the adjustment of the brake shoes. The handbrake can, however, be adjusted independently from the footbrake operated hydaulic system.
7 On vehicles exported to N. America, the hydraulic system is of dual circuit type and incorporates a tandem master cylinder. In the event of a leak or fault developing in ine circuit, then the brakes controlled by the second circuit will continue to operate.

2 Maintenance

1 At the weekly service intervals, the fluid level in the master cylinder reservoir should be checked.
2 On early models unscrew the cap and if the fluid level is below the mark inside the reservoir (just below the rim) top up with clean hydraulic brake fluid.
3 On later models with a translucent reservoir, the check can be carried out visually and if necessary the fluid topped up to the full mark.
4 The need for topping up should be very infrequent as the level drops very slowly to compensate for wear in the brake pads or lining.
5 A rapid drop in fluid level or the need for regular topping up will probably be due to a leak in the system which must be investigated immediately and rectified.
6 Always top up with clean fluid from a sealed tin. Brake fluid is hygroscopic (absorbs moisture from the air) and if not kept in a sealed tin its use in the hydraulic system will cause corrosion of the internal components of the system.
7 Provided a universal brake fluid of reputable make is used, it can be intermixed with any other fluid made by a different manufacturer.
8 Do not shake a brake fluid tin or it may become aerated in which case it will have to stand for 24 hours before use.
9 At the intervals specified in *Routine Maintenance* adjust the rear brakes. Check front brake pads and rear brake linings for wear.
10 Finally, remember that brake fluid damages paintwork.

3 Rear brakes – adjustment

1 Place chocks under the front wheels, jack-up the rear of the car, and ensure the handbrake is off.
2 A small square headed adjuster will be found on the rear at the bottom of each backplate. The edges of the adjuster are easily burred if an ordinary spanner is used. Use a square headed brake adjusting spanner if possible (photo).
3 Turn the adjuster a notch at a time until the wheel is locked. Then turn back the adjuster one notch or more so the wheel will rotate without binding.

3.2 Adjusting rear brake shoes

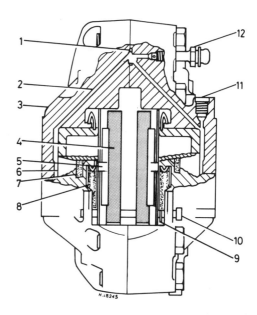

Fig. 9.1 Cutaway view of disc caliper (Sec 4 and 6)

1	O-ring	7 Piston seal
2	Fluid transfer channels	8 Dust excluder
3	Caliper body	9 Retaining clip
4	Pad	10 Brake pad retaining pin
5	Anti-squeal shim	11 Fluid inlet
6	Piston	12 Bleed nipple

4 Spin the wheel and apply the brakes hard to centralise the shoes. Recheck that it is not possible to turn the adjusting screw further without locking the shoe.
5 Sometimes brake adjusters seize up due to rust or corrosion. A smear of high melting point grease will keep them in good order. As the adjusters are retained by bolts, new ones can always be fitted.

4 Disc pads – inspection and renewal

1 Remove the front wheels and inspect the amount of friction material left on the friction pads. The pads must be renewed when the thickness of the material has worn down to the specified minimum thickness.

4.2a Disc brake pad retaining pins and clips

4.2b Disc pad retaining pins being withdrawn

2 Referring to Fig. 9.2 pull out the wire clips which secure the pad retaining pins in place, and remove the pins (photos).
3 The friction pads and anti-squeal shims can now be lifted from the caliper.
4 Carefully clean the recesses in the caliper in which the friction pad assemblies lie, and the exposed face of each piston from all traces of dirt and rust. Take care not to inhale the brake dust as this contains asbestos which is injurious to health.
5 The caliper pistons must now be pushed squarely into their cylinders in order to make room for the new thicker pads. Do this with a wide flat blade, but remember that this action will cause the fluid level in the master cylinder reservoir to rise. Anticipate this by syphoning out some fluid using an old battery hydrometer (clean) or a poultry basting syringe.
6 Fit new pads and refit the anti-squeal shims with the arrow towards the direction of rotation. Insert the pad retainer pins and secure them with the retainer clips.
7 Repeat the operations on the opposite brake as new pads must always be fitted in axle sets.
8 Depress the foot brake pedal hard several times to bring the pads up against the discs and then top up the fluid in the master cylinder reservoir to the correct level.

4.3 Disc pad with anti-squeal shim

5 Rear brake shoes – inspection and renewal

1 Remove the hub caps, loosen off the wheel nuts, then securely jack-up the car, and remove the road wheel. Ensure the handbrake is off.
2 Completely slacken off the brake adjustment and take out the two setscrews which hold the drum in place.
3 Remove the brake drum. If it proves obstinate tap the rim gently with a soft headed hammer. The shoes are now exposed for inspection (photo).
4 Brush or preferably vacuum away all brake dust from the drum and shoes, taking care not to inhale the dust which is injurious to health.
5 The brake linings should be renewed if they are so worn that the rivet heads are flush with the surface of the lining. If bonded linings are fitted they must be removed when the material has worn down to 0.079 in (2.0 mm).
6 Before removing the shoes, note and sketch if necessary, the position of the leading and trailing edges of the shoes. The brake shoes are not fitted symmetrically. Also note the position of the holes in the shoe webs in which the return springs engage. The springs are fitted inboard of the brake shoes.
7 Press in each brake shoe steady pin securing washer against the pressure of its spring.
8 Turn the head of the washer 90° so the slot will clear the securing bar on the steadypin and remove the spring and washer (photo).

5.3 Removing a brake drum

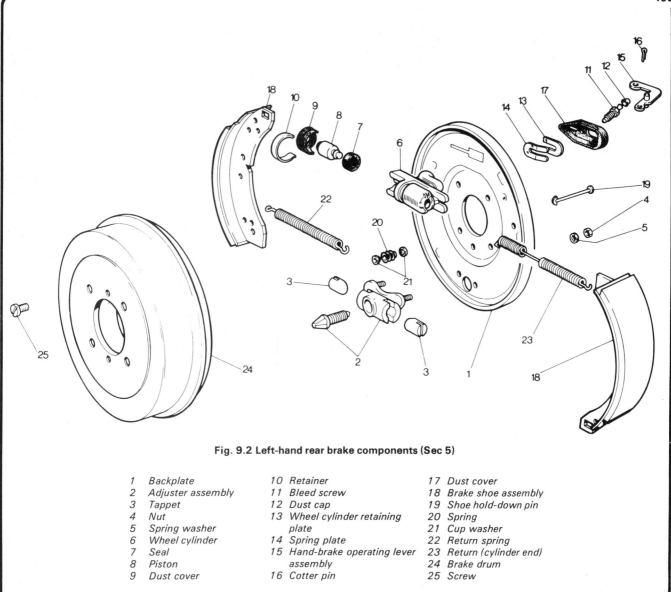

Fig. 9.2 Left-hand rear brake components (Sec 5)

1 Backplate
2 Adjuster assembly
3 Tappet
4 Nut
5 Spring washer
6 Wheel cylinder
7 Seal
8 Piston
9 Dust cover
10 Retainer
11 Bleed screw
12 Dust cap
13 Wheel cylinder retaining plate
14 Spring plate
15 Hand-brake operating lever assembly
16 Cotter pin
17 Dust cover
18 Brake shoe assembly
19 Shoe hold-down pin
20 Spring
21 Cup washer
22 Return spring
23 Return (cylinder end)
24 Brake drum
25 Screw

5.6 Rear brake assembly (LHS)

5.8 Turning steady pin cup washer

5.10a Shoe upper attachment

5.10b Shoe lower attachment

6.6a Removing caliper mounting bolts and lockplate

6.6b Removing caliper

9 Detach the shoes and return springs by pulling one end of the shoes away from the slot in the closed end of the brake cylinder.
10 Disengage the brake shoe from the return spring carefully noting the holes into which the spring fits and then remove the remaining shoe in similar fashion. Place rubber bands over the wheel cylinders to prevent any possibility of the pistons dropping out (photos).
11 Check that the pistons are free in their cylinders and that the rubber dust covers are undamaged and in position and that there are no hydraulic fluid leaks.
12 Prior to reassembly smear a trace of brake grease to all sliding surfaces. The shoes should be quite free to slide on the closed end of the cylinder and the piston anchorage point. It is vital that no grease or oil comes in contact with the brake drums or the brake linings.
13 Replacement is a straightforward reversal of the removal procedure, but make sure that the brake adjusters are backed right off before fitting the drum.
14 Adjust the brakes on completion.

6 Caliper – removal, overhaul and refitting

1 Jack up the car and remove the roadwheel.
2 Disconnect the hydraulic flexible hose. To do this, disconnect the union at the junction of the rigid and flexible brake line at the support bracket under the front wing. Cap the end of the rigid pipe immediately to prevent loss of fluid. A bleed nipple cap is useful for this.
3 Using two open-ended spanners, release and remove the end fitting nuts which secure the flexible hose to the support bracket.
4 The opposite end of the flexible hose may now be unscrewed from the caliper. Alternatively wait until the caliper is unbolted and removed to the bench.
5 Remove the disc brake friction pads and anti-squeal shims as previously described.
6 Unscrew the two caliper mounting bolts and lockwashers and remove the caliper assembly from the disc (photos).
7 Referring to Fig. 9.2 pull off the dust covers and remove the pistons from the caliper body. *On no account loosen the bolts which hold the two halves of the caliper together.*
8 The pistons are best ejected by applying air pressure from the tyre pump at the fluid entry hole in the caliper. Take care not to trap the fingers. Only low air pressure is required to start the pistons moving.
9 Very carefully remove the rubber piston sealing rings from their recesses in the caliper. The pistons, cylinders, and seals should be cleaned only with clean brake fluid or methylated spirit.
10 Examine the components carefully, renew the seals as a matter of course, and replace the caliper if cylinders or pistons are slightly grooved or otherwise worn.

Chapter 9 Braking system

11 Reassembly commences by carefully fitting new piston sealing rings into the recesses in the caliper cylinders, using the fingers only to manipulate them.
12 Fit the larger diameter lip of the rubber dust cover to the groove on the outside of the top of the cylinder.
13 Slide the pistons closed end first into the cylinders, with great care, and then fit the outer lip of the dust excluder into the groove in the outer end of the piston.
14 Fit the caliper over the disc, insert the two securing bolts, replace the anti-squeal shims and the pads. Reconnect the fllexible brake hose and bleed the system as described in Section 16.

7 Brake disc – inspection and renewal

1 Raise the front of the car to enable the roadwheel to be removed and support securely.
2 Remove the roadwheel, take out the disc pads.
3 Unscrew the caliper mounting bolts.
4 Pull the caliper off the disc and tie it up out of the way to avoid straining the hose.
5 Tap off the hub cap.
6 Extract the split pin now exposed. Unscrew the nut and remove it with the thrust washer from the stub axle.
7 Holding the hub level, carefully withdraw it complete with bearings and oil seal from the stub axle.
8 With the assembly secured in a vice, the disc and hub can be separated after unscrewing the four connecting bolts.
9 Examine the disc carefully. If the friction surfaces are deeply grooved or tiny cracks are evident, the disc must be renewed.
10 Refit by reversing the removal operation, but adjust the hub bearings in the following way before refitting the disc pads. Tighten the nut while turning the hub until a slight resistance can be felt. Release the nut and tighten it finger tight. The hub nut should now be unscrewed one flat to give the specified endfloat. This endfloat can be checked using a dial gauge or feeler blades.
11 Insert a new split pin, half fill the hub cap with multi-purpose grease and tap it on. Refit the disc brake pads.
12 Fit the roadwheel and lower the car to the ground.

8 Brake drum – inspection and renewal

1 Whenever the rear brake drums are removed to inspect the shoe linings (See Section 5) take the opportunity to check the drum for cracks or scoring or grooving. Uneven braking may be due to the drums being slightly distorted.
2 It is always best to fit new drums if the existing ones are defective in any way.

9 Rear wheel cylinder – overhaul without removal

1 If hydraulic fluid is leaking from one of the brake cylinders it will be necessary to dismantle the cylinder and replace the dust cover and

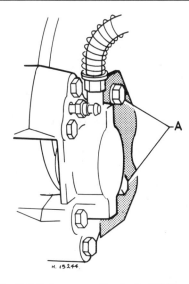

Fig. 9.3 Caliper mounting bolts (A) (Sec 6)

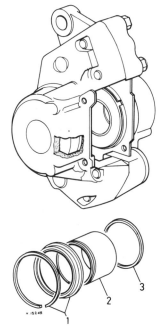

Fig. 9.4 Caliper piston components (Sec 6)

1 Retainer ring and dust excluder
2 Piston
3 Piston seal

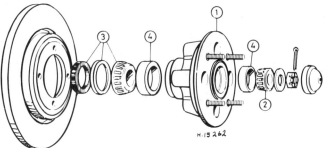

Fig. 9.5 Hub/disc components (Sec 7)

1 Hub
2 Outer bearing
3 Inner bearing and oil seal
4 Bearing tracks

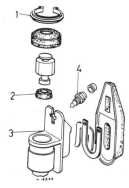

Fig. 9.6 Rear wheel cylinder components (Sec 8)

1 Dust excluder clip
2 Pistin seal
3 Cylinder body
4 Bleed nipple

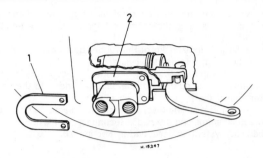

Fig. 9.7 Rear wheel cylinder attachment (Sec 10)

    1   Horseshoe plate   2   Spring plate

piston sealing rubber. If brake fluid is found running down the side of the wheel, or it is noticed that a pool of liquid forms alongside one wheel and the level in the master cylinder has dropped, and the hoses are all in good order, proceed as follows.

2   Remove the brake drums and brake shoes as described in Section 5.
3   Ensure that all the other wheels, and all the other brake drums are in place. Remove the piston assembly and the spring from the leaking cylinder by applying gentle pressure to the footbrake. Place a quantity of rag under the backplate or a tray to catch the hydraulic fluid as it pours out of the cylinder.
4   Inspect the inside of the cylinder for score marks. If any are found the cylinder and piston will require renewal together as an exchange assembly.
5   If the cylinder is sound thoroughly clean it out with fresh hydraulic fluid.
6   The old rubber seal will probably be swollen and visibly worn. Smear the new rubber seal with hydraulic fluid and reassemble in the cylinder the spring, seal and piston, and then the rubber boot. The seal must be fitted with its lip towards the bottom of the cylinder.
7   Replenish the brake fluid, replace the brake shoes and brake drum, and bleed the hydraulic system as described in Section 16.

10 Rear wheel cylinder – removal and refitting

1   Remove the left or right-hand brake drum and brake shoes as required, as described in Section 5.
2   Free the hydraulic pipe from the wheel cylinder at the union, and disconnect the handbrake cable clevis from its lever. Cap the end of the pipe to prevent loss of fluid.
3   Take off the dust excluder, the retaining plate and the spring clip, and remove the cylinder from the backplate.
4   On replacement smear the slot in the backplate and the cylinder neck with brake grease. The rest of the replacement process is a straightforward reversal of the removal sequence. Bleed the brakes on completion of reassembly.

11 Master cylinder (single type) – removal, overhaul and refitting

1   Working within the engine compartment, remove the split pin and clevis pin from the clevis fork which connects to the brake pedal (photos).
2   Disconnect the brake pipe from the master cylinder by unscrewing the union. Quickly cap the end of the pipe to prevent entry of dirt. A brake bleed nipple dust cap is useful for this. Allow the fluid to drain into a container.
3   Unscrew the two mounting bolts and lift the master cylinder from the bulkhead.
4   Tip the fluid from the reservoir and pull the dust excluder from the end of the cylinder and push it agnist the clevis fork.
5   Extract the circlip from the end of the cylinder and withdraw the pushrod and dished washer.
6   Prise up the tab on the spring thimble and pull the thimble from the piston.

11.1a Early type master cylinders

11.1b Later type master cylinders

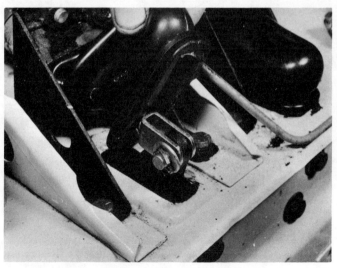

11.1c Exposing master cylinder pushrod clevis fork

Chapter 9 Braking system

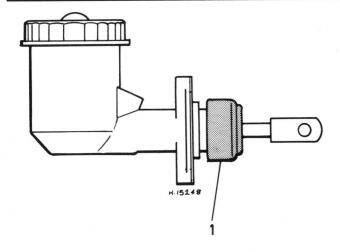

Fig. 9.8 Single circuit type master cylinder (Sec 11)

1   Dust excluder

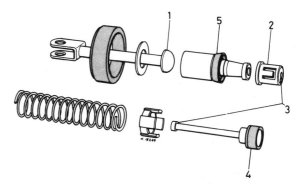

Fig. 9.9 Exploded view of single circuit type master cylinder (Sec 11)

1   Push rod   4   Valve seal
2   Thimble   5   Piston seal
3   Valve stem and keyhole slot

7   Release the valve stem from the key-hole slot in the thimble.
8   Remove the spring and then slide the valve and spacer along the valve stem.
9   Remove the valve seal from the valve stem.
10   Remove the seal from the piston.
11   Clean all components in brake fluid or methylated spirit, and examine for wear or damage. If the piston or cylinder is scored or corroded, renew the complete assembly. If these parts are in good condition, discard the seals and obtain a repair kit which will contain all the renewable items.
12   Reassembly is a reversal of dismantling, manipulate the seals into position using the fingers only. Dip the piston in clean brake fluid before inserting it into the cylinder. Make sure that the tab on the thimble is depressed with a pair of pliers as soon as it is fitted to the piston.
13   Refit the master cylinder by reversing the removal operations. Bleed the brakes on completion.

12 Master cylinder (tandem type) – removal, overhaul and refitting

1   This type of master cylinder is used on cars equipped with a dual circuit hydraulic system.
2   Working within the engine compartment, peel back the rubber boot to expose the pedal linkage and then remove the split pin and clevis pin. Pull off the rubber boot.
3   Disconnect the brake pipes from the master cylinder and cap the pipes to prevent entry of dirt. Allow the fluid from the master cylinder to drain into a container.
4   Unbolt the mounting bracket and remove the master cylinder/bracket assembly.
5   Unbolt the master cylinder from the mounting bracket.

Overhaul (earlier models)

6   Remove the fluid reservoir (four screws).
7   Unscrew the socket headed nut which retains the tipping valve in the master cylinder body.
8   Apply slight pressure to the end of the pushrod and extract the tipping valve.
9   Pull the dust excluding boot clear of the end of the master cylinder and extract the circlip.
10   Withdraw the pushrod.
11   Withdraw the primary and secondary piston assemblies.
12   Examine the piston and cylinder surfaces for scoring or corrosion. If these faults are evident, renew the complete master cylinder.
13   If these components are in good condition, separate the plungers and intermediate spring. Prise up the tab on the spring retainer and

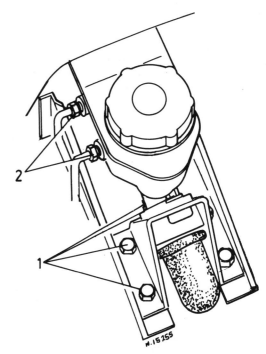

Fig. 9.10 Tandem type master cylinder mounting bracket bolts (1), brake pipe unions (2) (Sec 12)

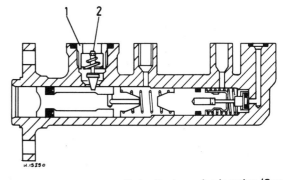

Fig. 9.11 Tandem master cylinder tipping valve location (Sec 12)

1   Retaining nut   2   Tipping valve

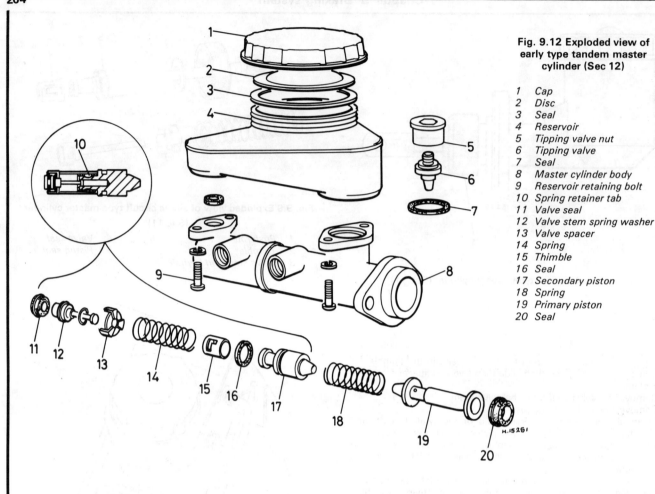

Fig. 9.12 Exploded view of early type tandem master cylinder (Sec 12)

1 Cap
2 Disc
3 Seal
4 Reservoir
5 Tipping valve nut
6 Tipping valve
7 Seal
8 Master cylinder body
9 Reservoir retaining bolt
10 Spring retainer tab
11 Valve seal
12 Valve stem spring washer
13 Valve spacer
14 Spring
15 Thimble
16 Seal
17 Secondary piston
18 Spring
19 Primary piston
20 Seal

Fig. 9.13 Exploded view of later type tandem master cylinder (Sec 12)

1 Reservoir
2 Sealing grommet
3 Pin
4 Dowel
5 Cylinder body
6 Spring
7 Retainer
8 Seal
9 Secondary piston
10 Spring
11 Retainer
12 Seal
13 Primary piston
14 Washer
15 Circlip
16 Dust excluder
17 Push rod clevis fork

remove the spring and centre valve sub-assembly from the secondary plunger.
14 Withdraw the valve spacer, spring washer and valve seal from the valve head.
15 Discard all seals and obtain a repair kit which will contain all the necessary renewable items.
16 Clean components in brake fluid or methylated spirit.
17 Reassembly is a reversal of dismantling, but manipulate the seals using the fingers only and dip the pistons in clean brake fluid before inserting them into the cylinder bore.
18 Remember to depress the tab on the spring retainer and tighten the tipping valve retaining nut (new seal fitted) to the specified torque.

Overhaul (later models)

19 This later type master cylinder can be identified by its reservoir which is a push-fit into rubber grommets, no fixing screws being used as on earlier models, just pins and clips.
20 Tip out the fluid from the reservoir, remove the securing clips and pins. Pull the reservoir from the master cylinder body.
21 Remove the sealing grommets.
22 Pull the dust excluding boot clear of the end of the cylinder body. Then extract the circlip and withdraw the push rod assembly.
23 Withdraw the primary piston assembly with spring.
24 Using a pair of long-nosed pliers, remove the dowel from the forward reservoir connecting hole.
25 Withdraw the secondary piston assembly and spring.
26 Examine the piston and cylinder surfaces for scoring or corrosion; if evident, renew the complete master cylinder. If these components are in good order, discard only the seals and obtain a repair kit which will contain all the necessary renewable items. The piston end seals are removable after extracting the spring from its recess and withdrawing the cap.
27 Reassembly is a reversal of removal. Manipulate the new seals into position using the fingers only. Dip the piston assemblies into clean brake fluid before inserting into the cylinder bores.
28 Refitting either type of master cylinder is a reversal of removal. Bleed the hydraulic system on completion as described in Section 16.

13 Pressure differential warning actuator (PDWA)

1 This device is fitted between the front and rear hydraulic circuits on cars with dual circuit systems.
2 Should either the front or rear brakes fail, the pressure drop on one side of the warning actuator causes a shuttle valve to move from its normal mid-position so actuating an electrical switch which brings on a warning light on the fascia.
3 The brake warning light is connected in series/parallel with the oil warning light. Thus the brake warning light comes on with the oil warning light when the ignition is turned on, and is extinguished when the engine is started. In this way it is possible to check that the brake **warning** bulb is working correctly, every time the ignition is turned on.
4 If the shuttle in the pressure differential actuator has moved, either because air has got into one of the circuits or because one of the circuits has failed, it will be necessary to centralise the shuttle, after the system has been bled and the faulty components replaced.
5 Fit a rubber tube to a bleed nipple on the opposite end of the car to the one that has just been bled, allowing one end of the tube to hang submerged in a clean jam jar containing a quantity of brake fluid.
6 Open the bleed nipple one turn, and turn on the ignition. The brake warning light will come on brightly but the oil warning light will not glow at all.
7 Slowly and gently depress the brake pedal until the brake warning light dims and the oil warning light comes on, at the same time a click can be felt at the pedal denoting that the shuttle has returned to its mid-position. Do not continue to press down further on the brake pedal or the shuttle will move to the other side of the valve.
8 Holding the brake pedal steady get a friend to tighten the bleed nipple. Check that after the brakes have been applied the warning actuator is working correctly.
9 In the event of a fault developing, the PDWA should not be dismantled as spares are not available. A complete assembly should be fitted. Removal is simply a matter of disconnecting the hydraulic pipelines and electrical connections from it and then unscrewing the mounting bolt. Take care to connect the brake pipes correctly and bleed the system on completion (See Section 16).

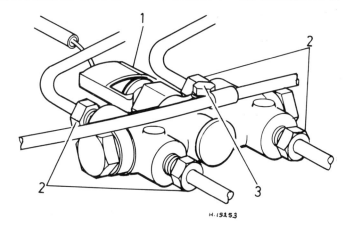

Fig. 9.14 PDWA (Sec 13)

1 Electrical connections   *3 Mounting bolt*
2 Union nuts

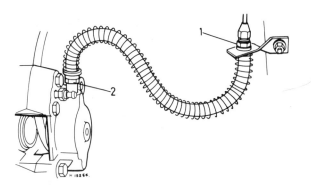

Fig. 9.15 Typical front flexible hose (Sec 14)

1 Bracket locknut   *2 End fitting at caliper*

14 Flexible hydraulic hoses – inspection and renewal

1 Inspect the condition of the flexible hydraulic hoses leading from the chassis mounted metal pipes to the brake backplates. If any are swollen, damaged, cut or chafed, they must be renewed.
2 If, when a hose is bent double with the fingers, tiny cracks open, this too is an indication that it has reached a dangerous condition.
3 Unscrew the metal pipe union nuts from the connection to the hose, and then holding the hexagon on the hose with a spanner, unscrew the attachment nut and washer.
4 The chassis end of the hose can now be pulled from the chassis mounting bracket and will be quite free.
5 Disconnect the flexible hose at the backplate by unscrewing it from the brake cylinder. When releasing the hose from the backplate, the chassis end must always be freed first.
6 Replacement is a straightforward reversal of the above procedure. A slight set may be given to a hose in order to provide greater clearance between it and a tyre or chassis, steering, or suspension component. This is achieved by releasing it from its bracket and twisting it by not more than one quarter turn in the appropriate direction. Hold the hose in the desired position while the bracket nuts are tightened.

15 Rigid brake pipes – inspection and renewal

1 Regularly check these pipes for damage or corrosion. The pipeline securing clips must always be pressed hard against the pipe to prevent rattle, chafing and subsequent rust after the plating has rubbed off. A

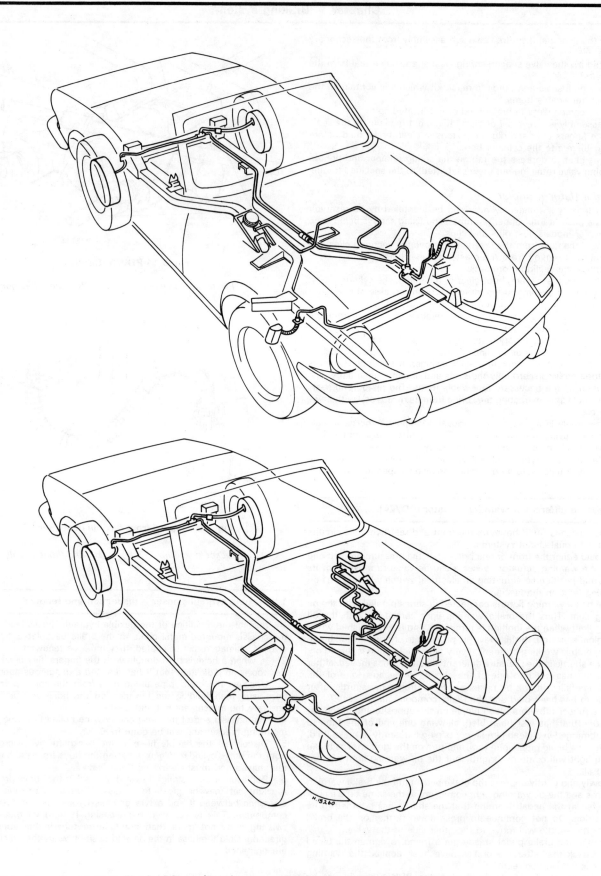

Fig. 9.16 Single (RHD) and dual (LHD) hydraulic circuit layouts (Sec 15)

Chapter 9 Braking system

split piece of plastic tubing makes a suitable insulator for the pipe at the clip locations.
2 New pipes can be made to pattern by many garages and motor supply stores and non-corrosive copper lines may be specified if preferred.

16 Brake hydraulic system – bleeding

1 On models with dual braking circuits the front brakes are operated by a separate system from the rear brakes.
2 If the master cylinder or PDWA has been disconnected and reconnected then the complete system (both circuits) must be bled.
3 If a component of one circuit only has been disturbed then only that particular circuit need be bled.
4 If the entire system is being bled, the sequence of bleeding should be carried out by starting at the bleed screw furthest from the master cylinder and finishing at the one nearest to it. Unless the pressure bleeding method is being used, do not forget to keep the fluid level in the master cylinder reservoir topped up to prevent air from being drawn into the system which would make any work done worthless.
5 Before commencing operations, check that all system hoses and pipes are in good condition with all unions tight and free from leaks.
6 Take great care not to allow hydraulic fluid to come into contact with the vehicle paintwork as it is an effective paint stripper. Wash off any spilled fluid immediately with cold water.

Bleeding – two man method

7 Gather together a clean glass jar and a length of rubber or plastic tubing which will be a tight fit on the brake bleed screws.
8 Engage the help of an assistant.
9 Push one end of the bleed tube onto the first bleed screw and immerse the other end in the glass jar which should contain enough hydraulic fluid to cover the end of the tube.
10 Open the bleed screw one half a turn and have your assistant depress the brake pedal fully then slowly release it. Tighten the bleed screw at the end of each pedal downstroke to obviate any chance of air or fluid being drawn back into the system.
11 Repeat the operation until clean hydraulic fluid, free from air bubbles, can be seen coming through into the jar.
12 Tighten the bleed screw at the end of a pedal downstroke and remove the bleed tube. Bleed the remaining screws in a similar way.

Bleeding – using one way valve kit

13 There are a number of one-man, one-way brake bleeding kits available from motor accessory shops. It is recommended that one of these kits is used wherever possible as it will greatly simplify the bleeding operation and also reduce the risk of air or fluid being drawn back into the system quite apart from being able to do the work without the help of an assistant.
14 To use the kit, connect the tube to the bleed screw and open the screw one half a turn.
15 Depress the brake pedal fully and slowly release it. The one-way valve in the kit will prevent expelled air from returning at the end of each pedal downstroke. Repeat this operation several times to be sure of ejecting all air from the system. Some kits include a translucent container which can be positioned so that the air bubbles can actually be seen being ejected from the system.
16 Tighten the bleed screw, remove the tube and repeat the operations on the remaining brakes.
17 On completion, depress the brake pedal. If it still feels spongy repeat the bleeding operations as air must still be trapped in the system.

Bleeding – using a pressure bleeding kit

18 These kits too are available from motor accessory shops and are usually operated by air pressure from the spare tyre.
19 By connecting a pressurised container to the master cylinder fluid reservoir, bleeding is then carried out by simply opening each bleed screw in turn and allowing the fluid to run out, rather like turning on a tap, until no air is visible in the expelled fluid.
20 By using this method, the large reserve of hydraulic fluid provides a safeguard against air being drawn into the master cylinder during

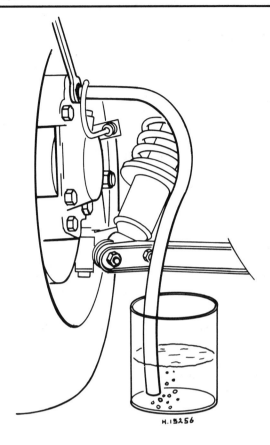

Fig. 9.17 Bleeding a disc caliper (Sec 16)

bleeding which often occurs if the fluid level in the reservoir is not maintained.
21 Pressure bleeding is particularly effective when bleeding complicated systems or when bleeding the complete system at time of routine fluid renewal.

All methods

22 When bleeding is completed, check and top up the fluid level in the master cylinder reservoir.
23 Check the feel of the brake pedal. If it feels at all spongy, air must still be present in the system and the need for further bleeding is indicated. Failure to bleed satisfactorily after repeating the bleeding operations may be due to worn master cylinder seals.
24 Discard brake fluid which has been expelled. It is almost certain to be contaminated with moisture, air and dirt making it unsuitable for further use.

17 Handbrake – adjustment

1 Raise the rear of the car, release the handbrake and turn both rear brake adjusters to lock the wheels.
2 Observe the relay lever which should be set at the angle shown in the diagram. If it is not, adjust the front handbrake cable as necessary to achieve the correct setting, by carrying out the following procedure:
3 Remove the arm rest, transmission cover and the carpet as described in Chapter 12.
4 With the handbrake linkage exposed, slacken the nut at the cable clevis fork. Rotate the cable threaded end fitting as necessary to obtain the specified angle of the relay lever.

Chapter 9 Braking system

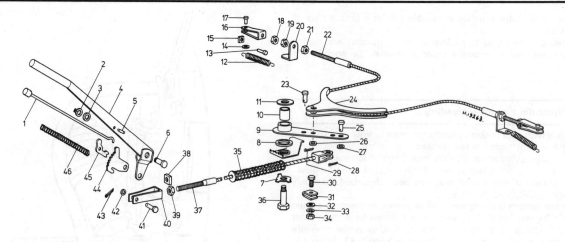

Fig. 9.18 Handbrake components (Sec 17)

1. Pawl release rod
2. Circlip
3. Washer
4. Handbrake lever
5. Pawl pivot pin
6. Lever pivot pin
7. Lock plate
8. Relay pivot seal
9. Relay lever
10. Bush
11. Felt seal
12. Return spring
13. Split-pin
14. Washer
15. Nut
16. Clevis fork
17. Clevis pin
18. Nut
19. Nut
20. Anchor plate for return spring
21. Nut
22. Rear (secondary) cable
23. Clevis pin
24. Equaliser
25. Clevis pin
26. Washer
27. Washer
28. Split-pin
29. Split-pin
30. Clamp bolt
31. Clamp
32. Washer
33. Spring washer
34. Nut
35. Spring
36. Pivot bolt
37. Front (primary) cable
38. Nut
39. Locknut
40. Clevis fork
41. Clevis pin
42. Washer
43. Split-pin
44. Ratchet
45. Pawl
46. Pawl spring

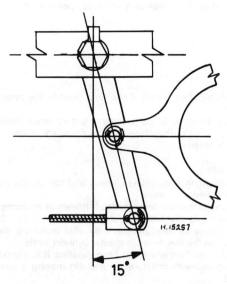

Fig. 9.19 Handbrake relay setting diagram (Sec 17)

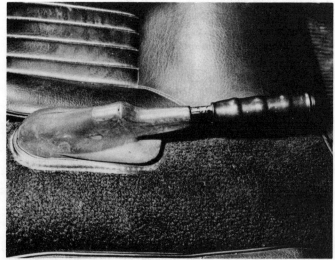

18.3 Handbrake lever

5 Working under the car, disconnect the brake cables from the levers on the brake backplates. This is carried out by extracting the split pins and clevis pins.
6 Now screw up the adjuster nuts at the clevis forks on the ends of both cables equally until the forks can be connected to the backplate levers without having to apply any force to the cable or fork in order to push the clevis pins into the holes.
7 Once this is done, fit the clevis pin washers, insert new split pins and tighten the clevis fork locknuts.
8 Slacken the brake shoe adjusters until the road wheel spins freely.
9 Lower the car to the ground.

18 Handbrake lever – removal and refitting

1 Remove the seats (see Chapter 12), release the brake.
2 Unbolt and remove the safety belt stalks from the transmission tunnel and the anchor plates from the floor.
3 Remove the carpet and pull off the hand grip from the lever. Remove the dust excluding boot (photo).
4 Refer to Chapter 12 and remove the armrest.
5 With the handbrake lever now exposed, extract the pivot pin circlip and then the pin.

Chapter 9 Braking system

6 Lift the handbrake lever sufficiently to be able to disconnect the cable clevis fork.
7 Refitting is a reversal of removal.

19 Handbrake pawl and ratchet – removal and refitting

1 Remove the handbrake lever as described in the preceding Section.
2 File off the end of the rivet which acts as the pawl pivot pin.
3 Depress the handbrake control plunger button and push out the pivot pin. Remove the pawl and ratchet.
4 Fitting the new pawl and ratchet is a reversal of dismantling. Make sure that the hooked end of the ratchet faces downwards and the pawl teeth face the teeth on the ratchet.
5 Rivet a new pivot pin in position.

20 Handbrake cables – renewal

Front
1 Remove the handbrake control lever as described in Section 26.
2 Disconnect the cable from the lever and then take the clevis fork from the front end of the cable.
3 Disconnect the cable from the relay lever by pulling out the split-pin and clevis pin.
4 Remove the cable.

Rear
5 Disconnect the equaliser from the relay lever by extracting the split-pin and clevis pin.
6 At the rear brake backplates, disconnect the cable return springs.
7 Disconnect the clevis forks from the backplate operating levers.
8 Withdraw the cable assembly from the car.
9 Refitting is a reversal of removal, but adjust the cables as described in Section 25.

21 Brake pedal – removal and refitting

1 The operations are as described for the clutch pedal in Chapter 5, Section 6.

22 Fault diagnosis – braking system

Symptom	Reason(s)
Pedal travels almost to floorboards before brakes operate	Brake fluid level too low Caliper leaking Master cylinder leaking (bubbles in master cylinder fluid) Brake flexible hose leaking Brake line fractured Brake system unions loose Rear brakes require adjustment
Brake pedal feels springy	New linings not yet bedded in Brake discs or drums badly worn or cracked Master cylinder securing nuts loose
Brake pedal feels spongy and soggy	Caliper or wheel cylinder leaking Master cylinder leaking (bubbles in master cylinder reservoir) Brake pipe, line or flexible hose leaking Unions in brake system loose Air in hydraulic system
Excessive effort required to brake car	Pad or shoe linings badly worn New pads or shoes recently fitted – not yet bedded-in Harder linings fitted than standard causing increase in pedal pressure Lining and brake drum contaminated with oil, grease or hydraulic fluid
Brakes uneven and pulling to one side	Linings and discs or drums contaminated with oil, grease or hydraulic fluid Tyre pressures unequal Radial ply tyres fitted at one end of the car only Brake caliper loose Brake pads or shoes fitted incorrectly Different type of linings fitted at each wheel Anchorages for front suspension or rear suspension loose Brake discs or drums badly worn, cracked or distorted
Brakes tend to bind, drag or lock-on	Air in hydraulic system Wheel cylinders seized Handbrake cables too tight

Chapter 10 Electrical system

Contents

Alternator – description, maintenance, precautions and testing	9
Alternator drivebelt – adjustment and renewal	10
Alternator – removal and refitting	11
Battery – maintenance	2
Battery – removal and refitting	3
Courtesy lamp switch – removal and refitting	19
Dynamo – maintenance	4
Dynamo – overhaul	7
Dynamo – removal and refitting	6
Dynamo – testing in the vehicle	5
Dynamo voltage regulator unit	8
Exterior lamps – bulb renewal	22
Exterior lamps – removal and refitting	23
Fault diagnosis – electrical system	46
Fuses and relays	15
General description	1
Headlamp beam alignment	21
Headlamp bulbs and sealed beam units – renewal	20
Horns	43
Instrument voltage stabiliser – removal and refitting	33
Interior lamps – bulb renewal	24
Key warning system (N. American models)	42
Master lighting switch – removal and refitting	18
Radio – installation	44
Radios and tape players – suppression of interference (general)	45
Seat belt warning system (N. America)	41
Service interval indicator (early N. American models) – removal and refitting	31
Speedometer – removal and refitting	26
Speedometer cable (complete) – removal and refitting	28
Speedometer cable (two section – N. America) – removal and refitting	29
Speedometer inner cable – renewal	27
Starter motor – description and testing in car	12
Starter motor – overhaul	14
Starter motor – removal and refitting	13
Steering column combination switch – removal and refitting	17
Tachometer – removal and refitting	25
Tachometer drive cable – removal and refitting	30
Temperature and fuel gauges – removal and refitting	32
Wheelboxes – removal and refitting	38
Windscreen washer system	40
Windscreen washer/wiper switch (early models) – removal and refitting	16
Windscreen wiper arms – removal and refitting	35
Windscreen wiper blades – removal and refitting	34
Windscreen wiper rack tubing – removal and refitting	39
Wiper motor (two speed) – removal, overhaul and refitting	37
Wiper motor and rack (single speed) – removal, overhaul and refitting	36

Specifications

System type .... 12V, positive earth (Mk I and Mk II) negative earth (Mk III and later). Dynamo or alternator and battery.

Battery .... 40 Ah at 20 hr rate

Dynamo .... Lucas C40-1
Maximum output .... 22 amps
No. of brushes 2
Minimum brush length .... $\frac{11}{32}$
Brush spring tension .... 22 to 25 ozs (0.62 to 0.71 kgs)
Field resistance .... 6 ohms

Voltage regulator/cut-out
Regulator/Control Box .... Lucas RB.340
Cut in voltage .... 12.6 to 13.4 volts
Drop off voltage .... 9.3 to 11.2 volts
Open circuit voltage settings .... 10°C (50°F) 14.9 to 15.5 volts
20°C (68°F) 14.7 to 15.3 volts
30°C (86°F) 14.5 to 15.1 volts
40°C (104°F) 14.3 to 14.9 volts
Reverse current .... 3.0 to 5.0 amps
Current regulator .... 22 + or – 1 amp

Alternator
Make/type	Lucas 15 ACR or 16 ACR
Polarity	Negative earth only
Minimum permissible length of brush to protrude from brush box when free	0.2 in (5.0 mm)
Control voltage	14.0 to 14.4V
Current	28A (15ACR) 34A (16ACR)
Drivebelt tension – total movement at mid-point of longest run	0.75 to 1.00 in (20 to 25 mm)

Starter motor
	Lucas M35G (early) M35J (later)
Brush wear limit	0.375 in (9.53 mm)

Windscreen wiper
	Lucas DR 3A single speed
Normal running current	2.7 to 3.4 amps
Drive to wheelboxes	Rack and cable
Armature endfloat	0.008 to 0.012 in (0.20 to 0.30 mm)
Armature resistance	0.29 to 0.352 ohms
Field resistance	8 to 9.5 ohms
Wiping speed	44 to 48 cycles per minute

Windscreen wiper motor
Make/type	Lucas 14 W (two speed)
Running current (connecting rod removed)	
Normal speed	1.5 amp
High speed	2.0 amp
Running speed:	
Normal speed	46 to 52 cycles/min
High speed	60 to 70 cycles/min
Armature endfloat	0.002 to 0.008 in (0.05 to 0.2 mm)
Minimum permissible brush length	0.125 in (3 mm)

Horns
	Lucas 9H – 12 volt
Maximum current consumption	$3\frac{1}{2}$ amps

Fuse Unit – Positive Earth Cars
	Located under left-hand side of fascia panel
No. of fuses	2 – 35 amps each. 1 in line fuse
No. of spare fuses	2

Fuse unit – Negative Earth Cars
	Located on the engine bulkhead
No. of fuses	3 – 35 amps each
No. of spare fuses	2

Bulbs
Vehicles with bulb type headlamps

	Wattage
Headlamps –	
R.H.D.	50/40
L.H.D.	36/36
L.H.D.	45/50
L.H.D.	45/50
L.H.D.	45/50
L.H.D.	35/35
Side lamps	6
Flashers	21
Stop/tail	21/6
Plate illumination	
Except USA	4
USA only	6
Panel and warning lamps	2.2

Vehicles with sealed beam headlamp units

	Wattage
Headlamps:	
LH Dip	60/45
RH Dip – USA	50/40
RH Dip – France	45/40
RH Dip – other markets	60/50
Front parking lamps	5
Front flasher lamps	21
Front marker lamps	4
Rear marker lamps	4
Tail/stop lamps	5/21
Rear flasher lamps	21
Reverse lamp	21

Plate illumination lamp	5
Instrument illumination	2.2
Courtesy light	2.2
Warning light	2.2
Seat belt warning light	2
EGR service indicator	2
Catalytic converter service indicator	2
Wash/wipe switch	2
Hazard warning	2
Brake warning	2

Torque wrench setting

	lbf ft	Nm
Alternator mounting bolts	20	27
Alternator adjuster link bolt	20	27
Starter motor bolts	30	41

1 General description

1 The electrical system fitted to all Spitfires is of the conventional 12-volt type. The major components consist of a twelve volt battery with the positive terminal earthed on Mk I and Mk II models, and the negative terminal earthed from March 1967, on the Spitfire Mk III; a control box, cut out and fuse unit.
2 Commencing with Mk III models, an alternator was fitted instead of a dynamo.

2 Battery – maintenance

1 Various designs of battery may be encountered, one with screw type vent plugs, the Autofil type or the latest low maintenance 'sealed for life' type.
2 With all batteries except the sealed type, the weekly maintenance consists of keeping the electrolyte level just above the separators by the addition of distilled or purified water. This water can be obtained by saving the ice condensate from your refrigerator.
3 To keep the battery terminals free from corrosion, smear them with petroleum jelly. Check the terminal clamps for tightness.
4 With an alternator, the need for topping-up a battery occurs only infrequently, but if regular topping-up is needed calling for excessive amounts of water, then the battery is probably being overcharged owing to a fault in the voltage control circuit.
5 The need for battery charging from a mains charger has virtually disappeared with the use of improved batteries and generators, but if the car is used only for very short journeys with frequent operation of the starter, then the use of a battery charger may be beneficial. Always disconnect the battery leads from the battery terminals before connecting the mains charger or cars equipped with an alternator.
6 Always keep a look out for corrosion of the battery mounting platform. This is evident when white fluffy deposits are seen. Clean it off, neutralize the metal surfaces with ammonia or sodium bicarbonate and apply some protective paint.
7 A hydrometer is useful for testing the condition of a battery. The specific gravity of the electrolyte should conform to the figures given in the following table. Any marked deviation in the specific gravity of one cell will indicate that the battery is failing and should be renewed at an early date.
8 Specific gravity is measured by drawing up into the body of a hydrometer sufficient electrolyte to allow the indicator to float freely. The level at which the indicator floats indicates the specific gravity.

Table A
Specific Gravity – Battery fully charged
1.268 at 100°F or 38°C electrolyte temperature
1.272 at 90°F or 32°c electrolyte temperature
1.276 at 80°F or 27°C electrolyte temperature
1.280 at 70°F or 21°C electrolyte temperature
1.284 at 60°F or 21°C electrolyte temperature
1.288 at 50°F or 10°C electrolyte temperature
1.292 at 40°F pr 4°C electrolyte temperature
1.296 at 30°F or –1.5°C electrolyte temperature

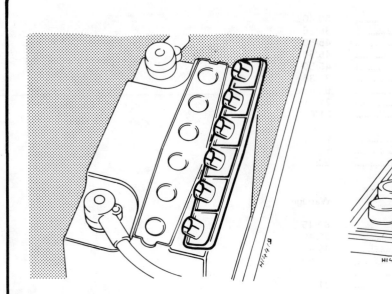

Fig. 10.1 Typical vent plug arrangement on battery (Sec 2)

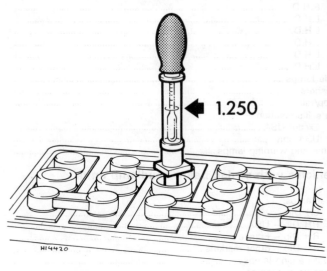

Fig. 10.2 Using a battery hydrometer (Sec 2)

Chapter 10 Electrical system

Table B
Specific Gravity — Battery fully discharged
1.098 at 100°F or 38°C electrolyte temperature
1.102 at 90°F or 32°C electrolyte temperature
1.106 at 80°F or 27°C electrolyte temperature
1.110 at 70°F or 21°C electrolyte temperature
1.114 at 60°F or 16°C electrolyte temperature
1.118 at 50°F or 10°C electrolyte temperature
1.122 at 40°F or 4°C electrolyte temperature
1.126 at 30°F or −1.5°C electrolyte temperature

3 Battery – removal and refitting

1 The battery is located on a recess in the engine compartment rear bulkhead.
2 Disconnect the earthed lead (positive early cars, negative later models).
3 Disconnect the remaining battery lead.
4 Release the battery clamp wing nuts, swing the retaining bar out of the way and remove the battery.
5 Refitting is a reversal of removal.

4 Dynamo – maintenance

1 Routine maintenance consists of checking the tension of the fan belt, and lubricating the dynamo rear bearing once every 12 000 miles (19 000 km).
2 The fan belt should be tight enough to ensure no slip between the belt and the dynamo pulley. If a shrieking noise comes from the engine when the unit is accelerated rapidly, it is likely that it is the fan belt slipping. On the other hand, the belt must not be too taut or the bearings will wear rapidly and cause dynamo failure or bearing seizure. Ideally $\frac{1}{2}$ in. (12.7 mm) of total free movement should be available at the fan belt midway between the fan and the dynamo pulley.
3 To adjust the fan belt tension slightly slacken the three dynamo retaining bolts, and swing the dynamo on the upper two bolts outwards to increase the tension, and inwards to lower it.
4 It is best to leave the bolts fairly tight so that considerable effort has to be used to move the dynamo; otherwise it is difficult to get the correct setting. If the dynamo is being moved outwards to increase the tension and the bolts have only been slackened a little, a long spanner acting as a lever placed behind the dynamo with the lower end resting against the block works very well in moving the dynamo outwards. Retighten the dynamo bolts and check that the dynamo pulley is correctly aligned with the fan belt.
5 Lubrication of the dynamo consists of inserting three drops of SAE 30 engine oil in the small oil hole in the centre of the commutator end bracket. This lubricates the rear bearing. The front bearing is pre-packed with grease and requires no attention.

5 Dynamo – testing in the vehicle

1 If, with the engine running no charge comes from the dynamo, or the charge is very low, first check that the fan belt is in place and is not slipping. Then check that the leads from the control box to the dynamo are firmly attached and that one has not come loose from its terminal.
2 The lead from the larger D terminal in the dynamo should be connected to the D terminal on the control box, and similarly the F terminals on the dynamo and control box should also be connected together. Check that this is so and that the leads have not been incorrectly fitted. Ensure that a good connection exists to control box terminal E.
3 Make sure none of the electrical equipment (such as the lights or radio) is on and then pull the leads off the dynamo terminals marked D and F, join the terminals together with a short length of wire.
4 Attach to the centre of this length of wire the negative clip of a 0-20 volts voltmeter and run the other clip to earth on the dynamo yoke. Start the engine and allow it to idle at approximately 750 rpm. At this speed the dynamo should give a reading of about 15 volts on the voltmeter. There is no point in raising the engine speed above a fast idle as the reading will then be inaccurate.
5 If no reading is recorded then check the brushes and brush connections. It a very low reading of approximately 1 volt is observed then the field winding may be suspect.
6 If a reading of between 4 to 6 amps is recorded it it likely that the armature winding is at fault.
7 If the voltmeter shows a good reading then with the temporary link still in position connect both leads from the control box to D and F on the dynamo (D to D and F to F). Release the lead from the D terminal at the control box end and clip one lead from the voltmeter to the end of the cable, and the other lead to a good earth. With the engine running at the same speed as previously, an identical voltage to that recorded at the dynamo should be noted on the voltmeter. If no voltage is recorded then there is a break in the wire. If the voltage is the same as recorded at the dynamo then check the F lead in similar fashion. If both readings are the same as at the dynamo then it will be necessary to test the control box.

6 Dynamo – removal and refitting

1 Slacken the two dynamo retaining bolts, and the nut on the sliding link, and move the dynamo in towards the engine so that the fan belt can be removed.
2 Disconnect the two leads from the dynamo terminals.
3 Remove the nuts from the sliding link bolt, and remove the two upper bolts. The dynamo is then free to be lifted away from the engine.
4 Replacement is a reversal of the above procedure. Do not finally tighten the retaining bolt and the nut on the sliding link until the fan belt has been tensioned correctly.

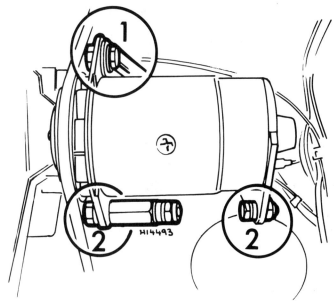

Fig. 10.3 Dynamo adjuster bolt (1) and mounting bolts (2) (Sec 6)

7 Dynamo – overhaul

1 Mount the dynamo in a vice and unscrew and remove the two through bolts from the commutator end bracket.
2 Mark the commutator end bracket and the dynamo casing so the end bracket can be replaced in its original position. Pull the end bracket off the armature shaft. Some versions of the dynamo may have a raised pip on the end bracket which locates in a recess on the edge of the casing. If so, marking the end bracket and casing is not necessary. A pip may also be found on the drive end bracket at the opposite end of the casing.
3 Lift the two brush springs and draw the brushes out of the brush holders.
4 Measure the brushes and if worn down $\frac{9}{32}$ in. or less unscrew the

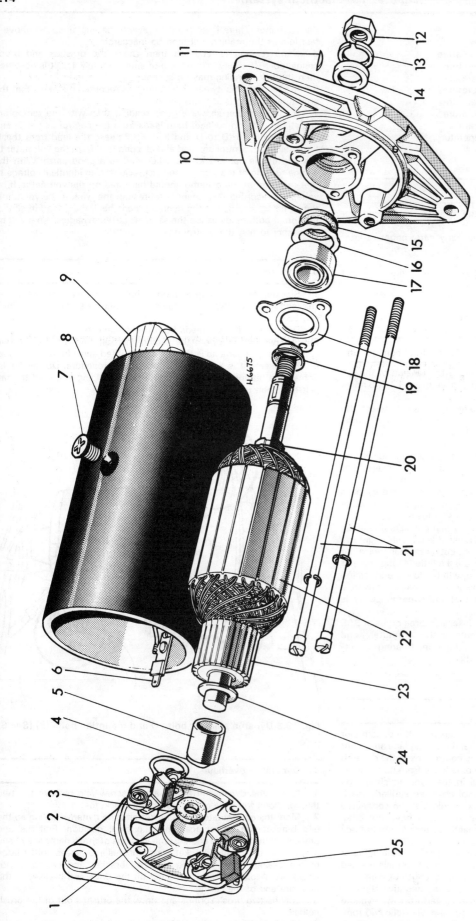

Fig. 10.4 Exploded view of dynamo (Sec 7)

1 Output terminal D
2 Commutator end bracket
3 Felt ring
4 Felt ring retainer
5 Bronze bush
6 Field terminal F
7 Pole shoe securing screws
8 Yoke
9 Field coils
10 Drive end bracket
11 Shaft key
12 Shaft nut
13 Lockwasher
14 Pulley spacer
15 Felt ring
16 Corrugated washer
17 Ballbearing
18 Bearing retaining plate
19 Shaft collar retaining cup
20 Shaft collar
21 Through bolts
22 Armature
23 Commutator
24 Thrustwasher
25 Brush

screws holding the brush leads to the end bracket. Take off the brushes complete with leads.

5 If no locating pip can be found, mark the drive end bracket and the dynamo casing so the drive end bracket can be replaced in its original position. Then pull the drive end bracket complete with armature out of the casing.

6 Check the condition of the ball bearing in the drive end plate by firmly holding the plate and noting if there is visible side movement of the armature shaft in relation to the end plate. If play is present the armature assembly must be separated from the end plate. If the bearing is sound there is no need to carry out the work described in the following two paragraphs.

7 Hold the armature in one hand (mount it carefully in a vice if preferred) and undo the nut holding the pulley wheel and fan in place. Pull off the pulley wheel and fan.

8 Next remove the Woodruff key from its slot in the armature shaft and also the bearing locating ring.

9 Place the drive end bracket across the open jaws of a vice with the armature downwards and gently tap the armature shaft from the bearing in the end plate with the aid of a suitable drift.

10 Carefully inspect the armature and check it for open or short circuited windings. It is a good indication of an open circuited armature when the commutator segments are burnt. If the armature has short circuited the commutator segments will be very badly burnt, and the overheated armature windings badly discoloured. If open or short circuits are suspected then test by substituting the suspect armature for a new one.

11 Check the resistance of the field coils. To do this, connect an ohmmeter between the field terminals and the yoke and note the reading on the ohmmeter which should be about 6 ohms. If the ohmmeter reading is infinity this indicates an open circuit in the field winding. If the ohmmeter reading is below 5 ohms this indicates that one of the field coils is faulty and must be replaced.

12 Field coil replacement involves the use of a wheel operated screwdriver, a soldering iron, caulking and riveting and this operation is considered to be beyond the scope of most owners. Therefore, if the field coils are at fault either purchase a rebuilt dynamo, or take the casing to a Triumph dealer or electrical engineering works for new field coils to be fitted.

13 Next check the condition of the commutator. If it is dirty and blackened as shown, clean it with a petrol dampened rag. If the commutator is in good condition the surface will be smooth and quite free from pits or burnt areas, and the insulated segments clearly defined.

14 If, after the commutator has been cleaned pits and burnt spots are still present, wrap a strip of glass paper round the commutator taking great care to move the commutator $\frac{1}{4}$ of a turn every ten rubs till it is thoroughly clean. Do not use emery cloth for this job.

15 In extreme cases of wear the commutator can be mounted in a lathe and with the lathe running at high speed, a very fine cut may be taken off the commutator. Then polish the commutator with glass paper. If the commutator has worn so that the insulators between the segments are level with the top of the segments, then undercut the insulators to a depth of $\frac{1}{32}$ in. (0.8 mm). The best tool to use for this purpose is half a hacksaw blade ground to the thickness of an insulator, and with the handle end of the blade covered in insulating tape to make it confortable to hold.

16 Check the bush bearing in the commutator end bracket for wear by noting if the armature spindle rocks when placed in it. If worn it must be renewed.

17 The bush bearing can be removed by a suitable extractor or by screwing a $\frac{5}{8}$ in (15.9 mm) tap four or five times into the bush. The tap complete with bush is then pulled out of the end bracket.

18 Note before fitting the new bush bearing that it is of the porous bronze type, and it is essential that it is allowed to stand in SAE 30 engine oil for at least 24 hours before fitting. In an emergency the bush can be immersed in hot oil (100°C) (212°F) for 2 hours.

19 Carefully fit the new bush into the endplate, pressing it in until the end of the bearing is flush with the inner side of the endplate. If available press the bush in with a smooth shouldered mandrel the same diameter as the armature shaft.

20 To renew the ball bearing fitted to the drive end bracket drill out the rivets which hold the bearing retainer plate to the end bracket and lift off the plate.

21 Press out the bearing from the end bracket and remove the corrugated and felt washers from the bearing housing.

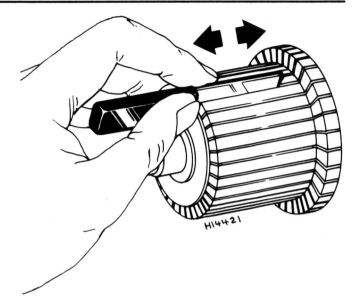

Fig. 10.5 Undercutting dynamo commutator insulation (Sec 7)

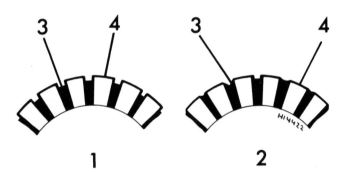

Fig. 10.6 Dynamo insulation undercut (Sec 7)

1 Correct
2 Incorrect
3 Insulation
4 Copper segments

22 Thoroughly clean the bearing housing, and the new bearing and pack with high metling-point grease.

23 Place the felt washer and corrugated washer in that order in the end bracket bearing housing.

24 Then fit the new bearing.

25 Gently tap the bearing into place with the aid of a suitable drift.

26 Replace the bearing plate and fit three new rivets.

27 Open up the rivets with the aid of a suitable cold chisel.

28 Finally peen over the open end of the rivets with the aid of a ball hammer.

29 Refit the drive end bracket to the armature shaft. Do not try and force the bracket on but with the aid of a suitable socket abutting the bearing tap the bearing in gently, so pulling the end bracket down with it.

30 Slide the spacer up the shaft and refit the Woodruff key.

31 Replace the fan and pulley wheel and then fit the spring washer and nut and tighten the latter. The drive bracket end of the dynamo is now fully assembled.

32 If the brushes are little worn and are to be used again then ensure that they are placed in the same holders from which they were removed. When refitting brushes, either new or old, check that they move freely in their holders. If either brush sticks, clean with a petrol moistened rag and if still stiff, lightly polish the sides of the brush with a very fine file until the brush moves quite freely in its holder.

33 Tighten the two retaining screws and washers which hold the wire leads to the brushes in place.

34 It is far easier to slip the end piece with brushes over the commutator if the brushes are raised in their holders as shown and held in this position by the pressure of the springs resting against their flanks.
35 Refit the armature to the casing and then the commutator end plate and screw up the two through bolts.
36 Finally, hook the ends of the two springs off the flanks of the brushes and onto their heads so the brushes are forced down into contact with the armature.

8 Dynamo voltage regulator unit

1 The control box comprises the voltage regulator and the cut-out. The voltage regulator controls the output from the dynamo depending on the state of the battery and the demands of the electrical equipment and ensures that the battery is not overcharged. The cut-out is really an automatic switch and connects the dynamo to the battery when the dynamo is turning fast enough to produce a charge. Similarly it disconnects the battery from the dynamo when the engine is idling or stationary so that the battery does not discharge through the dynamo.
2 Every 12 000 miles (19 000 km) disconnect the battery and check the cut-out and regulator contacts. If they are dirty or rough or burnt, place a piece of fine glass paper (DO NOT USE EMERY PAPER OR CARBORUNDUM PAPER) between the cut-out contacts, close them manually and draw the glass paper through several times.
3 Clean the regulator contacts in exactly the same way, but use emery paper or carborundum paper and not glass paper. Carefully clean both sets of contacts from all traces of dust with a rag moistened in methylated spirit.
4 If the battery is being undercharged check that the fan belt is not slipping and the dynamo is producing its correct output. Check the battery lead terminals for secureness on their posts. If the battery is being overcharged this points fairly definitely to an incorrectly set regulator.
5 Checking the action of the regulator and cut out is not difficult but must be completed as quickly as possible (not more than 30 seconds for each test) to avoid errors caused by heat of the coils. Essential test equipment comprises a 0-20 volt voltmeter and a moving coil – 40 to + 40 amp ohmmeter and an air temperature gauge. Also required is a special adjusting tool.
6 The regulator portion of the three bobbin type control box comprises the voltage regulator and the current regulator. The third bobbin at the B terminal end is the cut out.
7 To test the regulator take off the control box cover and slip a piece of thin card between the cut out points. Connect the voltmeter between control box terminal S and a good earth. Start and run the engine at about 3000 rpm. when a steady reading on the voltmeter should be given as shown in Table A.

Table A

Air Temperature	Type RB340 Open circuit voltage
10°C or 50°F	14.9 to 15.5
20°C or 68°F	14.7 to 15.3
30°C or 86°F	14.5 to 15.1
40°C or 104°F	14.3 to 14.9

8 If the reading fluctuates by more than 0.3 volts then it is likely that the contact points are dirty. If the reading is steady but incorrect turn the voltage adjustment cam clockwise with the special Lucas tool to increase the settings and anti-clockwise to lower it.
9 Stop the engine and then restart it, gradually increasing the speed. If the voltage continues to rise with a rise in engine speed this indicates short circuited or fused points or a faulty magnet coil. If this is the case the only remedy is to fit an exchange control box.
10 The dynamo should be able to provide 22 amps at 3000 rpm irrespective of the state of the battery.
11 To test the dynamo output take off the control box cover, and short out the voltage regulator contacts by holding them together with a bulldog clip.
12 Pull off the Lucar connectors from the control box terminals B and connect an ammeter reading to 40 volts to the two cables just disconnected to one of the B Lucar connectors.
13 Turn on all the lights and other electrical equipment and start the

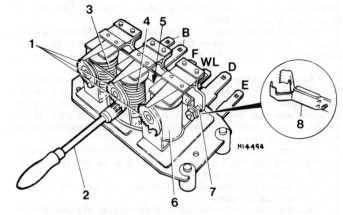

Fig. 10.7 Voltage control box (cover removed) (Sec 8)

1 Adjustment cams
2 Adjustment tool
3 Cut-out relay
4 Current regulator
5 Current regulator contacts
6 Voltage regulator
7 Voltage regulator contacts
8 Clip for use when testing output

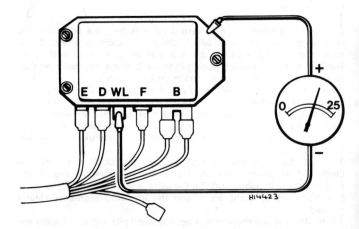

Fig. 10.8 Cut-in voltage checking circuit (Sec 8)

engine. At about 3000 rpm the dynamo should be giving between 21 and 23 amps as recorded on the ammeter. If the ammeter needle flickers it is likely that the contact points are dirty.
14 To increase the current turn the cam on top of the current regulator clockwise, and to lower it, anti-clockwise.
15 Check the voltage required to operate the cutout by connecting a voltage between the control box terminal WL and a good earth. Remove the control box cover, start the engine and gradually increase its speed until the cut-outs close. This should occur when the reading is between 12.7 to 13.3 volts.
16 If the reading is outside these limits turn the adjusting cam on the cut-out relay a fraction at a time clockwise to raise the voltage cut-in point and anti-clockwise to lower it.
17 To adjust the drop off voltage bend the fixed contact blade carefully. The adjustment to the cut-out should be completed within 30 seconds of starting the engine as otherwise heat build-up from the shunt coil will affect the readings.
18 If the cut-out fails to work, clean the contacts, and, if there is still no response, renew the cut-out and regulator unit.

9 Alternator – description, maintenance, precautions and testing

1 The alternator fitted to later models has an integral control unit.
2 The control unit contains sensitive components. The battery terminals must never be connected the wrong way round. (The cars

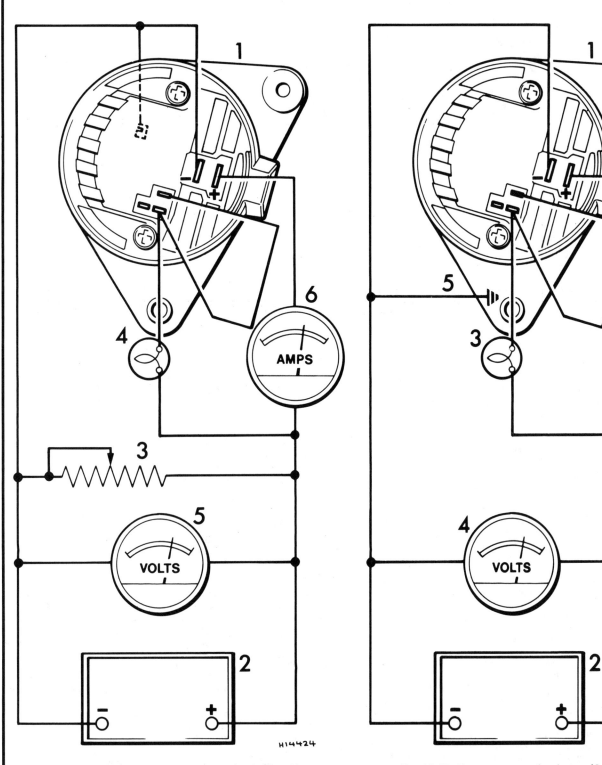

Fig. 10.9 Alternator current test circuit (Sec 9)

   1 Alternator   4 Test lamp
   2 Battery   5 Voltmeter
   3 Variable resistor 6 Ammeter

Fig. 10.10 Alternator control unit test (Sec 9)

   1 Alternator   4 Voltmeter
   2 Battery   5 Earth connection to
   3 Test lamp   alternator body

Chapter 10 Electrical system

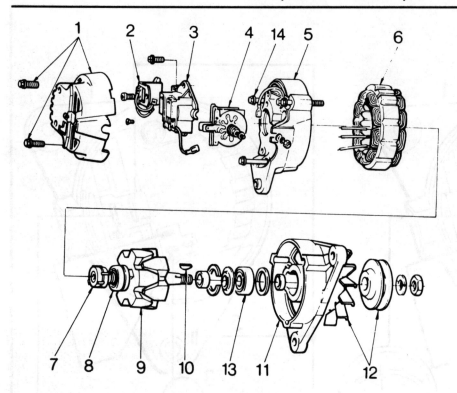

Fig. 10.11 Alternator main components (Sec 9)

1 End cover
2 Connector
3 Brush box
4 Rectifier pack
5 Slip ring end bracket
6 Stator
7 Slip rings
8 Slip ring end bearing
9 Rotor
10 Woodruff key
11 Drive end bracket
12 Fan and pulley
13 Drive end bearing
14 Tie bolt

are now all negative earth). When using electric welding equipment on the car the leads to the alternator should be disconnected.
3 The brushes work on simple slip rings instead of a commutator as in a dynamo, so last longer, and do not need cleaning very often.
4 To clean the brushes first remove the alternator (see Section 11) and clean the outside. Then remove the moulded end cover (two screws). Remove the brush box and control unit by disconnecting the Lucar connectors, and taking out three screws. Full stripping of the alternator requires the unsoldering of three leads from the rectifier pack to the stator. Do not do this unless you have experience of electronic work, or the diodes are likely to be damaged by the soldering heat.
5 The alternator output can be roughly checked with it in place on the engine. Insert an ammeter in series with the brown leads from the alternator and the terminal for them on the starter solenoid. The voltage rise when charging can be measured at the same point with a voltmeter from the terminal to earth. The alternator should have begun to charge even at idle speed.
6 More detailed tests of the alternator are outlined below.

Current test

(a) Check fan belt tension. Warm up the engine (and alternator).
(b) Disconnect leads to alternator. Remove the moulded end cover.
(c) Connect test circuit as shown in Fig. 10.10.
(d) Start the engine.
(e) The light should bo out at 620 engine rpm.
(f) Run the engine at 2500 rpm.
(g) Adjust the resistance to give a voltage of 14v.
(h) The current should now be approximately 28 amps (15 ACR) or 34 amps (16 ACR).
(i) If the output is down, clean the brushes and test again. If still down, a replacement alternator should be fitted.

Control unit test

(k) Having done the circuit test and stopped the engine, rewire to the test circuit shown in Fig. 10.11.
(l) Start up.
(m) The light should go out at 620 engine rpm.
(n) Run the engine at 2500 rpm. The voltmeter should give a steady reading between 14 to 14.7 volts.
(p) If it does not, and the current test has proved the alternator's ability to produce current, the control unit should be replaced.

10 Alternator drivebelt – adjustment and renewal

Refer to Chapter 2, Section 12.

11 Alternator – removal and refitting

1 Disconnect the battery negative lead.
2 Disconnect the wiring plug from the rear of the alternator and unclip the wiring harness from the clip on the adjuster link (photos).
3 Release the mounting ad adjuster link bolts, push the alternator in towards the engine and slip the drivebelt from the pulley.
4 Remove the mounting and adjuster link bolts and lift the alternator from the engine.
5 Refitting is a reversal of removal, adjust the drivebelt.

12 Starter motor – description and testing in car

1 The starter motor is of the inertia drive type.
2 The unit is mounted on the left-hand side of the engine endplate, and is held in position by two bolts which also clamp the bellhousing flange. The motor is of the four field coil, four pole piece type, and utilises four spring-loaded commutator brushes. Two of these brushes are earthed, and the other two are insulated and attached to the field coil ends.
3 If the starter motor fails to operate then check the condition of the battery by turning on the headlamps. If they glow brightly for several seconds and then gradually dim, the battery is in an uncharged condition.
4 If the headlamps glow brightly and it is obvious that the battery is in good condition then check the tightness of the battery wiring connections (and in particular the earth lead from the battery terminal to its connection on the bodyframe). Check the tightness of the connections at the relay switch and at the starter motor. Check the wiring with a voltmeter for breaks or shorts.
5 If the wiring is in order then check that the starter motor switch is operating. To do this press the rubber covered button in the centre of the relay switch under the bonnet. If it is working the starter motor will be heard to 'click' as it tries to rotate. Alternatively check it with a voltmeter.

Chapter 10 Electrical system

11.2a Alternator connecting plug

11.2b Alternator wiring harness clip

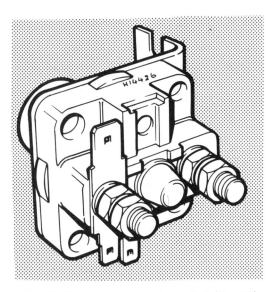

Fig. 10.12 Starter motor relay switch (Sec 12)

13.2 Starter motor terminal

6 If the battery is fully charged, the wiring in order, and the switch working and the starter motor fails to operate then it will have to be removed from the car for examination. Before this is done, however, ensure that the starter pinion has not jammed in mesh with the flywheel. Check by turning the square end of armature shaft with a spanner. This will free the pinion if it is stuck in engagement with the flywheel teeth.

13 Starter motor – removal and refitting

1 Disconnect the battery earth lead from the positive terminal (negative on later models).
2 Disconnect the heavy lead to the starter motor from the starter motor terminal (photo).
3 Undo and remove the two bolts which hold the starter motor in place and withdraw it upwards together with the distance piece.
4 Generally replacement is a straightforward reversal of the removal sequence. Check that the electrical cable is firmly attached to the starter motor terminal before fitting the starter motor in place. Also ensure that any packing washers originally fitted are replaced so as to give the correct out of mesh clearance between the stationary starter pinion and the flywheel ring gear of $\frac{3}{32}$ in. to $\frac{5}{32}$ in. (2.4 to 4.0 mm).

14 Starter motor – overhaul

1 With the starter motor on the bench, loosen the screw on the cover band and slip the cover band off. With a piece of wire bent into the shape of a hook, lift back each of the brush springs in turn and check the movement of the brushes in their holders by pulling on the flexible connectors. If the brushes are so worn that their faces do not rest against the commutator, or if the ends of the brush leads are exposed on their working face, they must be renewed.
2 If any of the brushes tend to stick in their holders then wash them with a petrol moistened cloth and, if necessary, lightly polish the sides of the brush with a very fine file, until the brushes move quite freely in their holders.
3 If the surface of the commutator is dirty or blackened, clean it with a petrol dampened rag. Secure the starter motor in a vice and check it by connecting a heavy gauge cable between the starter motor terminal and a 12-volt battery.
4 Connect the cable from the other battery terminal to earth in the starter motor body. If the motor turns at high speed it is in good order.
5 If the starter motor still fails to function or if it is wished to renew the brushes, then it is necessary to further dismantle the motor.
6 Lift the brush springs with the wire hook and lift all four brushes out of their holders one at a time.

Chapter 10 Electrical system

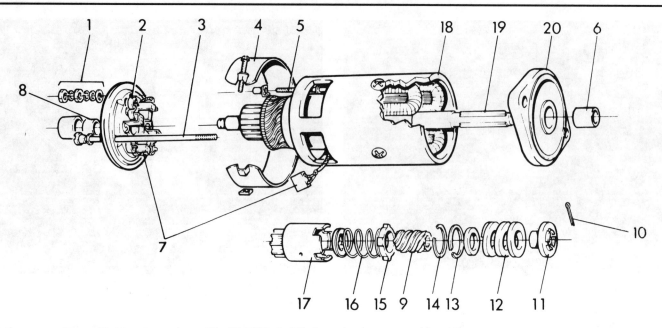

Fig. 10.13 Exploded view of starter motor (Sec 14)

1 Terminal nuts and washers
2 Brush spring
3 Through-bolt
4 Band cover
5 Terminal post
6 Bearing bush
7 Brushes
8 Bearing bush
9 Sleeve
10 Split pin
11 Shaft nut
12 Main spring
13 Retaining ring
14 Washer
15 Control nut
16 Restraining spring
17 Pinion and barrel
18 Yoke
19 Armature shaft
20 Driving end bracket

7 Remove the terminal nuts and washers from the terminal post on the commutator end bracket.
8 Unscrew the two through bolts which hold the end plates together and pull off the commutator end bracket. Also remove the driving end bracket which will come away complete with the armature.
9 At this stage if the brushes are to be renewed, their flexible connectors must be unsoldered and the connectors of new brushes soldered in their place. Check that the new brushes move freely in their holders as detailed above. If cleaning the commutator with petrol fails to remove all the burnt areas and spots, then wrap a piece of glass paper round the commutator and rotate the armature.
10 If the commutator is very badly worn, remove the drivegear as detailed in the following section. Then mount the armature in a lathe and with the lathe turning at high speed, take a very fine cut out of the commutator and finish the surface by polishing with glass paper. *Do not undercut the mica insulators between the commutator segments.*
11 With the starter motor dismantled, test the four field coils for an open circuit. Connect a 12-volt battery with a 12-volt bulb in one of the leads between the field terminal post and the tapping point of the field coils to which the brushes are connected. An open circuit is proved by the bulb not lighting.
12 If the bulb lights, it does not necessarily mean that the field coils are in order, as there is a possibility that one of the coils will be earthed to the starter yoke or pole shoes. To check this, remove the lead from the brush connector and place it against a clean portion of the starter yoke. If the bulb lights the field coils are earthing. Replacement of the field coils calls for the use of a wheel operated screwdriver, a soldering iron, caulking and riveting operations and is beyond the scope of the majority of owners. The starter yoke should be taken to a reputable electrical engineering works for new field coils to be fitted. Alternatively, purchase an exchange Lucas starter motor.
13 If the armature is damaged this will be evident after visual inspection. Look for signs of burning, discolouration, and for conductors that have lifted away from the commutator. Reassembly is a straightforward reversal of the dismantling procedure.
14 One of two types of starter motor drive may be fitted. Early cars make use of the drive shown in Fig. 10.15. To dismantle the earlier drive, extract the split pin from the shaft nut on the end of the starter drive.

15 Holding the squared end of the armature shaft at the commutator end bracket with a suitable spanner, unscrew the shaft nut which has a right-hand thread, and pull off the mainspring.
16 Slide the remaining parts with a rotary action off the armature shaft.
17 Reassembly is a straightforward reversal of the above procedure. Ensure that the split-pin is refitted. It is most important that the drivegear is completely free from grease and dirt. With the drivegear removed, clean all parts thoroughly in paraffin, dry and apply a little light oil to the screwed sleeve and shaft.
18 The later type of drive shown in Fig. 10.14 has a circlip at the retaining end of the shaft instead of a nut and split-pin.
19 To dismantle the later drive use a press to push the retainer clear of the circlip which can then be removed. The remainder of the dismantling sequence is the same as for the earlier type. On replacement use a suitable press to compress the spring and retainer sufficiently to allow a new circlip to be fitted to its groove on the shaft. Remove the press.
20 With the starter motor stripped down check the condition of the bushes. They should be renewed when they are sufficiently worn to allow visible side movement of the armature shaft.
21 The old bushes are simply driven out with a suitable drift and the new bushes inserted by the same method. As the bearings are of the phosphor bronze type it is essential that they are allowed to stand in SAE 30 engine oil for at least 24 hours before fitting.

Later models
22 The starter motor used on later models is the Lucas type M35J.
23 Instead of brushes being circumferentially around the commutator they are axially at its end.
24 To strip the starter remove the two bolts holding the drive end bracket to the yoke, and take off that bracket, withdrawing the armature. Remove the thrust washer from the brush end of the armature.
25 Undo the four small bolts holding on the commutator end bracket. Pull the bracket aside. Note the way the flexible cable from the field windings to the brushes is fitted. Then undo these brushes.
26 The brushes should be renewed if worn down below the specified minimum length. One pair should come already soldered to a terminal

Chapter 10 Electrical system

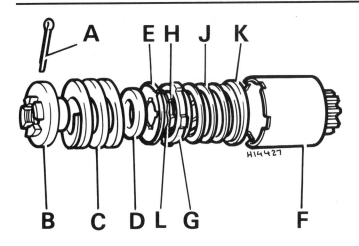

Fig. 10.14 Early type starter drive (Sec 14)

A Split-pin
B Castellated nut
C Main spring
D Buffer washer
E Retaining spring
F Pinion/barrel
G Control nut
H Sleeve
J Retaining coil spring
K Splined washer
L Corrugated washer

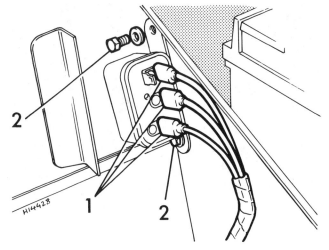

Fig. 10.15 Horn relay

1 Spade connection   2 Mounting bolts

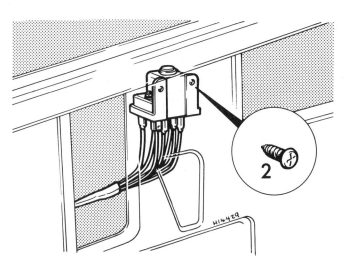

Fig. 10.16 Night dimmer relay (Sec 15)

2 Mounting screw

post, but if loose the soldering is straightforward. The field brushes should come attached to a common flexible lead. When cutting off the old brushes leave $\frac{1}{4}$ in (6 mm) of old lead to which to solder the new. The actual lead to the field winding is aluminium, and will not take solder.

27 The brushes for the shaft bearings can be renewed. They are porous self-lubricating ones. They must be soaked in engine oil for 24 hours before fitting. They must not be reamed or the porosity will be adversely affected. The old drive end bush is pressed out, and the new pressed in. To remove the commutator end bush drill out the two rivets in the end plate. Screw a $\frac{1}{2}$ in (12.7 mm) tap into the old bush as an extractor, and pull it out. The new bush is pressed in. Use a new plate, rivets and felt seal.

28 Do not undercut the commutator segments. The commutator may be lightly skimmed and then polished. Do not cut below the specified thickness.

15 Fuses and relays

Fuses – positive earth cars

1 Two 35 amp fuses are fitted in a Lucas 4FJ box (which also carries two spare fuses) mounted close to the cylindrical flasher unit under the fascia panel on the left-hand side of the car. In addition a line fuse is fitted adjacent to the fuse box.

2 The line fuse protects the horns and headlamp flasher circuits; the fuse fed by the red/green cable protects the sidelights; and the fuse fed by the white cable protects the instruments and ancillary items.

Fuses – negative earth cars

3 Three 35 amp fuses are fitted in a fuse box (which also carries two spare fuses) with a transparent cover on the front of the engine bulkhead (photo).

4 The top fuse which is fed by a white cable protects the following circuits – flasher lamps; stop lamps; windscreen wiper motor; heater motor; fuel gauge; reversing lamps; and the water temperature gauge.

5 The middle fuse which is fed by a red/green cable protects the rear lights, number plate light, and front sidelight circuits.

6 The bottom fuse which is fed by a brown cable protects the horn and headlamp flasher circuit.

Relays

7 Various relays may be fitted depending upon the particular vehicle specification.

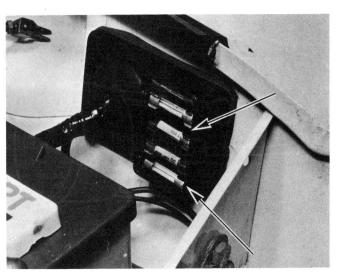

15.3 Negative earth fuse block. Spare fuses arrowed

15.8 Direction indicator flasher unit

Direction indicator flasher unit

8 This is located on a clip behind the fascia panel (photo).
9 Ultra rapid flashing or failure to flash at all may be due to a blown bulb or a poor earth at the lamp unit or bulb holder.

Hazard warning flasher unit

10 On earlier cars, this is located on the left-hand side of the bulkhead as an independent unit.
11 On later models, the flasher unit behind the fascia panel serves both the direction indicator and hazard warning circuits.

Horn relay

12 This relay is located on the engine compartment rear bulkhead close to the battery.

Night dimmer relay

13 This relay is to be found in earlier models, its purpose being to reduce the intensity of stop lamps and rear flasher lamps when the parking lamps are switched on so avoiding dazzle to following drivers.
14 The relay is located within the luggage compartment on the left-hand wheel arch panel. Before removing it, carefully identify the connecting wires for correct replacement.

Overdrive relay

15 On cars equipped with overdrive, the relay which controls the circuit is mounted together with the horn relay on the engine compartment rear bulkhead.
16 Always identify the connecting leads before disconnecting them from their particular relay.

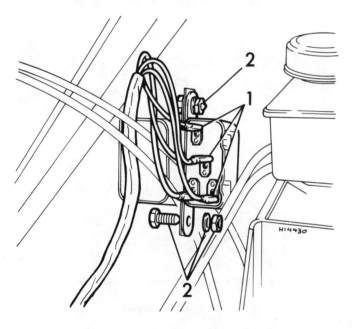

Fig. 10.17 Overdrive relay (Sec 15)

1 Spade terminals   2 Mounting bolts

16 Windscreen washer/wiper switch (early models) – removal and refitting

1 The wiper switch is fascia-mounted on earlier models.
2 Disconnect the battery.
3 Insert a probe into the hole in the underside of the control knob, depress the retainer and pull off the switch knob.
4 Unscrew the switch retaining bezel, lower the switch until the electrical leads and the washer pipes can be identified. Then disconnect them and remove the switch.
5 Refitting is a reversal of removal.

17 Steering column combination switch – removal and refitting

1 Refer to Chapter 11 and remove the steering column upper shrouds and the steering wheel.
2 Disconnect the battery.
3 Disconnect the wiring harness connecting plugs and pull the harness from the retaining clips.
4 Slacken the switch clamp screw and withdraw the switch complete with wiring harness (photo).
5 The individual switches may be renewed after unscrewing the mounting bolts.
6 Refitting is a reversal of removal but observe the following points:

(a) Make sure that the switch tongue is located in the column tube slot.
(b) Push the switch against the column tube before tightening the clamp screw.
(c) Before fitting the steering wheel, check that the direction indicator cancelling collar is correctly aligned (with the front roadwheels in the straight ahead position) and the steering wheel spokes correctly set.

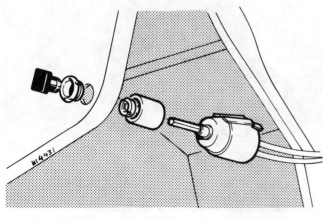

Fig. 10.18 Early type washer/wiper switch (Sec 16)

18 Master lighting switch – removal and refitting

1 Disconnect the battery.
2 Lower the fascia centre section as described in Chapter 12, Section 3).

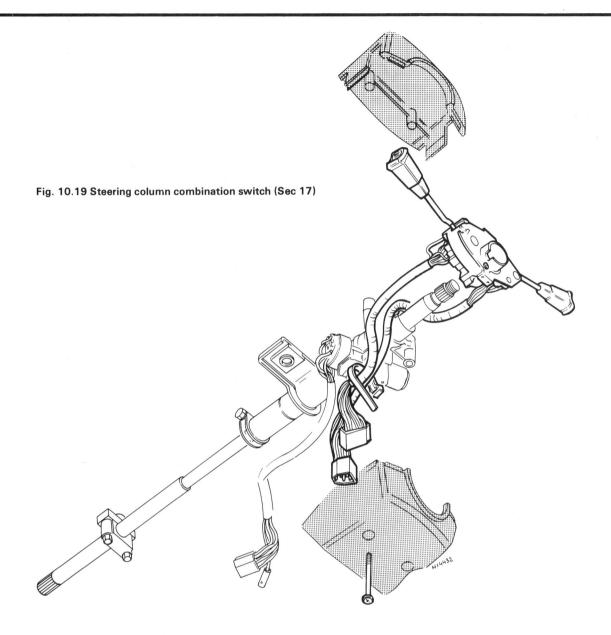

Fig. 10.19 Steering column combination switch (Sec 17)

17.4 Steering column combination switch

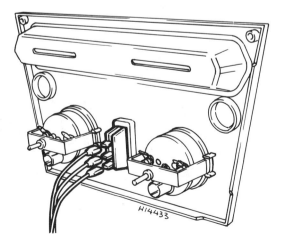

Fig. 10.20 Master lighting switch (Sec 18)

Chapter 10 Electrical system

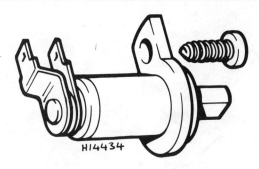

Fig. 10.21 Courtesy lamp switch (Sec 19)

3 Identify and then disconnect the three leads from the switch terminals.
4 Compress the retainers and push the switch from the panel.
5 Refitting is a reversal of removal.

19 Courtesy lamp switch – removal and refitting

1 Disconnect the battery.
2 Open the door and extract the screw which retains the courtesy lamp door plunger switch.
3 Withdraw the switch and disconnect the leads but take care not to allow the leads to slip into the body cavity.
4 Refitting is a reversal of removal. To prevent corrosion of the switch, smear it with petroleum jelly.

20 Headlamp bulbs and sealed beam units – renewal

Bulbs
1 Bulb type headlamps were fitted to Mk I, II and III models not destined for the N. American market.
2 To replace the bulb gently prise away the rim and remove the dust excluder.
3 Unscrew the three cross-headed securing screws. Remove the retaining rings and lamp unit. The bulb holder has a bayonet fitting into the back of the lamp unit.
4 Turn the bulb holder anti-clockwise to release it and remove the bulb.
5 Insert a new bulb and reverse the above procedure.

Sealed beam unit
6 Refer to Fig. 10.25 and remove the nut and two bolts which retain the headlamp surround. Lift off the surround (photo).
7 Extract the three retaining screws and lift off the lamp retaining rim. Take care not to unscrew the adjustment screws B in error (photo).
8 Working at the back of the lamp, pull off the wiring connector (photo).
9 Lift the sealed beam unit from the lamp body.
10 Refit by reversing the removal operations. Provided the beam adjusting screws have not been touched, the headlamp alignment should not have altered.

20.7 Removing headlamp surround

20.8 Headlamp rim screw (A) and adjuster screw (B)

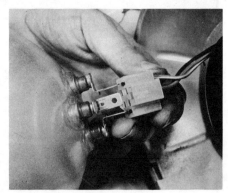

20.9 Pulling off headlamp wiring connector plug

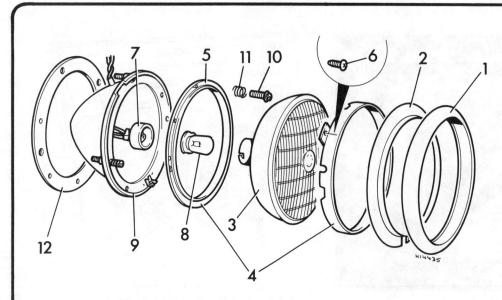

Fig. 10.22 Components of bulb type headlamp (Sec 20)

1 Rim
2 Dust excluder
3 Lamp unit
4 Retaining ring
5 Seating ring
6 Screw
7 Bulb holder
8 Bulb
9 Lamp body
10 Adjuster screw
11 Spring
12 Gasket

Chapter 10 Electrical system

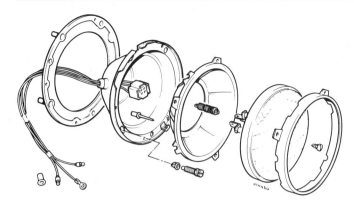

Fig. 10.23 Components of sealed beam type headlamp (Sec 20)

21 Headlamp beam alignment

It is preferable to leave this work to a service station equipped with optical beam setting equipment.

22 Exterior lamps – bulb renewal

Front parking/flasher lamps – Mk I, II and III
1 Separate lamps are fitted to these early models and access to the bulbs is obtained by prising the lamp rim and lens from the lips of the rubber lamp body (photo).

Later models
2 On these models the lamps are in the form of a combined unit.
3 The bulbs are accessible after unscrewing the rim, removing the two lenses and gasket (photo).
4 The bulbs are of bayonet fitting type (photo).

Rear lamps – Mk I, II and III
5 Extract the lens securing screw and remove the lens for access to the stop/tail twin filament bulb.
6 The bulb can be removed from the rear direction indicator lamp as described in paragraph 1 for the front lamp, as it is of similar type.

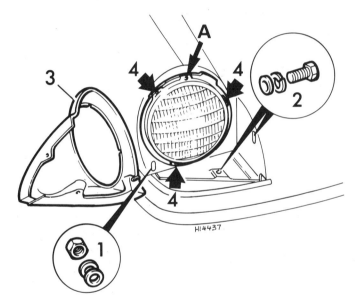

Fig. 10.24 Headlamp sealed beam removal points (Sec 20)

1 Nut and washers
2 Bolts
3 Headlamp surround
4 Rim securing screws
A Vertical adjuster screw

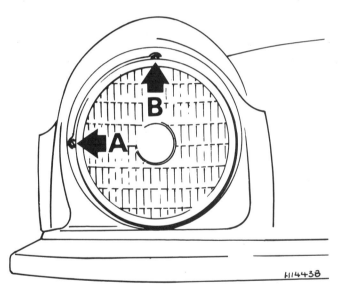

Fig. 10.25 Headlamp beam adjustment screws (Sec 21)

A Horizontal   B Vertical

22.1 Early front parking lamp

22.3 Later front parking/flasher lamp

22.4 Later front parking and flasher bulbs

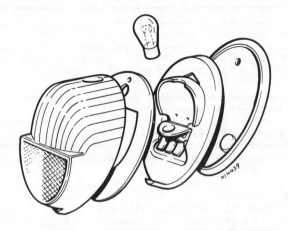

Fig. 10.26 Early type tail/stop lamp (Sec 22)

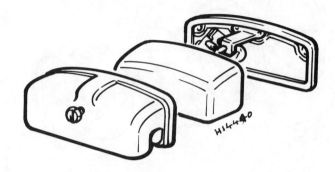

Fig. 10.27 Early type rear number plate lamp (Sec 22)

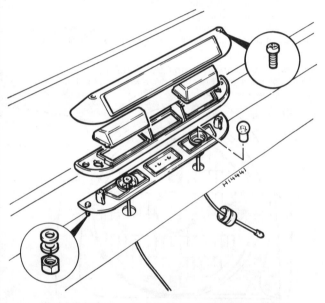

Fig. 10.28 Later type number plate lamp (Sec 22)

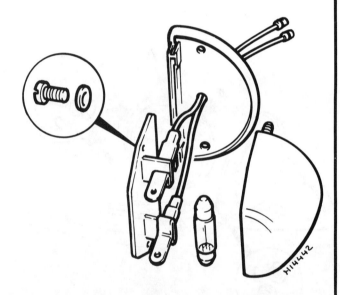

Fig. 10.29 Rear number plate lamp (N. America) (Sec 22)

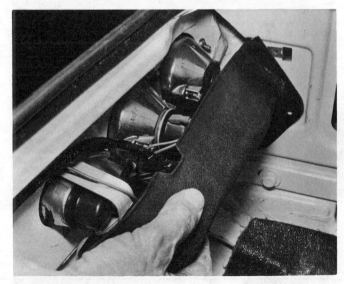

22.8 Rear lamp protective cover

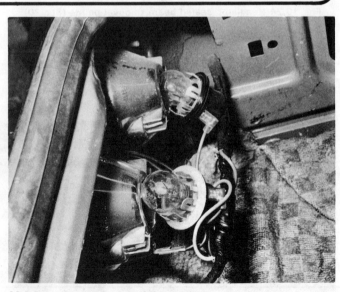

22.9 Rear lamp cluster bulb holders

Chapter 10 Electrical system

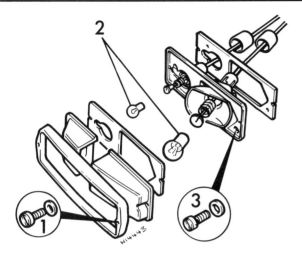

Fig. 10.30 Front parking/flasher lamp components (Sec 23)

1. Rim, lenses, gasket and screws
2. Bulbs
3. Lamp base screws

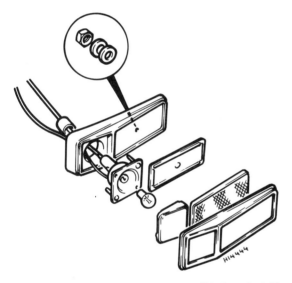

Fig. 10.31 Rear marker lamp components (N. America) (Sec 23)

Fig. 10.32 Rear lamp components (later models) (Sec 23)

Later models
7 Open the luggage boot.
8 Extract the screws and remove the lamp protective cover (photo).
9 Pull out the bulb holders and remove the bayonet fitting type bulbs (photo).
10 Renew the bulb with one of similar type and wattage. When refitting the bulb holder make sure that the wider spring tag enters the appropriate groove.

Rear number plate lamp (except N. America) – Mk I, II and III
11 Unscrew the lamp cover and remove it, taking care not to drop the lens.
12 Remove the bulb from its holder.

Later models
13 Disconnect the lamp leads from inside the luggage boot and pull the wires through the panel grommet.
14 Extract the screws and lift off the lamp cover.
15 Remove the lenses.
16 Remove the bayonet fitting type bulbs.

Rear number plate lamp (N. America)
17 Remove the two screws and withdraw the lens assembly.
18 Pull the festoon type bulb from its contacts.

Front and rear marker lamps (N. America)
19 The bulbs are accessible after prising the lamp retaining rims from the rubber lamp bases. Remove the lens and the bayonet type bulb.

23 Exterior lamps – removal and refitting

1 All lamps can be removed after disconnecting the electrical leads at the snap connectors and removing the lamp base securing nuts or screws.
2 Note the earth bonding wires on the front parking lamps (later models).
3 When refitting the lamp, check that the weathersealing gasket is in good condition.

24 Interior lamps – bulb renewal

Luggage boot lamp
1 The lamp bulb is of festoon type and removable after extracting the lamp fixing screws from the trim panel (photo).

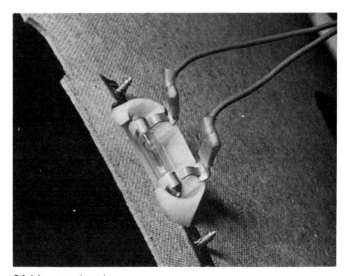

24.1 Luggage boot lamp

Warning, indicator and instrument panel lamps

2 These bulbs can be pulled from their holders once the fascia centre section has been lowered as described in Chapter 12, Section 37.

Footwell illuminating lamp

3 To remove the bulb simply unscrew it from its holder.

25 Tachometer – removal and refitting

1 Lower the fascia centre panel as described in Chapter 12, Section 37.

Mk I, II, III and IV

2 Lower the fascia centre panel as described in Chapter 12, Section 37.
3 Unscrew the knurled nuts, spring washers and remove the tachometer retaining clamp legs.
4 Pull out the bulb holder.
5 Unscrew the knurled ring which holds the drive cable to the rear of the tachometer.
6 Withdraw the tachometer from the panel.

1500 models

7 The tachometer on these cars is electrically driven. Operations are similar to those described in the preceding paragraphs, but of course there is no drive cable to disconnect.
8 Refitting is a reversal of removal.

26 Speedometer – removal and refitting

1 Lower the fascia centre panel as described in Chapter 12, Section 37.
2 Remove the tahometer as described in the preceding Section of this Chapter.
3 Depress the catch A and pull the drive cable from the speedometer head (Fig. 10.37).
4 Unscrew the trip reset knurled knob retaining nut from the lower edge of the fascia panel (photo).
5 Remove the clamp legs from the instrument, detach the earth wire and remove the speedometer.
6 Refitting is a reversal of removal.

27 Speedometer inner cable – renewal

1 Lower the fascia panel as described in Chapter 12, Section 37.
2 Remove the tachometer as previously described to give access to the rear of the speedometer.
3 Release the speedometer cable from the rear of the speedometer head.

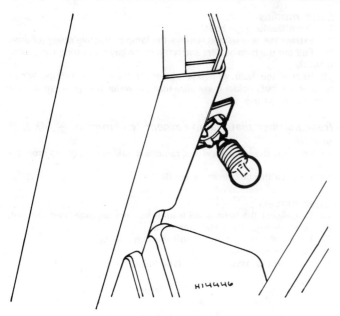

Fig. 10.33 Footwell lamp bulb (Sec 24)

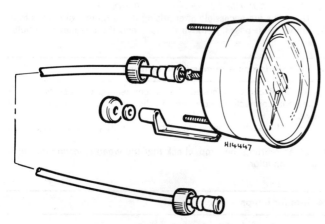

Fig. 10.34 Tachometer (cable driven type) (Sec 25)

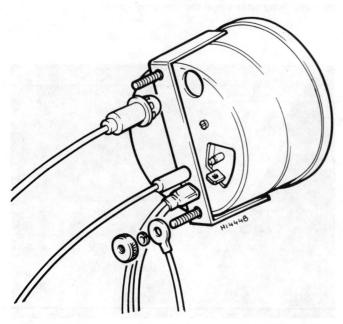

Fig. 10.35 Tachometer (electrically-operated type) (Sec 25)

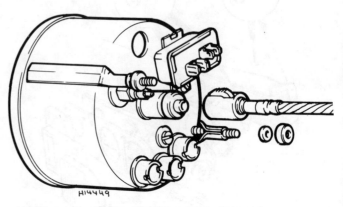

Fig. 10.36 Speedometer (Sec 26)

Chapter 10 Electrical system

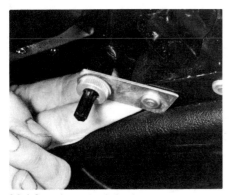

26.4 Speedometer trip knob

27.4 Speedometer inner and outer cables

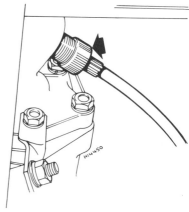

Fig. 10.37 Speedometer cable connection at gearbox (early models) (Sec 28)

Knurled ring – arrowed

4 Using a pair of long-nosed pliers withdraw the inner cable. If the cable has broken, the lower end of the outer cable will have to be disconnected from the gearbox and the lower section of inner cable withdrawn downwards (photo).
5 Apply multi-purpose grease to the new cable, but do not grease the upper few inches near the speedometer head.
6 Insert the cable using a twisting motion until the squared lower end is felt to engage in the speedometer pinion gear. Now withdraw the inner cable so that it projects by about 1.0 in (25.4 mm) from the outer cable.
7 Connect the cables to the speedometer and then refit all other removed items.

28 Speedometer cable (complete) – removal and refitting

Non-overdrive models
1 Disconnect the speedometer drive cable from the rear of the speedometer head as described in the preceding Section.
2 Working under the car, disconnect the cable from the gearbox. On early models the cable connection takes the form of a knurled ring while on later cars a forked clamp plate is used.
3 Withdraw the cable assembly through the body sealing grommet noting the cable routing.

Overdrive models
4 Disconnect the speedometer drive cable from the rear of the speedometer head as described in Section 27.
5 Remove the gearbox tunnel cover as described in Chapter 6.
6 Release the cable from the gearbox extension by unscrewing the knurled ring.
7 Withdraw the cable assembly through the body sealing grommet noting the cable routing.
8 Refitting is a reversal of removal.

29 Speedometer cable (two section – N. America) – removal and refitting

1 On early N. American models, the speedometer drive cable is in two sections.

Upper section
2 Remove the fascia centre section as described in Chapter 12, Section 37.
3 Remove the tachometer (see Section 25 of this Chapter).
4 Disconnect the speedometer drive cable from the speedometer.
5 Disconnect the speedometer cable by unscrewing the knurled nut from the extension above the EGR service interval indicator.

Lower section
6 Working under the car, disconnect the cable from the gearbox by removing the forked clamp plate.

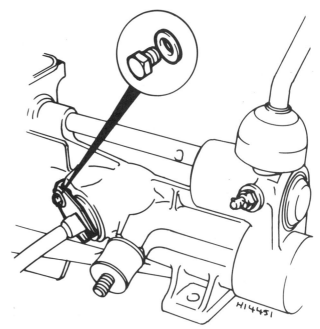

Fig. 10.38 Speedometer cable connection at gearbox (later models) (Sec 28)

Clamp plate and bolt – arrowed

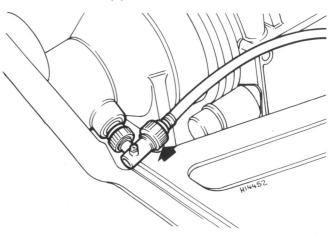

Fig. 10.39 Speedometer cable connection (overdrive models) (Sec 28)

Knurled ring – arrowed

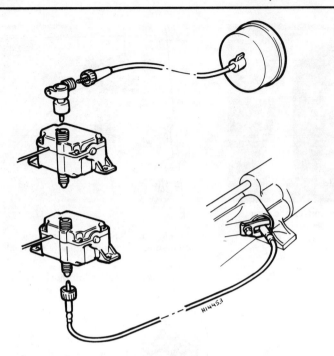

Fig. 10.40 Speedometer cables (two section – early N. American models) (Sec 29)

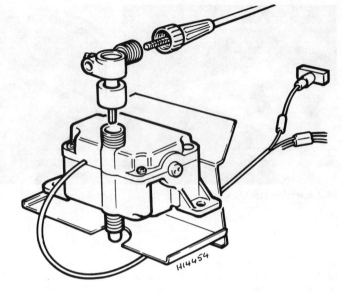

Fig. 10.41 Service interval indicator (EGR) on early N. American models (Sec 31)

7 Unscrew the knurled nut from the upper end of the cable.
8 Withdraw both cable sections noting their routing.
9 Refitting is a reversal of removal.

30 Tachometer drive cable – removal and refitting

On earlier models (up to Mk IV) with a tachometer cable-driven from the distributor, removal and refitting of the inner cable or complete cable assembly is very similar to the operations described for the speedometer drive cable in Sections 27 and 28.

31 Service interval indicator (early N. American models) – removal and refitting

EGR system
1 Lower the fascia centre panel as described in Chapter 12, Section 37.
2 Disconnect the two electrical leads (one green, one light green/white).
3 Disconnect the speedometer cable from the angled extension on top of the indicator housing.
4 Except on those vehicles equipped with a catalyst service interval indicator, release the speedometer cable from below the indicator housing.
5 Unbolt the indicator from the bulkhead and remove it, carefully easing the electrical leads through the grommet.

Catalytic converter
6 On Californian cars equipped with this service interval indicator, carry out the operations described in paragraphs 1 and 2 of this Section.
7 Unbolt the EGR indicator housing from the bulkhead and raise it carefully. Now remove the catalyst service interval indicator driveshaft and rubber sleeve.
8 Release the speedometer cable from below the catalyst indicator housing.
9 Unbolt the catalyst indicator housing from the bulkhead and remove it.
10 Refitting and reconnection are reversals of disconnection and removal.

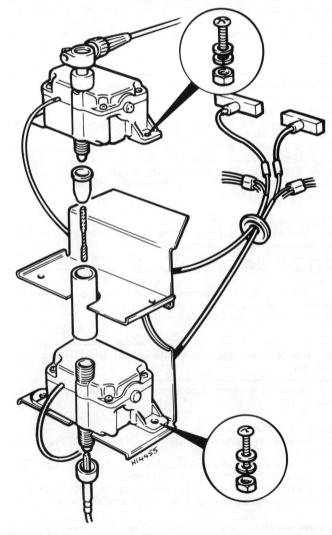

Fig. 10.42 Service interval indicator (catalytic converter) on early N. American models (Sec 31)

Chapter 10 Electrical system

32 Temperature and fuel gauges – removal and refitting

1 Lower the fascia centre panel as described in Chapter 12, Section 37.
2 Disconnect the bulb holders from the rear of the gauge by pulling them out.
3 Pull off the electrical leads.
4 Unscrew the knurled nut which secures the gauge fixing clamp.
5 Refitting is a reversal of removal.

33 Instrument voltage stabiliser – removal and refitting

1 This device is attached to the rear of the speedometer. An indication that the stabiliser has developed a fault will occur if both the coolant temperature and fuel contents gauges show incorrect readings at the same time (photo).
2 Access to the voltage stabiliser is obtained by removing the speedometer.

33.1 Instrument voltage regulator

34 Windscreen wiper blades – removal and refitting

1 The wiper blades should be renewed as soon as they become less effective.
2 Pull the wiper arm away from the glass until it locks.
3 Using the thumbnail, depress the small tag on the blade to arm connector to release the locking 'pip' and slide the blade from the arm (photo).
4 Rubber inserts or complete blades are available for replacement purposes.
5 Fit the blade by reversing the removal operations.

35 Windscreen wiper arms – removal and refitting

1 With the wipers in the parked position, mark the position of the arms in relation to the windscreen glass. Strips of masking tape stuck to the glass on either side of the blade are useful for this.
2 Pull the wiper arm from the splined spindle. If it is tight, prise it off by inserting a screwdriver between the wheelbox nut and the arm (photo).
3 Grease the splines before refitting. Do not push the arm fully home on the splines before checking the alignment against the screen.

34.3 Wiper blade connection to arm

36 Wiper motor and rack (single speed) – removal, overhaul and refitting

1 The wiper motor is always removed in unit with the flexible inner cable rack. First pull off the wiper arms and blades.
2 Unscrew the large nut which holds the rigid tubing to the wiper motor gearbox.
3 Pull off the three wires from their Lucar connectors on the end cover.
4 Undo the nuts and washers from the three bolts which hold the wiper motor gearbox in place and take off the motor complete with the inner cable rack. On some models it will be necessary to undo the bolts which hold the motor mounting bracket to the dash panel before access can be gained to the nuts and washers which hold the motor to the mounting bracket.
5 Mark the domed cover in relation to the flat gearbox lid, undo the four screws holding the gearbox lid in place and lift off the lid and domed cover.
6 Pull off the small circlip, and remove the limit switch wiper. The connecting rod and cable rack can now be lifted off. Take particular note of the spacer located between the final drive wheel and the connecting rod.
7 Undo and remove the two through bolts from the commutator end cover. Pull off the end cover.
8 Lift out the brushgear retainer and then remove the brushgear. Clean the commutator and brush gear and if worn fit new brushes. The resistance between adjacent commutator segments should be 0.34 to 0.41 ohm.

35.2 Removing wiper arm

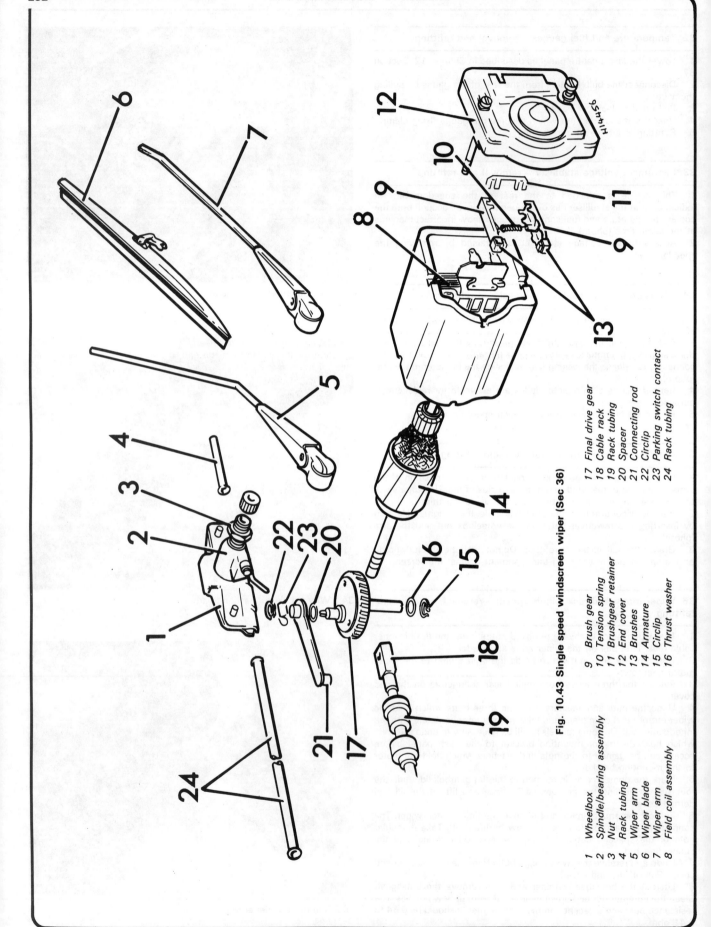

Fig. 10.43 Single speed windscreen wiper (Sec 36)

1 Wheelbox
2 Spindle/bearing assembly
3 Nut
4 Rack tubing
5 Wiper arm
6 Wiper blade
7 Wiper arm
8 Field coil assembly
9 Brush gear
10 Tension spring
11 Brushgear retainer
12 End cover
13 Brushes
14 Armature
15 Circlip
16 Thrust washer
17 Final drive gear
18 Cable rack
19 Rack tubing
20 Spacer
21 Connecting rod
22 Circlip
23 Parking switch contact
24 Rack tubing

Chapter 10 Electrical system

9 Carefully examine the internal wiring for signs of chafing, breaks or charring which would lead to a short circuit. Insulate or replace any damaged wiring.
10 Measure the value of the field resistance which should be between 12.8 to 14 ohms. If a lower reading than this is obtained it is likely that there is a short circuit and a new field coil should be fitted.
11 Renew the gearbox gear teeth if they are damaged, chipped or worn.
12 Reassembly is a straightforward reversal of the dismantling sequence, but ensure the following items are lubricated:

  (a) *Immerse the self aligning armature bearing in SAE 20 engine oil for 24 hours before assembly.*
  (b) *Oil the armature bearings in SAE 20 engine oil.*
  (c) *Soak the felt lubricator in the gearbox with SAE 20 engine oil.*
  (d) *Grease generously the worm wheel bearings, cross head, guide channel, connecting rod, crankpin, worm, cable rack and wheelboxes and the final gear shaft.*

13 Replacement is a straightforward reversal of the removal sequence. Lubricate the flexible inner cable with multi-purpose grease or similar to ensure smooth functioning in the rigid tubing.

37 Wiper motor (two speed) – removal, overhaul and refitting

1 Disconnect the battery and disconnect the electrical leads from the wiper motor at the wiring connector (photo).
2 Unscrew the union nut which holds the tubing to the wiper gearbox.
3 Remove the motor mounting strap and disconnect the earth lead.
4 Remove the gearbox cover (four screws).
5 Extract the crankpin spring clip and remove the thrust washer.

37.1 Two speed wiper motor

6 Remove the connecting rod and washer.
7 Detach the rack assembly and lift the wiper motor assembly from the engine compartment bulkhead.
8 Remove the circlip from the end of the final drive gearshaft and take off the thrust washer.
9 Withdraw the shaft and dished washer.
10 Remove the thrust screw.
11 Unscrew and remove the tie bolts.
12 Withdraw the cover and armature a small distance and then

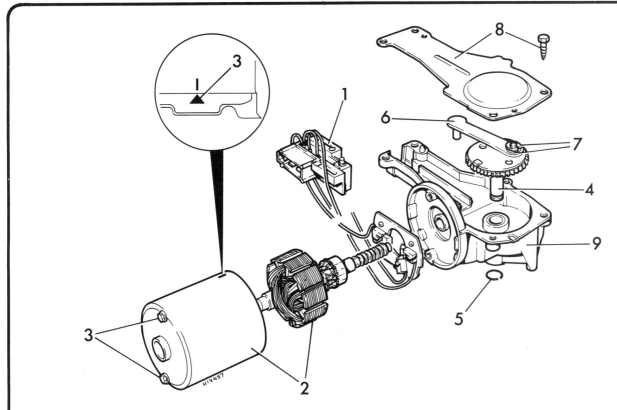

Fig. 10.44 Main components of two-speed wiper (Sec 37)

1 Limit switch
2 Armature and cover
3 Tie bolts (inset – alignment marks)
4 Final drive gear and shaft
5 Circlip
6 Connecting rod
7 Thrust washer and retaining clip
8 Gearbox cover and screw
9 Gear case

withdraw a little more until the brushes drop clear of the commutator. Do not allow them to become contaminated with grease.
13 Pull the armature from the cover. Magnetic resistance will be felt as this is done.
14 Remove the brush assembly (three screws).
15 Lift the limit switch up and slide it out sideways to release the spring clip. The limit switch and brush assembly are joined by wires.
16 Renew parts as necessary and commence reassembly by fitting the brush assembly and the limit switch.
17 Lubricate the cover bearing and saturate the cover bearing felt washer with engine oil.
18 Locate the armature in the cover.
19 Lubricate the self-aligning bearing with engine oil and insert the armature shaft through the bearing. Hold the three brushes back to clear the commutator.
20 Seat the cover against the gearbox and then turn the cover until the triangular and dash marks are in alignment. Fit the tie bolts.
21 Fit the thrust screw. If the screw is of the non-adjustable type, check the armature endfloat in the following way. Place a feeler gauge between the end of the armature shaft and the thrust screw. Push the armature towards the cover. The end-float should be as specified. Adjustment may be made by adding shims under the thrust screw head or by reducing the thickness of the thrust screw head by taking metal from the underside.
22 If an adjustable type thrust screw is used, then adjust the end-float by releasing the screw locknut and turning the screw in until resistance is felt. Now unscrew it one quarter of a turn and lock it.
23 Apply oil to the final drive gear bushes, and grease to the cam. The dished washer should be fitted so that the concave side is towards the final drive gear.
24 Pack grease around the gears and into the cross-head guide channel and oil the final drive gear crankpin.
25 Insert the connecting rod, washer, spring clip and fit the gearbox cover.

38.2 Wiper wheelbox mounting plate

38 Wheelboxes – removal and refitting

Left-hand wiper wheelbox
1 Remove the windscreen wiper motor and rack as previously described.
2 Unscrew the two nuts from the wheelbox mounting plate (photo).
3 Pull the rack tubing ends aside.
4 Working outside the car below the windscreen, remove the wiper arm spindle cover nut, spacer and seal.
5 Withdraw the wheelbox.

Right-hand wiper wheelbox
6 Disconnect the battery.
7 Remove the windscreen wiper motor and rack as previously described.
8 Remove the parcels shelf from the passenger side (refer to Chapter 12).
9 Unclip the windscreen demister hose from the heater casing on the driver's side.
10 Unscrew the two nuts from the wheelbox mounting plate.
11 Pull the rack tubing ends aside.
12 Remove the wiper arm spindle cover nut, spacer and seal and then withdraw the wheelbox.
13 Refitting is a reversal of removal.

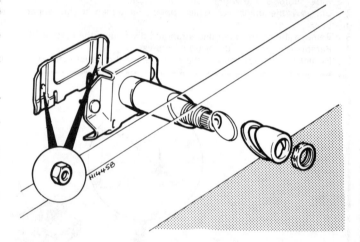

Fig. 10.45 Wiper arm wheelbox components (Sec 38)

39 Windscreen wiper rack tubing – removal and refitting

This can be withdrawn from behind the fascia panel after the wiper motor and rack assemblies have been withdrawn and the nuts released from the wheelbox mounting plates as described in Sections 36, 37 or 28 as applicable.

40 Windscreen washer system

1 On earlier models, the electric washer pump was combined with the fascia-mounted wiper switch (see Section 16).
2 On later models with a steering column stalk switch, the washer

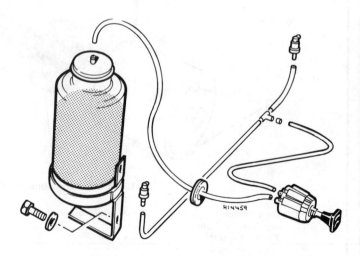

Fig. 10.46 Early type windscreen washer assembly (Sec 40)

Chapter 10 Electrical system

40.2 Washer pump

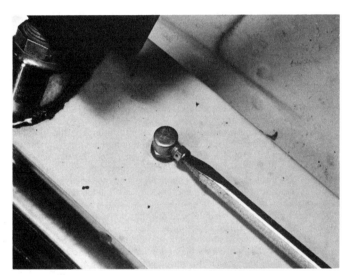

40.4 Adjusting a washer jet nozzle

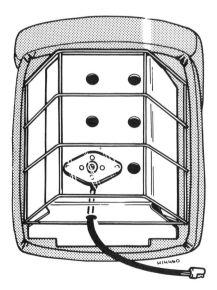

Fig. 10.47 Weight sensitive seat belt switch (Sec 41)

pump is mounted on the engine compartment rear bulkhead near the washer fluid reservoir (photo).
3 Keep the reservoir well filled with clean water to which a proprietary cleaning fluid has been added. In winter, proprietary windscreen washer antifreeze should be used. Never add cooling system antifreeze or damage to the paintwork will result.
4 The washer jet nozzles may be adjusted with a screwdriver to obtain the most satisfactory wash pattern on the windscreen glass (photo).

41 Seat belt warning system (N. America)

1 This system fitted to later models, gives audible and visual warnings when the ignition is switched on, but the seat belts have not been coupled.
2 The main components of the system consist of the following:

(a) Weight sensitive seat switch
(b) Seat buckle switch
(c) Warning lamp
(d) Warning buzzer

42 Key warning system (N. American models)

1 This consists of a buzzer designed to give an audibe warning if the driver's door is opened without first having removed the ignition key.
2 The buzzer is in fact the same unit as used for the seat belt warning system.

43 Horns

1 The horns consist of a high note and low note unit.
2 The horns require no attention other than keeping the terminals free from corrosion and the leads securely connected.

44 Radio – installation

1 A position for installation of a radio /cassette player is provided just below the fascia centre panel.
2 Most equipment is of standard size and will fit the space allocated once the blanking panel has been removed.
3 Some radios will have mounting brackets provided together with instructions; others will need to be fitted using drilled and slotted metal strips, bent to form mounting brackets – these strips are available from most accessory stores. The unit must be properly earthed, by fitting a separate earth lead between the casing of the radio and the vehicle frame.
4 Use the radio manufacturer's instructions when wiring the radio into the vehicle's electrical system. If no instructions are available, refer to the relevant wiring diagram to find the 'feed' connection in the vehicle's wiring circuit. A 1-2 amp 'in-line' fuse must be fitted in the radio's 'feed' wire – a choke may also be necessary (see next Section).
5 The type of aerial used, and its fitted position is a matter of personal preference. In general the taller the aerial, the better the reception. It is best to fit a fully retractable aerial – espeically, if a mechanical car-wash is used or if you live in an area where cars tend to be vandalised. In this respect electric aerials which are raised and lowered automatically when switching the radio on or off, are convenient, but are more likely to give trouble than the manual type.
6 When choosing a site for the aerial the following points should be considered:

(a) The aerial lead should be as short as possible bearing in mind that the aerial can only be mounted on a rear wing.
(b) The aerial must be mounted as far away from the distributor and HT leads as possible.
(c) The part of the aerial which protrudes beneath the mounting point must not foul the roadwheels, or anything else.
(d) If possible the aerial should be positioned so that the coaxial lead does not have to be routed through the engine compartment.

(e) The plane of the panel on which the aerial is mounted should not be so steeply angled that the aerial cannot be mounted vertically (in relation to the 'end-on' aspect of the vehicle). Most aerials have a small amount of adjustment available.

7 Having decided on a mounting position, a relatively large hole will have to be made in the panel. The exact size of the hole will depend upon the specific aerial being fitted although, generally, the hole required is of approximately ¾ inch diameter. On metal bodied cars, a 'tank cutter' of the relevant diameter is the best tool to be used for making the hole. This tool needs a small diameter pilot hole drilled through the panel, through which the tool clamping bolt is inserted. When the hole has been made, the raw edges should be de-burred with a file and then painted, to prevent corrosion.

8 Fit the aerial according to the manufacturer's instructions. If the aerial is very tall, or if it protrudes beneath the mounting panel for a considerable distance, it is a good idea to fit a stay between the aerial and the vehicle frame. This stay can be manufactured from the slotted and drilled metal strips previously mentioned. The stay should be securely screwed or bolted in place. For best reception it is advisable to fit an earth lead between the aerial body and the vehicle frame.

9 It will probably be necessary to drill one or two holes through bodyworn panels in order to feed the aerial lead into the interior of the vehicle. Where this is the case ensure that the holes are fitted with rubber grommets to protect the cable, and to stop possible entry of water.

10 Due to the restricted amount of space within the car, the speaker will have to be mounted in the fascia centre support trim panel or on the parcels shelf or within the doors.

11 Take great care not to damage the speaker dipahragm whilst doing this. It is a good idea to fit a gasket between the speaker frame and the mounting panel in order to prevent vibration – some speakers will already have such a gasket fitted.

12 To fit recessed-type speakers in the front doors first check that there is sufficient room to mount a speaker in each door without it fouling the latch or window winding mechanism. Hold the speaker against the skin of the door, and draw a line around the periphery of the speaker. With the speaker removed draw a second cutting-line within the first, to allow enough room for the entry of the speaker back but at the same time providing a broad seat for the speaker flange. When you are sure that the cutting-line is correct, drill a series of holes around its periphery. Pass a hacksaw blade through one of the holes and then cut through the metal between the holes until the centre section of the panel falls out.

13 De-burr the edges of the hole and then paint the raw metal to prevent corrosion. Cut a corresponding hole in the door trim panel, ensuring that it will be completely covered by the speaker grille. Now drill a hole in the door edge and a corresponding hole in the door surround. These holes are to feed the speaker leads through, to fit grommets. Pass the speaker leads through the door trim, door skin and out through the holes in the side of the door and door surround. Refit the door trim panel and then secure the speaker to the door using self-tapping screws. *Note: If the speaker is fitted with a shield to prevent water dripping on it, ensure that this shield is at the top.*

14 By now you will have several yards of additional wiring in the car; use PVC tape to secure this wiring out of harm's way. Do not leave electrical leads dangling. Ensure that all new electrical connections are properly made (wires twisted together will not do) and completely secure. The radio should now be working, but before you pack away your tools it will be necessary to 'trim' the radio to the aerial. Follow the radio manufacturer's instructions regarding this adjustment.

Cassette players

15 Fitting instructions for cassette players are generally similar to those for fitting a radio. Tape players are not usually prone to electrical interference like radios – although it can occur – so positioning is not so critical. If possible the player should be mounted on an 'even keel'. Also, it must be possible for a driver wearing a seatbelt to reach the unit in order to change, or turn over, tapes.

45 Radios and tape players – suppression of interference (general)

To eliminate buzzes and other unwanted noises costs very little and is not as difficult as sometimes thought. With a modicum of common sense and patience, and following the instructions in the following paragraphs, interference can be virtually eliminated.

The first cause for concern is the generator. The noise this makes over the radio is like an electric mixer and the noise speeds up when the engine is revved. (To prove the point, remove the fanbelt and try it). The remedy for this is simple; connect a $1.0\,\mu F - 3.0\,\mu F$ capacitor between earth (probably the bolt that holds down the generator base) and the *large* (B+) terminal on the alternator. This is most important, for if you connect it to the small terminal, you will probably damage the alternator permanently.

A second common cause of electrical interference is the ignition system. Here a $1.0\,\mu F$ capacitor must be connected between earth and the 'SW' or '+' terminal on the coil. This may stop the tick-tick-tick sound that comes over the speaker. Next comes the spark itself. The ignition HT leads are of suppressed type and no further action is required. Do not fit plug suppressor caps or cut the leads to fit in-line suppressors.

At this stage it is advisable to check that the radio is well earthed, also the aerial, and to see that the aerial plug is pushed well into the set and that the radio is properly trimmed (see preceding Section). In addition, check that the wire which supplies the power to the set is as short as possible and does not wander all over the car. It is a good idea to check that the fuse is of the correct rating. For most sets this will be about 1 to 2 amps.

At this point, the more usual causes of interference have been suppressed. If the problem still exists, a look at the cause of

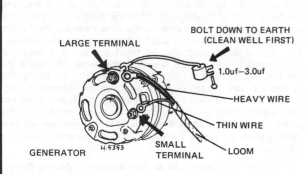

Fig. 10.48 Correct way to connect a suppressor to the generator; alternator shown (Sec 45)

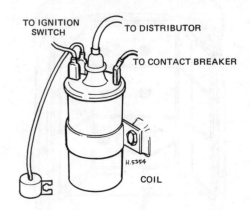

Fig. 10.49 Correct way to connect a suppressor to the ignition coil (Sec 45)

Chapter 10 Electrical system

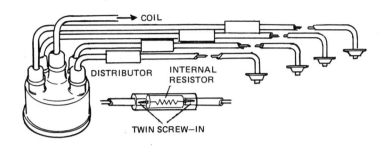

Fig. 10.50 Ignition HT lead suppressors (Sec 45)

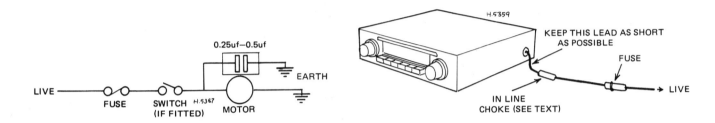

Fig. 10.51 Correct method of suppressing electric motors (Sec 45)

Fig. 10.52 Location of in-line choke on radio feed wire (Sec 45)

interference may help to pinpoint the component generating the stray electrical discharges.

The radio picks up electromagnetic waves in the air. Some are made by regular broadcasters and some, which we do not want, are made by the car itself. The home-made signals are produced by stray electrical discharges floating around in the car. Common producers of these signals are electrical motors, ie the windscreen wipers, electric screen washers, electric window winders, heater fan or an electric aerial if fitted. Other sources of interference are electric fuel pumps, flashing turn signals and instruments. Turn signals are not normally suppressed. In recent years, radio manufacturers have included in the line (live) of the radio, in addition to the fuse, an 'in-line' choke. If your installation lacks one of these, put one in (Fig. 10.52).

All the foregoing components are available from radio shops or accessory shops. For a transistor radio, a 2A choke should be adequate. If you have an electric clock fitted, this should be suppressed by connecting a 0.5 μF capacitor directly across it as shown for a motor in Fig. 10.51.

If after all this you are still experiencing radio interference, first assess how bad it is, for the human ear can filter out unobtrusive unwanted noises quite easily. But if you are still adamant about eradicating the noise, then continue.

As a first step, a few experts seem to favour a screen between the radio and the engine. This is OK as far as it goes, literally! The whole set is screened anyway and if interference can get past that then a small piece of aluminium is not going to stop it.

A more sensible way of screening is to discover if interference is coming down the line. First, take the live lead; interference can get between the set and the choke (hence the reason for keeping the wires short). One remedy here is to screen the wire and this is done by buying screened wire and fitting that. The loudspeaker lead could be screened also to prevent pick-up getting back to the radio although this is unlikely.

Without doubt, the worst source of radio interference comes from the ignition HT leads, even if they have been suppressed. The ideal way of suppressing these is to slide screening tubes over the leads themselves. As this is impractical, we can place an aluminium shield over the majority of the lead areas.

Now for the really difficult cases, here are a few tips to try out. Where metal comes into contact with metal, an electrical disturbance is caused which is why good clean connections are essential. To remove interference due to overlapping or butting panels, you must bridge the join with a wide braided earth strap (like that from the frame to the engine/transmission). The most common moving parts that could create noise and should be strapped are, in order of importance:

(a) Silencer-to-frame
(b) Exhaust pipe-to-engine block and frame
(c) Air cleaner-to-frame
(d) Front and rear bumpers-to-frame
(e) Steering column-to-frame
(f) Bonnet and boot lids-to-frame

These faults are most pronounced when the engine is idling or labouring under load. Although the moving parts are already connected with nuts, bolts, etc, these do tent to rust and corrode, this creating a high resistance interference source.

If you have a 'ragged' sounding pulse when mobile, this could be wheel or tyre static. This can be cured by buying some anti-static powder and sprinkling inside the tyres.

If the interference takes the shape of a high pitched screeching noise that changes its note when the car is in motion and only comes now and then, this could be related to the aerial, especially if it is of the telescopc or whip type. This source can be cured quite simply by pushing a small rubber ball on top of the aerial (yes really!) as this breaks the electric field before it can form; but it would be much better to buy yourself a new aerial of a reputable brand. If, on the other hand, you are getting a loud rushing sound every time you brake, then this is brake static. This effect is most prominent on hot dry days and is cured only by fitting a special kit, which is quite expensive.

In conclusion, it is pointed out that it is relatively easy, and therefore cheap, to eliminate 95 per cent of all noise, but to eliminate the final 5 per cent is time and money consuming. It is up to the individual to decide it if is worth it. Please remember also, that you cannot get a concert hall performance out of a cheap radio.

Finally, at the beginning of this Section are mentioned tape players; these are not usually affected by interference but in a very bad case, the best remedies are the first three suggestions plus using a 3 – 5 amp choke in the 'live' line, and in difficult cases screen the live and speaker wires.

Note: *If your car is fitted with electronic ignition, then it is not recommended that either the spark plug resistors or the ignition coil capacitor be fitted as these may damage the system. Most electronic ignition units have built in suppression and should, therefore, not cause interference.*

46 Fault diagnosis – electrical system

Symptom	Reason(s)
Starter motor fails to turn engine	
No electricity at starter motor	Battery discharged
	Battery defective internally
	Battery terminal leads loose or earth lead not securely attached to body
	Loose or broken connections in starter motor circuit
	Starter motor switch or solenoid faulty
Electricity at starter motor; faulty motor	Starter motor pinion jammed in mesh with flywheel ring gear
	Starter brushes badly worn, sticking, or brush wires loose
	Commutator dirty, worn or burnt
	Starter motor armature faulty
	Field coils earthed
Starter motor turns engine very slowly	
Electrical defects	Battery in discharged condition
	Starter brushes badly worn, sticking or, brush wires loose
	Loose wires in starter motor circuit
Starter motor operates without turning engine	
Dirt or oil on drive gear	Starter motor pinion sticking on the screwed sleeve
Mechanical damage	Pinion or flywheel gear teeth broken or worn
Starter motor noisy or excessively rough engagement	
Lack of attention or mechanical damage	Pinion or flywheel gear teeth broken or worn
	Starter drive main spring broken
	Starter motor retaining bolts loose
Battery will not hold charge for more than a few days	
Wear or damage	Battery defective internally
	Electrolyte level too low or electrolyte too weak due to leakage
	Plate separators no longer fully effective
	Battery plates severely sulphated
Insufficient current flow to keep battery charged	Dynamo/alternator belt slipping
	Battery terminal connections loose or corroded
	Dynamo or alternator not charging properly
	Short in lighting circuit causing continual battery drain
	Regulator unit not working correctly
Ignition light fails to go out, battery runs flat in a few days	
Generator not charging	Fan belt loose and slipping, or broken
	Brushes worn, sticking, broken, or dirty
	Brush spring weak of broken
	Commutator or slip rings dirty, worn or burnt
Ignition light fails to come on but battery remains charged	
Generator charging	Generator warning light control defective

Failure of individual electrical equipment to function correctly is dealt with alphabetically, item-by-item, under the headings listed below:

Fuel gauge	
Fuel gauge gives no reading	Faulty voltage stabiliser
	Fuel tank empty!
	Electric cable between tank sender unit and gauge earthed or loose
	Fuel gauge case not earthed
	Fuel gauge supply cable interrupted
	Fuel gauge unit broken
Fuel gauge registers full all the time	Electrical cable between tank unit and gauge broken or disconnected
	Faulty voltage stabiliser
Horn	
Horn operates all the time	Horn push either earthed or stuck down
	Horn cable to horn push earthed
Horn fails to operate	Blown fuse
	Cable or cable connection loose, broken or disconnected
	Horn has an internal fault

Chapter 10 Electrical system

Symptom	Reason(s)
Horn emits intermittent or unsatisfactory noise	Cable connections loose Horn incorrectly adjusted
Lights	
Lights do not come on	If engine not running, battery discharged Light bulb filament burnt out or bulbs broken Wire connections loose, disconnected or broken Light switch shorting or otherwise faulty
Lights come on but fade out	If engine not running, battery discharged
Lights give very poor illumination	Lamp glasses dirty Lamps badly out of adjustment Electrical wiring connections faulty
Lights work erratically – flashing on and off, especially over bumps	Lights not earthing properly Contacts in light switch faulty
Wipers	
Wiper motor fails to work	Blown fuse Wire connections loose, disconnected, or broken Brushes badly worn Armature worn or faulty Field coils faulty
Wiper motor works very slowly and takes excessive current	Commutator dirty, greasy, or burnt Drive linkage bent or unlubricated Wheelbox spindle binding or damaged Armature bearings dry or unaligned

'Wiring diagrams overleaf'

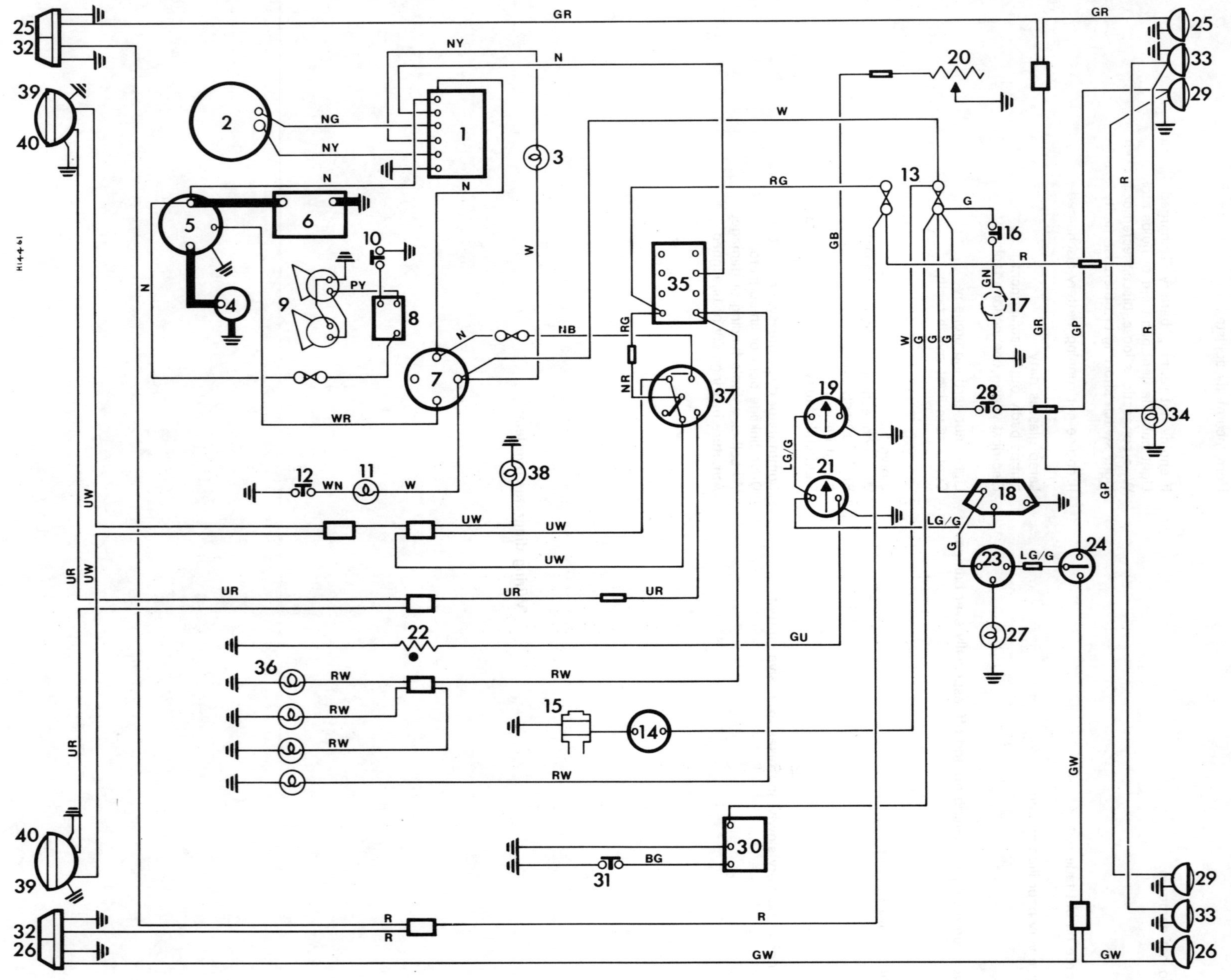

Fig. 10.53 Wiring diagram for Spitfire Mk I and II

Key to wiring diagram for Spitfire Mk I and II

1 Control box
2 Generator
3 Ignition warning lamp
4 Starter motor
5 Starter solenoid
6 Battery
7 Ignition/starter switch
8 Horn fuse
9 Horns
10 Horn push
11 Oil warning lamp
12 Oil pressure switch
13 Fuse unit
14 Ignition coil

15 Distributor
16 Heater blower switch
17 Heater blower motor*
18 Voltage stabiliser
19 Fuel gauge
20 Fuel tank unit
21 Temperature gauge
22 Temperature sender
23 Flasher unit
24 Turn signal switch
25 LH indicator lamps
26 RH indicator lamps
27 Indicator monitor

28 Brake/stop lamp switch
29 Brake/stop lamps
30 Wiper motor
31 Wiper motor switch
32 Front parking lamps
33 Rear lamps
34 Plate illumination lamps
35 Lighting switch
36 Instrument lighting
37 Steering column light switch
38 Main beam warning lamp
39 Headlamp main beams
40 Headlamp dipped beams

*Special accessory

Wire colour codes

B	Black	M	Medium	S	Slate
D	Dark	N	Brown	U	Blue
G	Green	P	Purple	W	White
K	Pink	R	Red	Y	Yellow
L	Light				

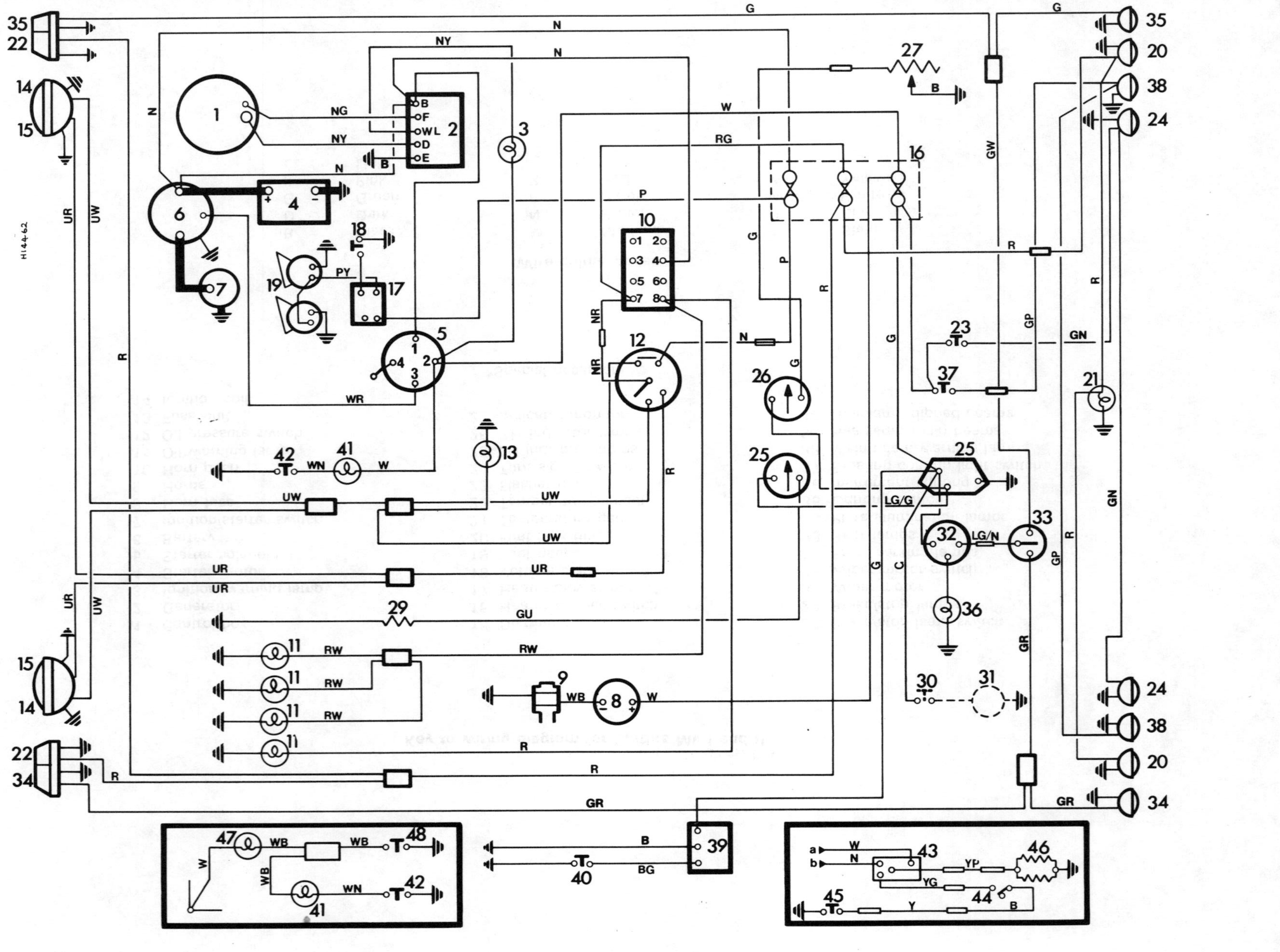

Fig. 10.54 Wiring diagram for Spitfire Mk III negative earth RHD and LHD

Key to wiring diagram for Spitfire Mk III negative earth RHD and LHD

1. Generator
2. Control box
3. Ignition warning light
4. Battery
5. Ignition/starter switch
5A. Ignition/starter switch radio supply connector
6. Starter solenoid
7. Starter motor
8. Ignition coil
9. Distributor
10. Light switch
11. Instrument illumination
12. Column light switch
13. Main beam warning light
14. Main beam
15. Dip beam
16. Fuse assembly
17. Horn relay
18. Horn push
19. Horn
20. Rear lamp
21. Plate illumination lamp
22. Front parking lamp
23. Reverse lamp switch
24. Reverse lamp
25. Voltage stabiliser
26. Fuel gauge
27. Fuel tank unit
28. Temperature gauge
29. Temperature sender
30. Heater switch (if fitted)
31. Heater motor (if fitted)
32. Flasher unit
33. Flasher switch
34. LH indicator
35. RH indicator
36. Flasher warning light
37. Stop lamp switch
38. Stop lamp
39. Wiper motor
40. Wiper switch
41. Oil pressure warning light
42. Oil pressure switch
43A. Overdrive relay (if fitted)
44A. Overdrive column switch (if fitted)
45A. Overdrive gearbox switch
46A. Overdrive solenoid
a. From ignition/starter switch – connector 2
b. From ignition/starter switch – connector 1
47. Brake line failure
48. Brake line failure switch

Wire colour codes

N	Brown	G	Green	Y	Yellow
U	Blue	LG	Light Green	S	Slate
R	Red	W	White	B	Black
P	Purple				

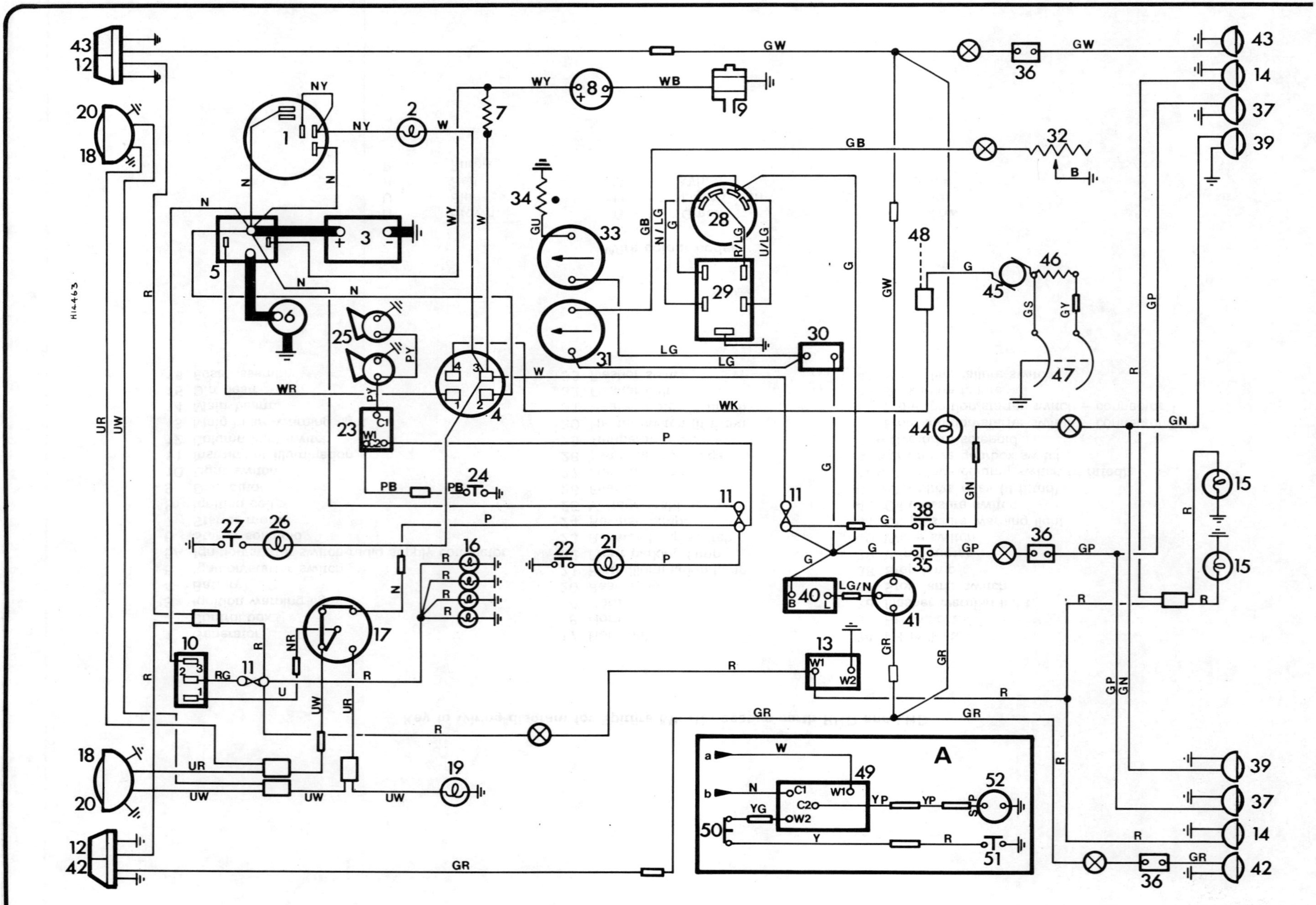

Fig. 10.55 Wiring diagram for RHD Spitfire Mk IV

Key to Wiring diagram for RHD Spitfire Mk IV

1 Alternator
2 Ignition warning light
3 Battery
4 Ignition/starter switch
5 Starter solenoid
6 Starter motor
7 Ballast resistor
8 Ignition coil (6 volt)
9 Distributor
10 Light switch
11 Fuse
12 Front parking lamp
13 Night dimming relay winding
14 Rear lamp
15 Number plate lamp
16 Instrument lamps
17 Column light switch
18 Dip beam
19 Main beam warning light
20 Main beam
21 Courtesy light
22 Door switch
23 Horn relay
24 Horn push
25 Horn
26 Oil pressure warning light
27 Oil pressure switch
28 Wiper switch
29 Wiper motor
30 Voltage stabiliser
31 Fuel gauge
32 Fuel tank
33 Temperature gauge
34 Temperature sender unit
35 Stop lamp switch
36 Night dimming relay
37 Stop lamp
38 Reverse lamp switch
39 Reverse lamp
40 Indicator flasher unit
41 Indicator switch
42 LH indicator lamp
43 RH indicator lamp
44 Indicator warning light
45 Heater motor
46 Heater rheostat
47 Heater switch
48 Radio supply
A Overdrive (optional extra)
49 Overdrive relay
50 Overdrive gearbox switch
51 Overdrive gear lever switch
52 Overdrive solenoid
a From ignition/starter switch – terminal 3
b From ignition/starter switch – terminal 2

Wire colour codes

B	Black	N	Brown	U	Blue
G	Green	O	Orange	W	White
K	Pink	P	Purple	Y	Yellow
LG	Light green	S	Slate	R	Red

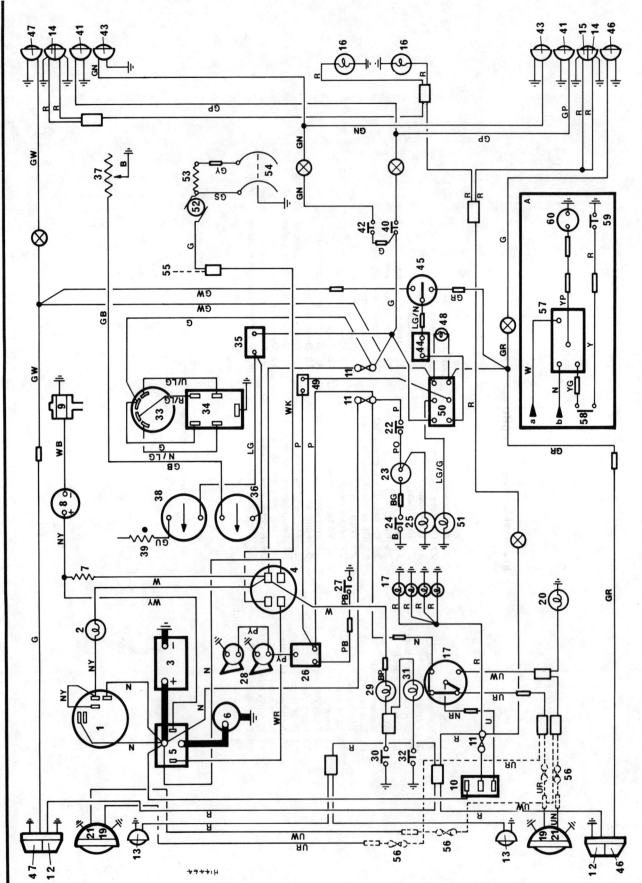

Fig. 10.56 Wiring diagram for LHD Spitfire Mk IV to 1971

Key to wiring diagram for LHD Spitfire Mk IV to 1971

1. Alternator
2. Ignition warning light
3. Battery
4. Ignition/starter switch
5. Starter solenoid
6. Starter motor
7. Ballast resistor
8. Ignition coil (6 volt)
9. Distributor
10. Light switch
11. Fuse
12. Front parking lamp
13. Front marker lamp
14. Rear marker lamp
15. Tail lamp
16. Number plate lamp
17. Instrument lamps
18. Column light switch
19. Dip beam
20. Main beam warning light
21. Main beam
22. Door switch
23. Buzzer
24. Key switch
25. Key light
26. Horn relay
27. Horn push
28. Horn
29. Brake warning light
30. Brake line failure switch
31. Oil pressure warning light
32. Oil pressure switch
33. Wiper switch
34. Wiper motor
35. Voltage stabiliser
36. Fuel gauge
37. Fuel tank unit
38. Temperature gauge
39. Temperature sender unit
40. Stop lamp switch
41. Stop lamp
42. Reverse lamp switch
43. Reverse lamp
44. Indicator flasher unit
45. Indicator switch
46. LH indicator lamp
47. RH indicator lamp
48. Indicator warning lamp
49. Hazard flasher unit
50. Hazard warning switch
51. Hazard warning light
52. Heater motor
53. Heater rheostat
54. Heater switch
55. Radio supply
56. Line fuse (25 amp) (Italy only)
A. Overdrive (optional extra)
57. Overdrive relay
58. Overdrive gearbox switch
59. Overdrive gear lever switch
60. Overdrive solenoid
a. From ignition/starter switch – terminal 3
b. From ignition/starter switch – terminal 2

Wire colour codes

B	Black	N	Brown	U	Blue
G	Green	O	Orange	W	White
K	Pink	P	Purple	Y	Yellow
LG	Light green	S	Slate	R	Red

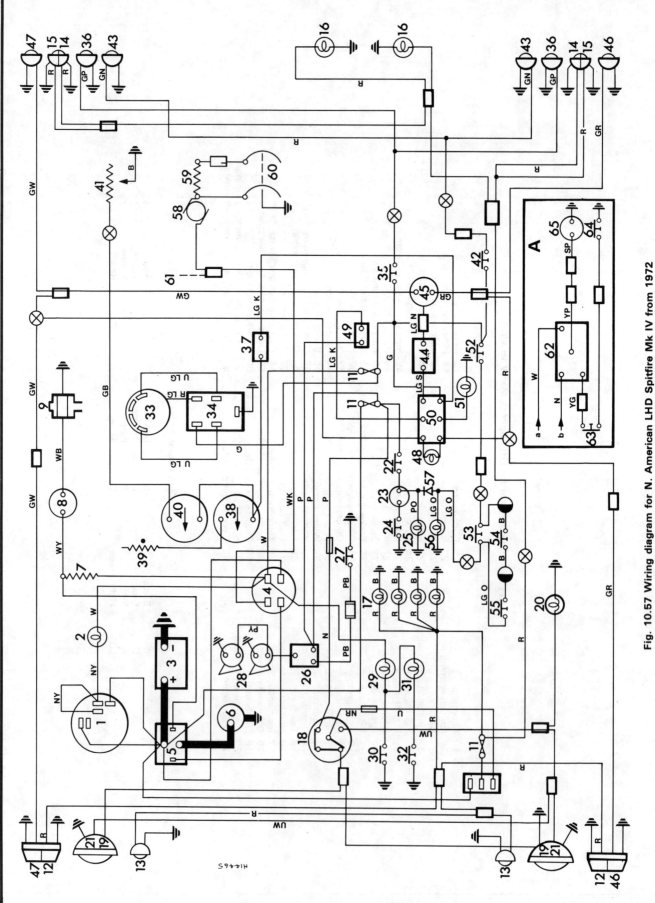

Fig. 10.57 Wiring diagram for N. American LHD Spitfire Mk IV from 1972

Key to wiring diagram for N. American LHD Spitfire Mk IV from 1972

1 Alternator
2 Ignition warning light
3 Battery
4 Ignition/starter switch
5 Starter solenoid
6 Starter motor
7 Ballast resistor
8 Ignition coil (6 volt)
9 Distributor
10 Light switch
11 Fuse
12 Front parking light
13 Front marker lamp
14 Rear marker lamp
15 Tail lamp
16 Number plate lamp
17 Instrument lamps
18 Column light switch
19 Dip beam
20 Main beam warning light
21 Main beam
22 Door switch
23 Buzzer
24 Key switch
25 Key light
26 Horn relay
27 Horn push
28 Horn
29 Brake warning light
30 Brake failure switch
31 Oil pressure warning light
32 Oil pressure switch
33 Wiper switch
34 Wiper motor
35 Stop lamp switch
36 Stop lamp
37 Voltage stabiliser
38 Temperature gauge
39 Temperature sender unit
40 Fuel gauge
41 Fuel tank unit
42 Reverse lamp switch
43 Reverse lamp
44 Indicator flasher unit
45 Indicator switch
46 LH indicator lamp
47 RH indicator lamp
48 Indicator warning lamp
49 Hazard flasher unit
50 Hazard warning switch
51 Hazard warning light
52 Belt warning gearbox switch
53 Driver's belt switch
54 Passenger's seat switch
55 Passenger belt switch
56 Belt warning light
57 Diode
58 Heater motor
59 Heater rheostat
60 Heater switch
61 Radio supply
A Overdrive (optional extra)
62 Overdrive relay
63 Overdrive gearbox switch
64 Overdrive gear lever switch
65 Overdrive solenoid

Wire colour codes

B	Black	N	Brown	U	Blue
G	Green	O	Orange	W	White
K	Pink	P	Purple	Y	Yellow
LG	Light green	S	Slate	R	Red

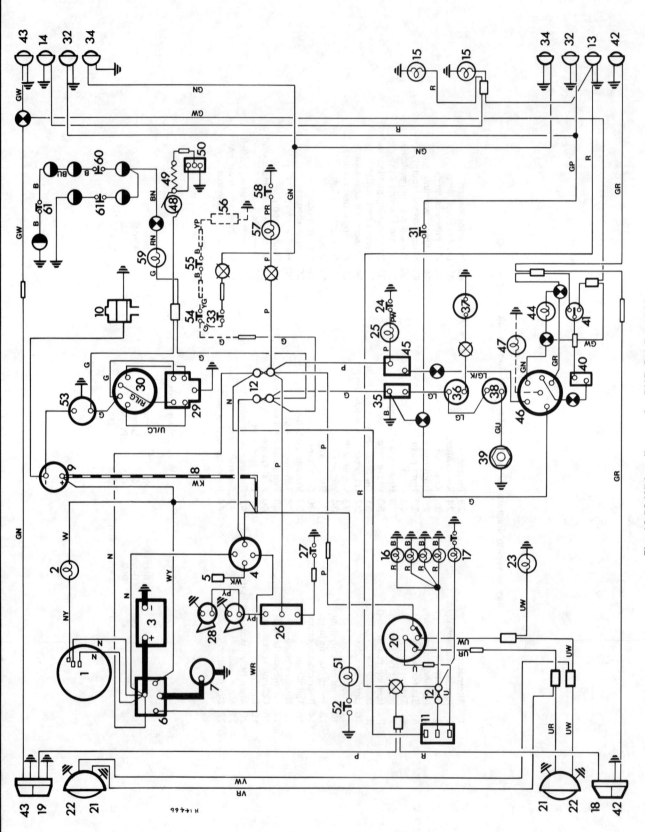

Fig. 10.58 Wiring diagram for RHD Spitfire 1500

Key to wiring diagram for RHD Spitfire 1500

1 Alternator
2 Ignition warning light
3 Battery
4 Ignition/starter switch
5 Wiper/washer switch
6 Starter solenoid
7 Starter motor
8 Ballast resistor wire
9 Coil
10 Distributor
11 Light switch
12 Fuse
13 LH tail lamp
14 RH tail lamp
15 Number plate lamp
16 Instrument lamps
17 Map light
18 LH front parking lamp
19 RH front parking lamp
20 Column light switch
21 Dip beam
22 Main beam
23 Main beam warning light
24 Door switch
25 Courtesy lamp
26 Horn relay
27 Horn switch
28 Horn
29 Wiper/washer switch
30 Wiper motor
31 Stop lamp switch
32 Stop lamp
33 Reversing lamp switch
34 Reversing lamp
35 Voltage stabiliser
36 Fuel gauge
37 Fuel tank unit
38 Temperature gauge
39 Temperature sender unit
40 Indicator flasher unit
41 Indicator switch
42 LH indicator repeater lamp
43 RH indicator repeater lamp
44 Indicator warning light
45 Hazard flasher unit
46 Hazard warning switch
47 Hazard flasher warning light
48 Heater motor
49 Heater resistor
50 Heater switch
51 Oil pressure warning light
52 Overdrive gearbox switch
53 Tachometer
54 Overdrive gearbox switch
55 Overdrive selector switch
56 Overdrive solenoid
57 Luggage compartment lamp
58 Luggage compartment switch
59 Seat belt warning light
60 Seat belt switch
61 Seat sensor switch

Wire colour codes

B	Black	N	Brown	U	Blue
G	Green	O	Orange	W	White
K	Pink	P	Purple	Y	Yellow
LG	Light green	S	Slate	R	Red

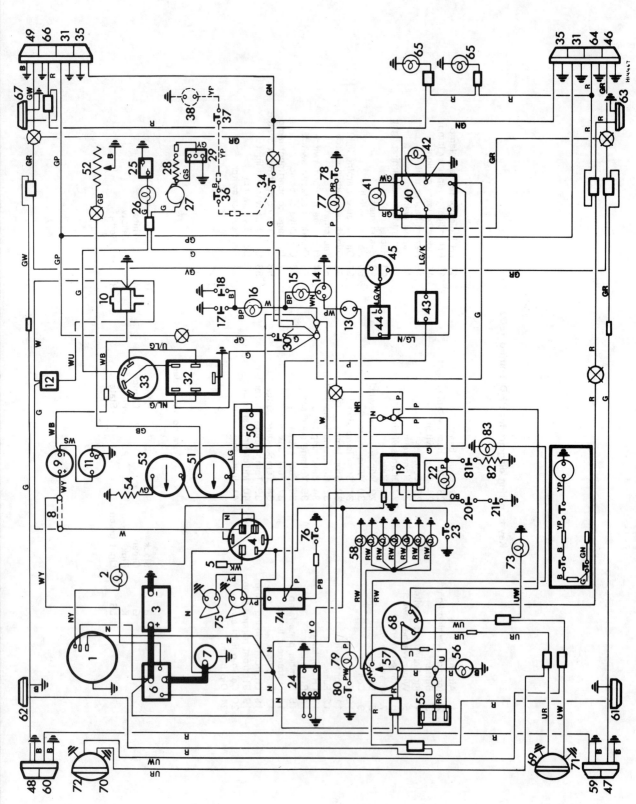

Fig. 10.59 Wiring diagram for N. American LHD Spitfire 1500 up to 1978

Key to wiring diagram for N. American LHD Spitfire 1500 up to 1978

1. Alternator
2. Ignition warning light
3. Battery
4. Ignition/starter switch
5. Radio supply
6. Starter solenoid
7. Starter motor
8. Ballast resistor wire
9. Coil (6 volt)
10. Distributor
11. Tachometer
12. Drive resistor
13. Anti-run-on valve
14. Oil pressure switch
15. Oil pressure warning light
16. Dual brake warning light
17. Dual brake switch
18. Handbrake switch
19. Seat belt timer
20. LH door switch
21. Ignition key contacts
22. Seat belt warning light
23. Buckle switch
24. Starter relay
25. Service interval instrument (SII)
26. SII warning light
27. Two-speed heater motor
28. Heater resistor
29. Two-speed heater switch
30. Stop lamp switch
31. Stop lamp
32. Two-speed heater motor
33. Two-speed wiper motor
34. Reversing lamp switch
35. Reversing lamp
36. Overdrive switch (optional)
37. Gearbox switch (optional)
38. Overdrive solenoid
39. Seat belt neutral switch
40. Hazard warning switch
41. Indicator warning light
42. Hazard warning light
43. Hazard flasher unit
44. Indicator flasher unit
45. Indicator switch
46. LH rear indicator lamp
47. LH front indicator lamp
48. RH front indicator lamp
49. RH rear indicator lamp
50. Voltage stabiliser
51. Fuel gauge
52. Tank unit
53. Temperature gauge
54. Temperature sender unit
55. Light switch
56. Map reading light
57. Panel rheostat
58. Panel illumination
59. LH front parking lamp
60. RH front parking lamp
61. LH front side marker
62. RH front side marker
63. LH rear side marker
64. LH tail lamp
65. Number plate lamp
66. RH tail lamp
67. RH rear side marker
68. Headlamp flash/dip switch
69. LH headlamp (dip)
70. RH headlamp (dip)
71. LH headlamp (main beam)
72. RH headlamp (main beam)
73. Main beam warning light
74. Horn relay
75. Horn
76. Horn switch
77. Luggage compartment lamp
78. Luggage compartment lamp switch
79. Ignition switch illumination
80. LH door switch
81. Cigarette lighter switch
82. Cigarette lighter element
83. Cigarette lighter illumination

Wire colour codes

B	Black	N	Brown	U	Blue
G	Green	O	Orange	W	White
K	Pink	P	Purple	Y	Yellow
LG	Light green	S	Slate	R	Red

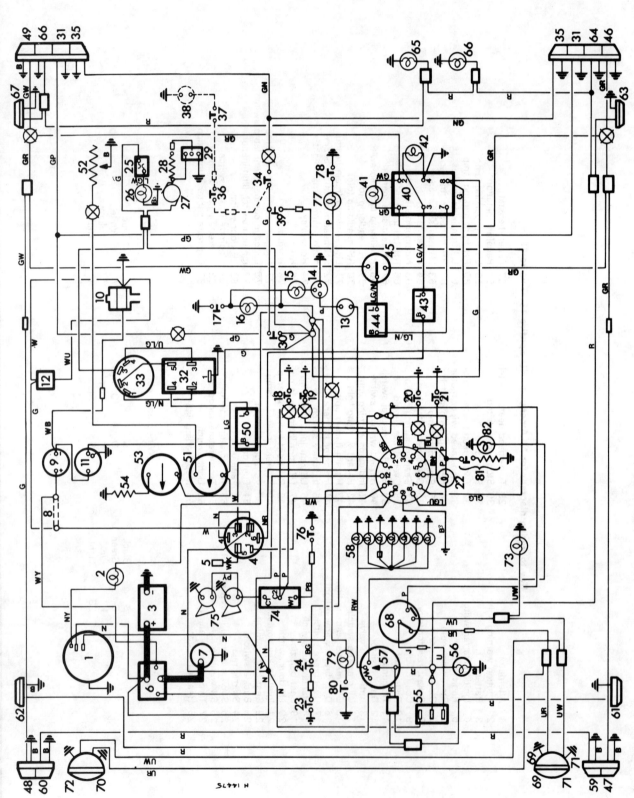

Fig. 10.60 Wiring diagram for N. American LHD Spitfire 1500

Key to wiring diagram for N. American LHD Spitfire 1500

1. Alternator
2. Ignition warning light
3. Battery
4. Ignition/starter switch
5. Radio supply
6. Starter solenoid
7. Starter motor
8. Ballast resistor wire
9. Coil (6 volt)
10. Distributor
11. Tachometer
12. Drive resistor
13. Anti-run-on valve
14. Oil pressure switch
15. Oil pressure warning light
16. Dual brake warning light
17. Dual brake switch
18. RH seat switch
19. RH seat belt switch
20. LH seat switch
21. LH seat belt switch
22. Seat belt warning light
23. Ignition key contacts
24. LH door switch
25. Service interval instrument (SII)
26. SII warning light
27. Two-speed heater motor
28. Heater resistor
29. Two-speed heater switch
30. Stop lamp switch
31. Stop lamp
32. Two-speed wiper switch
33. Two-speed wiper motor
34. Reversing lamp switch
35. Reversing lamp
36. Overdrive switch (optional)
37. Gearbox switch (optional)
38. Overdrive solenoid (optional)
39. Seat belt neutral switch
40. Hazard switch
41. Indicator warning light
42. Hazard warning light
43. Hazard flasher unit
44. Indicator flasher unit
45. Indicator switch
46. LH rear indicator lamp
47. LH front indicator lamp
48. RH front indicator lamp
49. RH rear indicator lamp
50. Voltage stabiliser
51. Fuel gauge
52. Tank unit
53. Temperature gauge
54. Temperature sender unit
55. Light switch
56. Map light
57. Panel rheostat
58. Panel lamps
59. LH front parking lamp
60. RH front parking lamp
61. LH front side marker
62. RH front side marker
63. LH rear side marker
64. LH tail lamp
65. Number plate lamp
66. RH tail lamp
67. RH rear side marker
68. Headlamp flash/dip switch
69. LH headlamp (dip)
70. RH headlamp (dip)
71. LH headlamp (main beam)
72. RH headlamp (main beam)
73. Main beam warning light
74. Horn relay
75. Horn
76. Horn switch
77. Luggage compartment lamp
78. Luggage compartment lamp switch
79. Ignition switch illumination
80. LH door switch
81. Cigar lighter
82. Cigarette lighter illumination (where applicable)

Wire colour codes

B	Black	N	Brown	U	Blue
G	Green	O	Orange	W	White
K	Pink	P	Purple	Y	Yellow
LG	Light green	S	Slate	R	Red

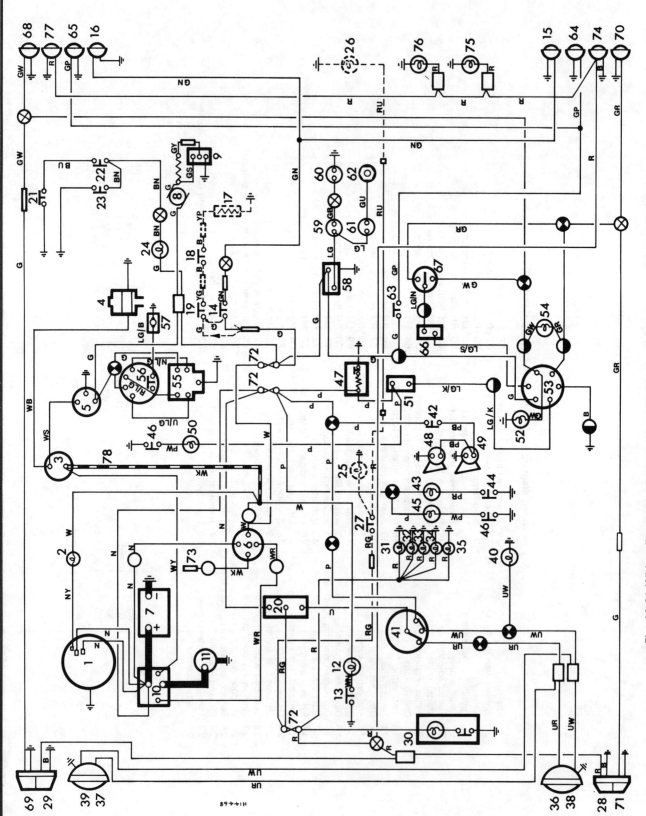

Fig. 10.61 Wiring diagram for RHD Spitfire 1500 from commission number FH100020

Key to wiring diagram for RHD Spitfire 1500 from commission number FH100020

1. Alternator
2. Ignition warning light
3. Ignition coil
4. Distributor
5. Tachometer
6. Ignition switch
7. Battery
8. Two-speed heater motor
9. Two-speed heater switch
10. Starter solenoid
11. Starter motor
12. Oil pressure warning light
13. Oil pressure switch
14. Reverse lamp switch
15. LH reverse lamp
16. RH reverse lamp
17. Overdrive solenoid
18. Overdrive in-out switch
19. Gearbox inhibitor switch
20. Headlamp switch
21. LH seat switch
22. LH belt switch
23. RH belt switch
24. Seat belt warning light
25. Rear fog warning light
26. Rear fog light
27. Rear fog light switch
28. LH sidelamp
29. RH sidelamp
30. Map light
31. Fuel gauge illumination
32. Temperature gauge illumination
33. Speedometer illumination
34. Tachometer illumination
35. Cigar lighter illumination
36. LH dip beam
37. RH dip beam
38. LH main beam
39. RH main beam
40. Main beam warning light
41. Main, dip beam and flasher switch
42. Horn push
43. Boot lamp
44. Boot lamp switch
45. LH interior light
46. Door switch
47. Cigar lighter
48. RH horn
49. LH horn
50. RH interior light
51. Hazard flasher unit
52. Hazard warning light
53. Hazard warning switch
54. Indicator warning light
55. Two-speed wiper motor
56. Two-speed wiper switch
57. Screen washer motor
58. Voltage stabiliser
59. Fuel gauge
60. Tank gauge
61. Temperature gauge
62. Temperature sender unit
63. Brake lamp switch
64. LH brake lamp
65. RH brake lamp
66. Flasher unit
67. Indicator switch
68. RH rear indicator lamp
69. RH front indicator lamp
70. LH rear indicator lamp
71. LH front indicator lamp
72. Fuses
73. Radio supply
74. LH tail lamp
75. LH number plate lamp
76. RH number plate lamp
77. RH tail lamp
78. Ballast resistor wire

Wire colour codes

B	Black	N	Brown	U	Blue
G	Green	O	Orange	W	White
K	Pink	P	Purple	Y	Yellow
LG	Light green	S	Slate	R	Red

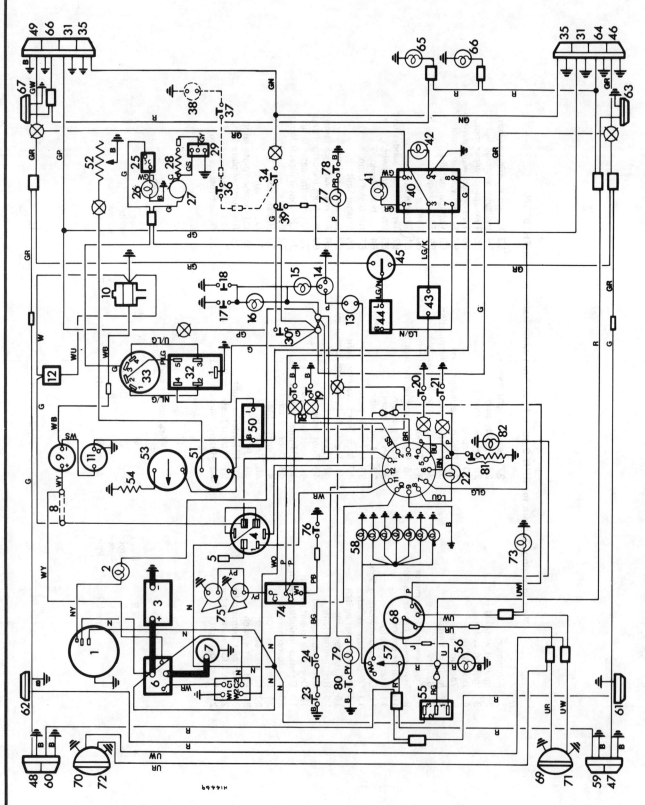

Fig. 10.62 Wiring diagram for N. American LHD Spitfire 1500 up to commission number FM32409

Key to wiring diagram for N. American LHD Spitfire 1500 up to commission number FM32409

1. Alternator
2. Ignition warning light
3. Battery
4. Ignition/starter switch
5. Starter solenoid
6. Starter motor
7. Ballast resistor wire
8. Coil (6 volt)
9. Distributor
10. Tachometer
11. Drive resistor
12. Anti-run-on valve
13. Oil pressure switch
14. Oil pressure warning light
15. Dual brake warning light
16. Dual brake switch
17. RH seat switch
18. RH belt switch
19. LH seat switch
20. LH belt switch
21. Ignition key contacts
22. LH door switch
23. Service interval instrument (SII)
24. SII warning light
25. Two-speed heater motor
26. Heater resistor
27. Two-speed heater switch
28. Stop lamp switch

Note: numbering in key: 1 Alternator, 2 Ignition warning light, 3 Battery, 4 Ignition/starter switch, 5 Radio supply, 6 Starter solenoid, 7 Starter motor, 8 Ballast resistor wire, 9 Coil (6 volt), 10 Distributor, 11 Tachometer, 12 Drive resistor, 13 Anti-run-on valve, 14 Oil pressure switch, 15 Oil pressure warning light, 16 Dual brake warning light, 17 Dual brake switch, 18 RH seat switch, 20 RH belt switch, 21 LH seat switch, 22 LH belt switch, 23 Ignition key contacts, 24 LH door switch, 25 Service interval instrument (SII), 26 SII warning light, 27 Two-speed heater motor, 28 Heater resistor

29. Two-speed heater switch
30. Stop lamp switch
31. Stop lamp
32. Two-speed wiper switch
33. Two-speed wiper motor
34. Reverse lamp switch
35. Reverse lamp
36. Overdrive switch (optional)
37. Gearbox switch (optional)
38. Overdrive solenoid (optional)
39. Seat belt neutral switch
40. Hazard warning switch
41. Indicator warning light
42. Hazard warning light
43. Hazard flasher unit
44. Indicator flasher unit
45. Indicator switch
46. LH rear indicator lamp
47. LH front indicator lamp
48. RH front indicator lamp
49. RH rear indicator lamp
50. Voltage stabiliser
51. Fuel gauge
52. Tank unit
53. Temperature gauge
54. Temperature sender unit
55. Light switch

56. Map light
57. Panel rheostat
58. Panel lamps
59. LH front parking lamp
60. RH front parking lamp
61. LH front side marker
62. RH front side marker
63. LH rear side marker
64. LH tail lamp
65. Number plate lamp
66. RH tail lamp
67. RH rear side marker
68. Headlamp flash/dip switch
69. LH dip beam
70. RH dip beam
71. LH main beam
72. RH main beam
73. Main beam warning light
74. Horn relay
75. Horn
76. Horn switch
77. Boot lamp
78. Boot lamp switch
79. Ignition switch illumination
80. LH door switch
81. Cigar lighter
82. Cigar lighter illumination

Wire colour codes

B	Black	N	Brown	U	Blue
G	Green	O	Orange	W	White
K	Pink	P	Purple	Y	Yellow
LG	Light green	S	Slate	R	Red

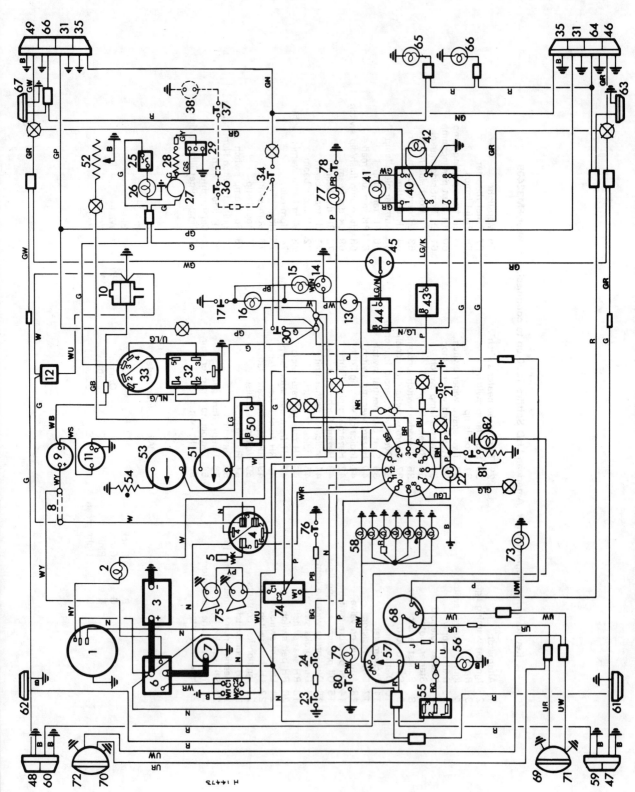

Fig. 10.63 Wiring diagram for N. American LHD Spitfire 1500 from commission number FM32410 to FM40000

Key to wiring diagram for N. American LHD Spitfire 1500 from commission number FM32410 to FM40000

1. Alternator
2. Ignition warning light
3. Battery
4. Ignition/starter switch
5. Radio supply
6. Starter solenoid
7. Starter motor
8. Ballast resistor wire
9. Coil (6 volt)
10. Distributor
11. Tachometer
12. Drive resistor
13. Anti-run-on valve
14. Oil pressure switch
15. Oil pressure warning light
16. Dual brake warning light
17. Dual brake switch
21. LH belt switch
22. Seat belt warning light
23. Ignition key contacts
24. LH door switch
25. Service interval instrument (SII)
26. SII warning light
27. Two-speed heater motor
28. Heater resistor
29. Two-speed heater switch
30. Stop lamp switch
31. Stop lamp
32. Two-speed wiper switch
33. Two-speed wiper motor
34. Reverse lamp switch
35. Reverse lamp
36. Overdrive switch (optional)
37. Gearbox switch (optional)
38. Overdrive solenoid (optional)
40. Hazard warning switch
41. Indicator warning light
42. Hazard warning light
43. Hazard flasher unit
44. Indicator flasher unit
45. Indicator switch
46. LH rear indicator lamp
47. LH front indicator lamp
48. RH front indicator lamp
49. RH rear indicator lamp
50. Voltage stabiliser
51. Fuel gauge
52. Tank unit
53. Temperature gauge
54. Temperature sender unit
55. Light switch
56. Map light
57. Panel rheostat
58. Panel lamps
59. LH front parking lamp
60. RH front parking lamp
61. LH front side marker
62. RH front side marker
63. LH rear side marker
64. LH tail lamp
65. Number plate lamp
66. RH tail lamp
67. RH rear side marker
68. Headlamp flash/dip switch
69. LH dip beam
70. RH dip beam
71. LH main beam
72. RH main beam
73. Main beam warning light
74. Horn relay
75. Horn
76. Horn switch
77. Boot lamp
78. Boot lamp switch
79. Ignition switch illumination
80. LH door switch
81. Cigar lighter
82. Cigar lighter illumination

Wire colour codes

B	Black	N	Brown	U	Blue
G	Green	O	Orange	W	White
K	Pink	P	Purple	Y	Yellow
LG	Light green	S	Slate	R	Red

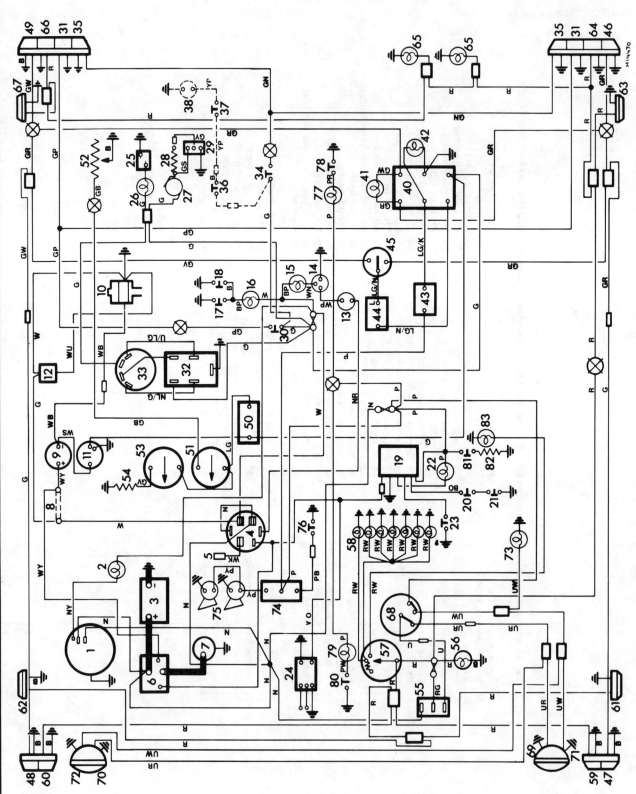

Fig. 10.64 Wiring diagram for N. American LHD Spitfire 1500 from commission number FM40001 to FM60000

Key to wiring diagram for N. American LHD Spitfire 1500 from commission number FM40001 to FM60000

1. Alternator
2. Ignition warning light
3. Battery
4. Ignition/starter switch
5. Radio supply
6. Starter solenoid
7. Starter motor
8. Ballast resistor wire
9. Coil (6 volt)
10. Distributor
11. Tachometer
12. Drive resistor
13. Anti-run-on valve
14. Oil pressure switch
15. Oil pressure warning light
16. Dual brake switch
17. Dual brake warning light
18. Handbrake switch
19. Seat belt timer
20. LH door switch
21. Ignition key contacts
22. Seat belt warning light
23. Buckle switch
24. Starter relay
25. Service interval instrument (SII)
26. SII warning light
27. Two-speed heater motor
28. Heater resistor
29. Two-speed heater switch
30. Stop lamp switch
31. Stop lamp
32. Two-speed wiper switch
33. Two-speed wiper motor
34. Reverse lamp switch
35. Reverse lamp
36. Overdrive switch (optional)
37. Gearbox switch (optional)
38. Overdrive solenoid (optional)
40. Hazard warning switch
41. Indicator warning light
42. Hazard warning light
43. Hazard flasher unit
44. Indicator flasher unit
45. Indicator switch
46. LH rear indicator lamp
47. LH front indicator lamp
48. RH front indicator lamp
49. RH rear indicator lamp
50. Voltage stabiliser
51. Fuel gauge
52. Tank unit
53. Temperature gauge
54. Temperature sender unit
55. Light switch
56. Map light
57. Panel rheostat
58. Panel illumination
59. LH front parking lamp
60. RH front parking lamp
61. LH front side marker
62. RH front side marker
63. LH rear side marker
64. LH tail lamp
65. Number plate lamp
66. RH tail lamp
67. RH rear side marker
68. Headlamp flash/dip switch
69. LH dip beam
70. RH dip beam
71. LH main beam
72. RH main beam
73. Main beam warning light
74. Horn relay
75. Horn
76. Horn switch
77. Boot lamp
78. Boot lamp switch
79. Ignition switch illumination
80. LH door switch
81. Cigar lighter switch
82. Cigar lighter element
83. Cigar lighter illumination

Wire colour codes

B	Black	N	Brown	U	Blue
G	Green	O	Orange	W	White
K	Pink	P	Purple	Y	Yellow
LG	Light green	S	Slate	R	Red

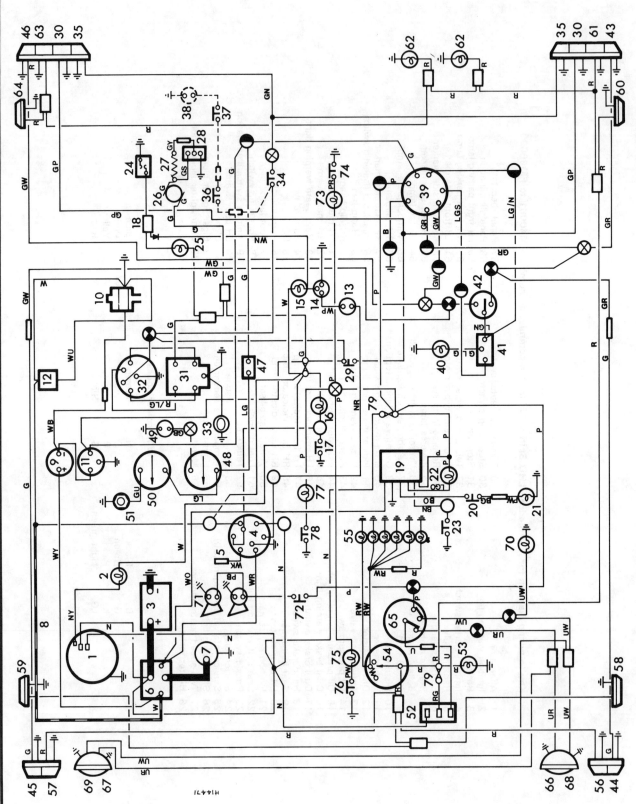

Fig. 10.65 Wiring diagram for N. American LHD Spitfire 1500 from commission number FM60001 to FM95000

Key to wiring diagram for N. American LHD Spitfire 1500 from commission number FM60001 to FM95000

1. Alternator
2. Ignition warning light
3. Battery
4. Ignition/starter switch
5. Radio supply
6. Starter solenoid
7. Starter motor
8. Ballast resistor wire
9. Coil (6 volt)
10. Distributor
11. Tachometer
12. Drive resistor
13. Anti-run-on valve
14. Oil pressure switch
15. Oil pressure warning light
16. Dual brake warning light
17. Dual brake switch
18. Diode
19. Buzzer unit
20. LH door switch
21. Ignition key contacts
22. Seat belt warning light
23. Buckle switch
24. Service interval instrument (SII)
25. SII warning light
26. Two-speed heater motor
27. Heater resistor
28. Two-speed heater switch
29. Stop lamp switch
30. Stop lamp
31. Two-speed wiper switch
32. Two-speed wiper motor
33. Windscreen washer motor
34. Reverse lamp switch
35. Reverse lamp
36. Overdrive switch (optional)
37. Gearbox switch (optional)
38. Overdrive solenoid (optional)
39. Hazard warning switch
40. Indicator warning light
41. Hazard flasher unit
42. Indicator switch
43. LH rear indicator lamp
44. LH front indicator lamp
45. RH front indicator lamp
46. RH rear indicator lamp
47. Voltage stabiliser
48. Fuel gauge
49. Tank unit
50. Temperature gauge
51. Temperature sender unit
52. Light switch
53. Map light
54. Panel rheostat
55. Panel illumination
56. LH front parking lamp
57. RH front parking lamp
58. LH front side marker
59. RH front side marker
60. LH rear side marker
61. LH tail lamp
62. Number plate lamp
63. RH tail lamp
64. RH rear side marker
65. Headlamp flash/dip switch
66. LH dip beam
67. RH dip beam
68. LH main beam
69. RH main beam
70. Main beam warning light
71. Horn
72. Horn switch
73. Boot lamp
74. Ignition switch illumination
75. Boot lamp switch
76. LH door switch
77. Passenger light
78. RH door switch
79. Fuses

Wire colour codes

B	Black	N	Brown	U	Blue
G	Green	O	Orange	W	White
K	Pink	P	Purple	Y	Yellow
LG	Light green	S	Slate	R	Red

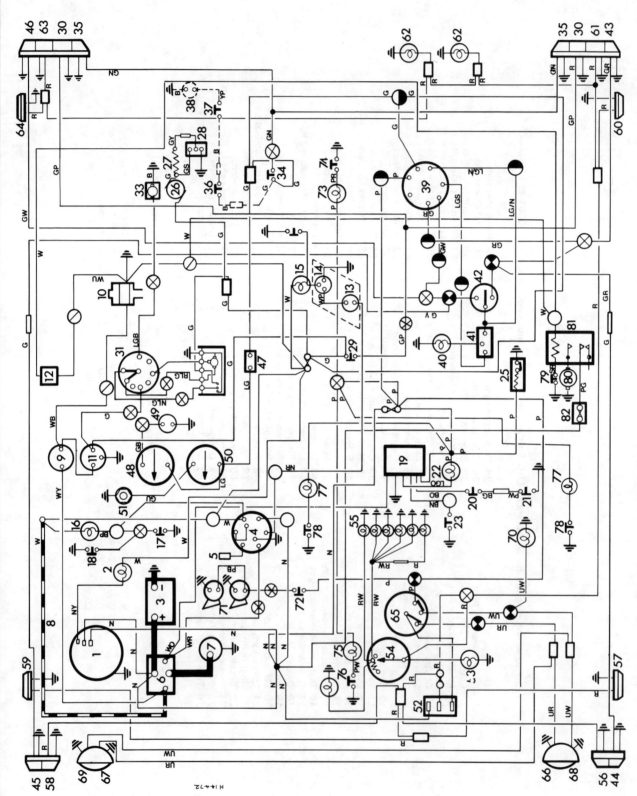

Fig. 10.66 Wiring diagram for N. American LHD Spitfire 1500 from commission number FM95001

Key to wiring diagram for N. American LHD Spitfire 1500 from commission number FM95001

1. Alternator
2. Ignition warning light
3. Battery
4. Ignition starter switch
5. Radio supply
6. Starter solenoid
7. Starter motor
8. Ballast resistor wire
9. Coil (6 volt)
10. Distributor
11. Tachometer
12. Drive resistor
13. Anti-run-on valve
14. Oil pressure switch
15. Oil pressure warning light
16. Dual brake warning light
17. Dual brake switch
18. Handbrake warning light switch
19. Buzzer unit
20. LH door switch
21. Ignition key contacts
22. Seat belt warning light
23. Buckle switch
24. Service interval instrument (SII)
25. Cigar lighter
26. Two-speed heater motor
27. Heater resistor
28. Two-speed heater switch
29. Stop lamp switch
30. Stop lamp
31. Two-speed wiper switch
32. Two-speed wiper motor
33. Windscreen washer motor
34. Reverse lamp switch
35. Reverse lamp
36. Overdrive switch (optional)
37. Gearbox switch (optional)
38. Overdrive solenoid (optional)
39. Hazard warning switch
40. Indicator warning light
41. Hazard/flasher unit
42. Indicator switch
43. LH rear indicator lamp
44. LH front indicator lamp
45. RH front indicator lamp
46. RH rear indicator lamp
47. Voltage stabiliser
48. Fuel gauge
49. Tank unit
50. Temperature gauge
51. Temperature sender unit
52. Light switch
53. Map light
54. Panel rheostat
55. Panel lamps
56. LH front parking lamp
57. LH front side marker
58. RH front parking lamp
59. RH front side marker
60. LH rear side marker
61. LH tail lamp
62. Number plate lamp
63. RH tail lamp
64. RH rear side marker
65. Headlamp flash/dip switch
66. LH dip beam
67. RH dip beam
68. LH main beam
69. RH main beam
70. Main beam warning light
71. Horn
72. Horn switch
73. Boot lamp
74. Ignition switch illumination
75. Boot lamp switch
76. LH door switch
77. Passenger light
78. RH door switch
79. Thermo switch
80. Electric radiator fan
81. Electric radiator fan relay
82. Electric radiator fan fuse

Wire colour codes

B	Black	N	Brown	U	Blue
G	Green	O	Orange	W	White
K	Pink	P	Purple	Y	Yellow
LG	Light green	S	Slate	R	Red

Chapter 11 Suspension and steering

Contents

Fault diagnosis	30
Front anti-roll bar – removal and refitting	5
Front coil spring/shock absorber – removal and refitting	6
Front hub bearing – adjustment	3
Front hub bearings – renewal	4
Front suspension lower wishbone – removal and refitting	11
Front suspension stub axle carrier – removal and refitting	9
Front suspension upper balljoint – renewal	8
Front suspension upper wishbone – removal and refitting	10
Front trunnion – removal, overhaul and refitting	7
General description	1
Maintenance and inspection	2
Rack-and-pinion backlash – adjustment	21
Rack-and-pinion steering gear – removal and refitting	23
Radius rod – removal and refitting	17
Rear hubs – removal and refitting	13
Rear roadspring – overhaul	15
Rear roadspring – removal and refitting	14
Rear shock absorbers – removal and refitting	12
Rear suspension vertical link – removal and refitting	16
Rear wheel alignment	28
Roadwheels and tyres	29
Steering angles and front wheel alignment	27
Steering column flexible coupling – removal and refitting	22
Steering column – removal, overhaul and refitting	25
Steering gear – overhaul	24
Steering lock/ignition switch – removal and refitting	26
Steering rack bellows – renewal	19
Steering wheel – removal and refitting	20
Tie-rod end balljoints – renewal	18

Specifications

Suspension type .... Independent on all four wheels, coil springs and telescopic shock absorbers at front, transverse leaf spring and telescopic shock absorbers at rear. Front anti-roll bar

Steering type .... Rack-and-pinion

Front hub bearing endfloat .... 0.003 to 0.005 in (0.08 to 0.13 mm)

Front wheel alignment
Vehicle carrying fuel, oil and water, but no occupant
Camber .... 2° to 4° positive
Castor .... 3° to 5° positive
King-pin inclination .... 4° 45' to 5° 45'
Toe-in .... $\frac{1}{16}$ to $\frac{1}{8}$ in (1.59 to 3.18 mm)

Rear wheel alignment
Vehicle carrying fuel, oil and water, but no occupant
Camber .... 0° to 2° negative
Toe-out .... $\frac{1}{32}$ to $\frac{3}{32}$ in (0.79 to 2.38 mm)

Turning circle (between kerbs) .... 24 ft (7.3 m)

Steering wheel diameter
Early models .... 14.5 in (368 mm)
Later models .... 13.5 in (343 mm)

Number of turns of wheel, lock-to-lock .... $3\frac{3}{4}$

Lubrication
Steering lower swivels .... Hypoid gear oil, viscosity SAE 90EP (Duckhams Hypoid 90S)
Steering rack .... General purpose grease (Duckhams LB 10)
Wheel hubs .... General purpose grease (Duckhams LB 10)

Tyre pressures

	Front	Rear
Mk I and II	18 lbf/in² (1.2 bar)	24 lbf/in² (1.7 bar)
Mk III:		
Cross-ply	18 lbf/in² (1.2 bar)	24 lbf/in² (1.7 bar)
Radial	21 lbf/in² (1.4 bar)	26 lbf/in² (1.8 bar)
Mk IV and 1500	21 lbf/in² (1.4 bar)	26 lbf/in² (1.8 bar)

Chapter 11 Suspension and steering

Torque wrench settings

	lbf ft	Nm
Front suspension		
Anti-roll bar clamp bolts	4	6
Anti-roll bar end link	34	46
Balljoint to upper wishbone	20	27
Balljoint taper pin to stub axle carrier	38	52
Brake caliper mounting bolt	65	88
Shock absorber lower mounting	46	63
Wishbone pivot bracket bolts	32	44
Stub axle to carrier	65	88
Tie-rod end balljoint to steering arm	32	44
Tie-rod end locknuts	38	52
Steering arm to stub axle carrier	65	88
Trunnion to lower wishbone	45	61
Wishbone pivot bracket bolts to chassis	25	34
Rear suspension		
Shock absorber lower mounting nut	38	52
Shock absorber upper mounting bolt	48	65
Radius arm bracket to body	32	44
Radius arm pivot bolts	32	44
Roadspring eye bolts	48	65
Roadspring clamp plate nuts	20	27
Vertical link to trunnion	48	65
Steering		
Steering column to lower bracket	9	12
Flexible coupling pinch bolts	14	19
Rack mounting nuts	16	22
Column safety clamp grubscrew	20	27
Column safety clamp bolts	9	12
Steering wheel nut	34	46
Roadwheels	48	65

1 General description

Front suspension

1 All four wheels are suspended independently on all models. At the front the suspension consists of two pairs of wishbones, the inner end of each pair being attached to and hinging on the chassis side members. The upper and lower suspension wishbones are fitted with nylon and rubber bushes and no greasing is therefore necessary.

2 Each stub axle carries the hub and road wheel and is attached by a nut and washer to the stub axle carrier. Stub axle carriers join each pair of wishbones at their outer ends. At the top of each carrier a balljoint and casing is held by two nuts and bolts between the front and rear arms of the wishbone. At the bottom of each carrier a bronze trunnion is held by a bolt, self-locking washer and nut between the outer ends of the lower wishbone.

3 A combined coil spring and telescopic shock absorber unit is fitted between the outer end of the lower wishbone assembly and the bracket on the chassis frame which also supports the inner ends of the top wishbone. An anti-roll bar is fitted between the lower wishbones on each side.

Rear suspension

4 At the rear, independent suspension is provided by a single transverse leaf spring dowelled and attached in its centre to the top of the differential unit housing to which it is clamped by a plate. At each end of the spring, two vertical links mounted on the spring eye carry the trunnion housing for the hub bearing and the swing axle shaft bearing assembly.

5 On the front face of the two vertical links a flexibly bushed radius arm is fitted to a bracket. The front end of the arm is fitted to a bracket on the chassis crossmember. In this way fore and aft movement of the rear suspension is firmly controlled. Telescopic rear shock absorbers are fitted between the bottom of the vertical links and a bracket on the chassis frame.

Steering

6 The rack-and-pinion steering gear is held in place behind the radiator and above the chassis frame by inverted U clamps at each end of the rack housing. Tie-rods from each end of the steering gear operate the steering arms via exposed and rubber gaiter enclosed balljoints. The upper splined end of the helically toothed pinion protrudes from the rack housing and engages with the splined end of the steering column. The pinion spline is grooved and the steering column is held to the pinion by a clamp bolt which partially rests in the pinion groove.

2 Maintenance and inspection

1 At the intervals specified in Routine Maintenance, remove the plug from the front suspension lower steering swivel and substitute a grease nipple (photo).

2.1 Grease nipple on front suspension steering swivel

2.3a Rear hub trunnion grease plug

2.3b Rear hub trunnion grease nipple

2.4 Grease nipple on steering rack

2 Give several strokes of a grease gun filled with EP90 oil **not grease**. Refit the plug.
3 Also at the specified intervals, remove the plug from the inside of the rear hub trunnion and substitute a grease nipple. Give several strokes of a grease gun using multi-purpose grease. Refit the plug (photos).
4 On earlier models, the rack-and-pinion steering gear is fitted with a grease plug. At the specified intervals, remove the plug and a grease gun filled with multi-purpose grease. Refit the plug (photo).
5 The most important check which must be carried out regularly to the suspension and steering is to inspect it visually.
6 To check for wear in the outer balljoints of the tie-rods place the car over a pit, or lie on the ground looking at the balljoints, and get a friend to turn the steering wheel from side to side. Wear is present if there is play in the joints.
7 To check for wear in the nylon and rubber bushes jack-up the front of the car until the wheels are clear of the ground. Hold each wheel in turn, at the top and bottom and try to rock it. If the wheel rocks, continue the movement at the same time inspecting the suspension to determine where play exists.
8 If the movement occurs between the wheel and the brake backplate then providing the wheel is on tightly the hub bearings will require replacement. The ball-pin between the top suspension arms may be worn and movement here will be clearly seen, as will movement at the outer end of the lower wishbones.
9 How well the shock absorbers function can be checked by bouncing the car at each corner. After each bounce the car should return to its normal ride position within $\frac{1}{2}$ to $\frac{3}{4}$ up-and-down movements. If the car continues to move up-and-down in decreasing amounts it means that the shock absorbers are worn and must be replaced.
10 Excessive play in the steering gear will lead to wheel wobble, and can be confirmed by checking if there is any lost movement between the end of the steering column and the rack. Rack-and-pinion steering is normally very accurate and lost motion in the steering gear indicates that the car has covered a considerable mileage or that lubrication has not been maintained.
11 If backlash develops it is quite possible to take up the wear which will have occurred between the teeth on the rack and the pinion by adjusting the pinion damper.
12 The outer balljoints at either end of the tie-rods are the most likely items to wear first, followed by the rack balljoints at the inner end of the tie-rods.
13 At the rear end, bangs, clonks and squeaks can arise from a variety of sources and to determine the exact point of trouble is not difficult as long as a methodical and thorough check is made.
14 Start by checking the radius arm bushes for play, as they are of rubber they are fairly easy to replace. The damper bushes may also have worn and should be checked by heavy bouncing. Wear in the spring eye bush is unusual but can happen after high mileages.
15 Perhaps the most common source of trouble is worn bushes in the trunnion housing. If with the trunnion housing bolt firmly tightened the trunnion housing can be seen to be moving loosely between the vertical links new bushes will have to be fitted.
16 Check for splits in the steering gear flexible bellows and check the front wheel alignment at regular intervals as specified in Routine Maintenance.

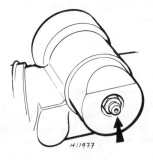

Fig. 11.1 Location for steering rack grease nipple (arrowed) (Sec 2)

3 Front hub bearings – adjustment

1 Raise the front of the car and remove the roadwheel.
2 Remove the disc pads (Chapter 9).
3 Tap off the grease cap and extract the split-pin. On some models, the grease cap has a tapped hole so that a screw may be inserted to force the cap off (photo).
4 Tighten the nut finger tight and then back it off one flat. This should give the correct endfloat as specified. This can be measured using a dial gauge or feeler blades.
5 Insert a new split-pin, half fill the grease cap with grease and refit it.
6 Refit the disc pads, the roadwheel and lower the jack.

4 Front hub bearings – renewal

1 Raise the front of the car and remove the roadwheel.
2 Unbolt and remove the caliper, tying it up out of the way to avoid straining the brake flexible hose.
3 Remove the grease cap (photo).
4 Extract the split-pin and then take off the castellated nut and thrust washer.
5 Withdraw the hub.
6 Prise out the oil seal from the inboard end of the hub and remove the tapered roller bearing races (photo).
7 Using a brass drift, drive out the bearing tracks (photo).
8 Clean out the old grease from the hub and fit the new bearing tracks.
9 Half fill the hub with multi-purpose grease, fit the bearing roller races, well greased and tap a new oil seal squarely into position.
10 Fit the hub to the stub axle taking care not to damage the felt oil seal which should have been sparingly greased (photo).
11 Fit the thrust washer and the nut. Then tighten the nut while the hub is turned until a slight resistance to rotation can be felt. This is to settle the bearings (photo).

3.3 Method of removing grease cap

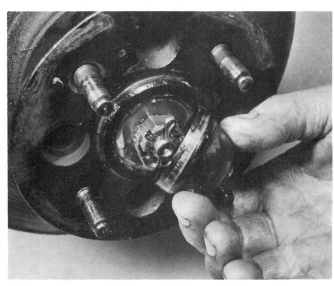

4.3 Removing grease cap

4.6 Front hub oil seal

4.7 Front hub bearing track

4.10 Fitting front hub

4.11 Fitting front hub thrust washer

272   Chapter 11 Suspension and steering

4.14 Inserting a new front hub split-pin

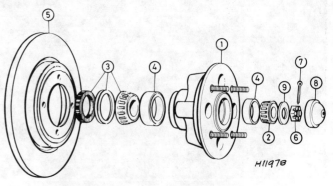

Fig. 11.2 Front hub components (Sec 4)

1 Hub
2 Outer tapered roller bearing
3 Inner tapered rubber bearing and seal
4 Outer bearing track
5 Disc
6 Castellated nut
7 Split pin
8 Grease cap
9 Thrust washer

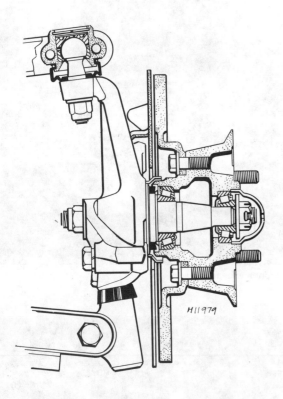

Fig. 11.3 Sectional view of front hub (Sec 4)

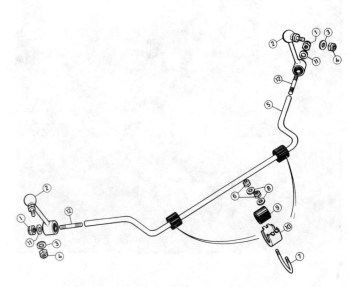

Fig. 11.4 Anti-roll bar components (Sec 5)

1 Self-locking nut
2 Link
3 Plain washer
4 Self-locking nut
5 Anti-roll bar
6 Self-locking nut
7 U bolt
8 Plain washer
9 Flexible bush
10 Clamp
11 Plain washer
12 Stud

12 Now unscrew the nut and retighten it using finger pressure only. Back the nut off one flat.
13 This should provide the correct hub endfloat as specified. This can be measured using a dial gauge or feeler blades.
14 Insert a new split-pin, half fill the grease cap with grease and refit it (photo).
15 Fit the caliper, the roadwheel and lower the car to the ground.

5 Front anti-roll bar – removal and refitting

1 Jack-up the front of the car and place stands under the chassis for safety. Undo the nyloc nuts and washers from the ends of the links.
2 The anti-roll bar is held to the chassis frame at two points by two U bolts and clamps. Undo the nuts and washers from the ends of the bolts and pull out the clamps. Remove the anti-roll bar from the car.
3 Replacement is a straightforward reversal of the removal sequence. Do not tighten down the nuts fully until the car is on the ground and the suspension in its normal laden position.

6 Front coil spring/shock absorber – removal and refitting

1 Loosen the nuts on the road wheel, jack-up the front of the car and support it on stands, and take off the wheel. Place the wheel nuts in the hub cap for safe keeping.
2 Open the bonnet and then undo the nuts which hold the ends of the anti-roll bar to the lower wishbone. Undo and remove the three

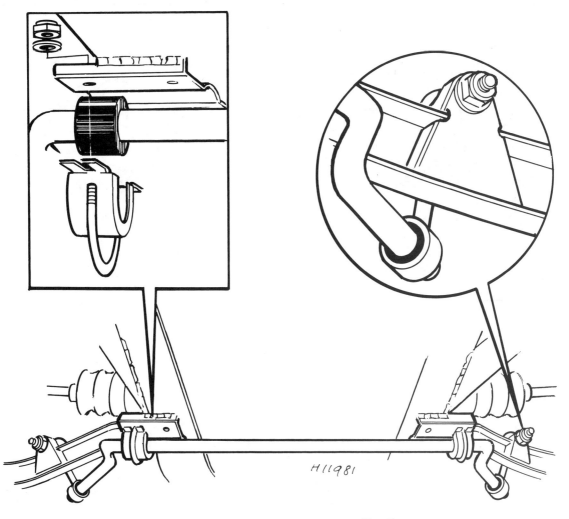

Fig. 11.5 Front anti-roll bar attachment (Sec 5)

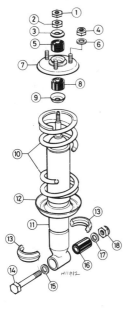

Fig. 11.6 Front coil spring/shock absorber (Sec 6)

1 Locknut
2 Nut
3 Washer
4 Self-locking nut
5 Rubber cushion
6 Washer
7 Spring upper pan
8 Rubber cushion
9 Washer
10 Coil spring
11 Shock absorber
12 Spring lower pan
13 Collets
14 Mounting bolt
15 Washer
16 Flexible bush
17 Washer
18 Self-locking nut

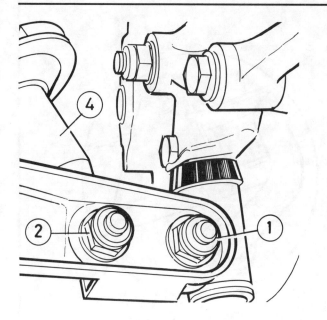 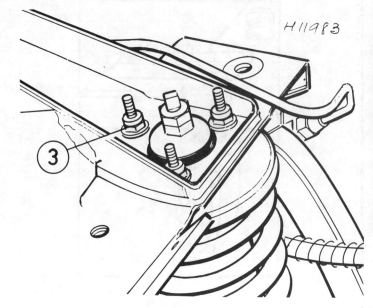

Fig. 11.7 Front shock absorber upper and lower mountings (Sec 6)

1 Trunnion nut
2 Shock absorber nut
3 Shock absorber upper mounting nut
4 Shock absorber

nuts and washers which hold the upper spring pan in place on the chassis sub frame (photo).
3 From the lower end of the shock absorber undo and remove the nut, bolt and washers which hold the shock absorber eye in place between the outer ends of the wishbone arms.
4 Slightly lift the brake drum and wishbones and pull out the coil spring/shock absorber units from between the lower suspension arms.
5 To remove the spring from the shock absorber, a suitable spring compressor must be used. Do not attempt this work using makeshift tools as the tension of the spring will make it a very dangerous operation.
6 Once the upper mounting flange is released from spring loading, the nut and locknut which secure the shock absorber to the flange can be removed.
7 Remove the mounting rubber cushions, the seats and the flange from the shock absorber rod.
8 Very slowly and carefully release the spring from the compressor tool.
9 Reassembly and refitting are reversals of removal and dismantling.

7 Front trunnion – removal, overhaul and refitting

1 Raise the front of the car and support securely.
2 Remove the roadwheel.
3 Remove the hub (Section 4).
4 Remove the bolt which secures the steering arm and disc shield to the stub axle carrier. Remove the disc shield.
5 Remove the trunnion securing nut and bolts.
6 Slacken the shock absorber lower mounting nut and bolt.
7 Remove the trunnion from the lower wishbone. Unscrew the trunnion from the stub axle carrier.
8 With the trunnion removed, take off the two washers and dust seals.
9 Press out the spacer sleeve.
10 Take out the nylon bearings and the cupped washers.
11 Reassembly is a reversal of dismantling.
12 When screwing on the trunnions to the stub axle remember that the left-hand trunnion has a left-hand thread. The right-hand trunnion can be identified by the reduced diameter at its lower end.
13 Screw the trunnion onto the stub axle carrier finger tight and then unscrew it just enough to enable full lock to be obtained in both

6.2 Front coil spring upper attachment

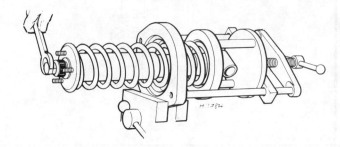

Fig. 11.8 Typical roadspring compressor (Sec 6)

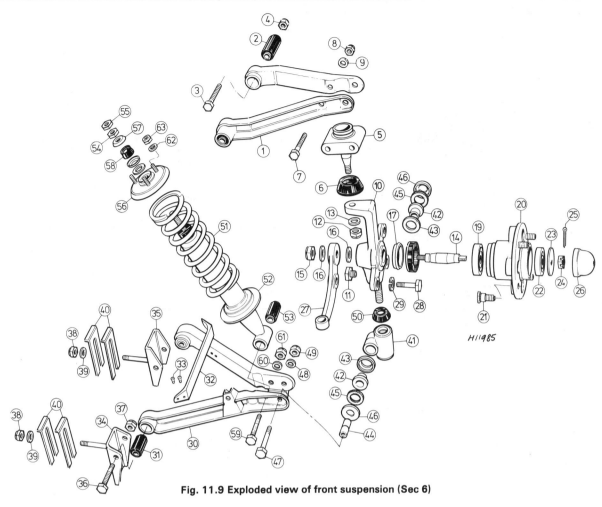

Fig. 11.9 Exploded view of front suspension (Sec 6)

1 Upper wishbone arm	17 Water shield	33 Rivet	49 Self-locking nut
2 Flexible bush	18 Oil seal	34 Pivot bracket	50 Oil seal
3 Pivot bolt	19 Inner bearing	35 Pivot bracket	51 Coil spring
4 Self-locking nut	20 Hub	36 Pivot bolt	52 Shock absorber
5 Balljoint	21 Roadwheel stud	37 Self-locking nut	43 Flexible bush
6 Dust excluder	22 Outer bearing	38 Self-locking nut	54 Nut
7 Bolt	23 Thrust washer	39 Washer	55 Locknut
8 Self-locking nut	24 Castellated nut	40 Shim	56 Spring upper plate
9 Washer	25 Split-pin	41 Trunnion	57 Washer
10 Stub axle carrier	26 Grease cap	42 Bearing	58 Rubber cushion
11 Grease plug	27 Steering arm	43 Water shield	59 Bolt
12 Self-locking nut	28 Bolt	44 Distance piece	60 Washer
13 Washer	29 Spring washer	45 Seal	61 Self-locking nut
14 Stub axle	30 Lower wishbone	46 Water shield	62 Washer
15 Self-locking nut	31 Flexible bush	47 Bolt	63 Self-locking nut
16 Washer	32 Strut	48 Washer	

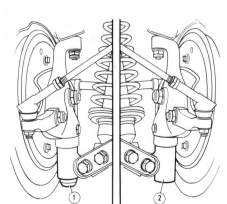

Fig. 11.10 Front trunnions (Sec 7)

1 Right-hand
2 Left-hand

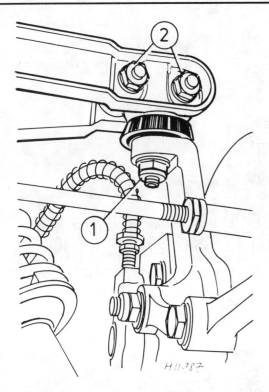

Fig. 11.11 Front suspension upper balljoint attachment (Sec 8)

1 Ball pin nut   2 Balljoint mounting nuts

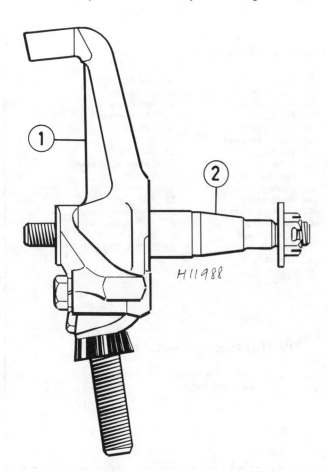

Fig. 11.12 Diagrammatic view of stub axle carrier (1) and stub axle (2) (Sec 9)

directions with the trunnion in its correct alignment to receive the wishbone.

14 Reassembly and refitting are reversals of removal. Tighten all nuts and bolts to the specified torque and oil the trunnion on completion as described in Section 2.

8 Front suspension upper balljoint – renewal

1 Raise the front of the car and remove the roadwheel.
2 Unscrew the nut which holds the balljoint to the stub axle carrier.
3 Using a suitable balljoint extractor, release the balljoint from the stub axle carrier.
4 Unbolt the balljoint from the wishbone arm and remove it.
5 Fitting the balljoint is a reversal of removal.

9 Front suspension stub axle carrier – removal and refitting

1 Raise the front of the car and remove the roadwheel.
2 Unbolt the brake caliper and tie it up out of the way to avoid straining the flexible hose.
3 Remove the hub and brake disc (refer to Section 4).
4 Unbolt and remove the disc shield and the steering arm.
5 Disconnect the upper balljoint from the stub axle carrier as described in Section 8.
6 Slacken the shock absorber lower mounting bolt.
7 Unbolt the lower swivel from the wishbone.
8 Remove the stub axle carrier.
9 If the stub axle must be separated from the carrier, a press will be required.
10 Refitting is a reversal of removal.

10 Front suspension upper wishbone – removal and refitting

1 Raise the front of the car and remove the roadwheel.
2 Remove the coil spring/shock absorber as described in Section 6.
3 Unbolt the balljoint from the outboard end of the upper wishbone (photo).
4 Unbolt the inboard ends of the upper wishbone from the support brackets and lift the wishbone from the car.
5 Refitting is a reversal of removal, but do not tighten the inboard pivot bolts of the wishbone to the specified torque until the weight of the car is again on its roadwheels.

11 Front suspension lower wishbone – removal and refitting

1 Raise the front of the car and support securely.
2 Disconnect the anti-roll bar from the lower wishbone.

10.3 One side of the front suspension

Chapter 11 Suspension and steering

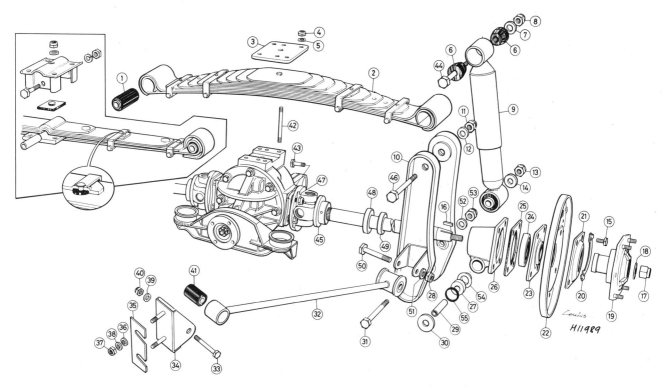

Fig. 11.13 Exploded view of rear suspension (Sec 12)

1 Flexible bush	16 Woodruff key	30 Dust seal	44 Bolt
2 Leaf spring (Mk I, II and III)	17 Nut	31 Bolt	45 Driveshaft
3 Clamp plate	18 Washer	32 Radius arm	46 Bolt
4 Nut	19 Hub	33 Bolt	47 Nut
5 Washer	20 Lockplate	34 Radius arm bracket	48 Flinger
6 Flexible bush	21 Grease retainer	35 Shim	49 Seal
7 Washer	22 Brake backplate	36 Washer	50 Bolt
8 Nut	23 Seal housing	37 Nut	51 Washer
9 Shock absorber	24 Bearing	38 Washer	52 Washer
10 Vertical link	25 Gasket	39 Washer	53 Nut
11 Nut	25 Gasket	40 Nut	54 Dust seal
12 Washer	26 Trunnion housing	41 Flexible bush	55 Rubber seal
13 Nut	27 Nylon bush	42 Stud	Inset – Later type roadspring (Mk IV and 1500)
14 Washer	28 Nut	43 Bolt	
15 Bolt	29 Steel bush		

3 Unbolt the trunnion and the shock absorber lower mounting from the outboard end of the wishbone.

4 Unbolt the inboard ends of the wishbone from its support brackets and lift the wishbone from the car, retaining any shims (see Section 27).

5 Refitting is a reversal of removal, but do not tighten the inboard pivot bolts to the specified torque until the weight of the car is again on its roadwheels. Make sure that the camber adjusting shims are returned to their original positions.

12 Rear shock absorbers – removal and refitting

1 Remove the hub cap, loosen the wheel nuts and jack-up the rear of the car. Undo the nuts and take off the roadwheel. Support the rear of the chassis on suitable stands.

2 Place the jack under the crossmember joining the two halves of the vertical links, and raise the jack so the load is taken off the shock absorber.

3 Undo and remove the nut and bolt from the upper eye of the shock absorber where it is attached to the chassis. Then remove the nut and washer from the vertical link crossmember extension and pull the shock absorber off.

4 Replacement is a straightforward reversal of the removal sequence. Always fit new rubber bushes to the top eye of the shock absorber and bleed the shock absorber before fitment by operating it vertically over its full travel. Hold the shock absorber vertically while fitting it.

13 Rear hubs – removal and refitting

1 As explained in Chapter 7, Section 5, due to the exceptionally tight fit of the rear hub on the driveshaft, a special tool (S4221A) or other heavy duty extractor will be required to remove it.

2 Chapter 7, Section 5 also describes overhaul of the rear hub in detail.

14 Rear roadspring – removal and refitting

Mk I, II and III models

1 Loosen the rear wheel securing nuts, jack-up the rear of the car and place support blocks or stands under the chassis frame. Remove the road wheels.

2 Disconnect the flexible hydraulic pipes from the chassis bracket as described in Chapter 9 and disconnect the handbrake cable by

which hold the cover in place over the centre of the spring and then undo the six nuts and washers from the studs which hold the spring in place.
7 Lift off the spring clamp plate and undo the three rear studs from the differential casing.
8 The road spring can now be pulled away from the side of the car.
9 In general replacement is a straightforward reversal of the removal sequence, but note the following points.

(a) The spring must be fitted with the edge marked 'front' towards the front of the car
(b) The studs are fitted with their shorter threaded ends into the differential casing
(c) To prevent any water entering the car through the spring access cover smear the edge with sealant, and with the cover in place smear the joint liberally with mastic.
(d) Do not tighten fully the spring eye and damper mounting nuts and bolts until the car is on the ground in its static laden condition. This is to allow the rubber bushes to assume their normal working positions

Mk IV and 1500 models

10 The rear spring on the Mk IV and 1500 models has fewer leaves. It can be removed without dismantling the brakes or propeller shaft as for earlier cars.
11 Jack-up the rear of the car, put blocks under the chassis and lower it onto them. Note that you will need about a four feet space beside the car to pull the spring out.
12 Remove both rear wheels.
13 Remove the nuts and washers from the bolts holding the spring eyes to the suspension vertical links.
14 At this stage there is some load from the suspension on the bolts through the spring eyes, and they will anyway be stiff with dirt. Try to remove them. Put an ordinary nut (not the self locking one) on the threads to protect them, and tap the bolts part way out. At this stage just try gently tapping the bolts out with a punch.
15 The load on the spring can be eased by disconnecting the shock absorbers. Remove the nut from the bottom end of the shock absorber. Take the weight of the suspension off it with the jack under the bottom of the vertical link. Try and pull the shock absorber off its bottom pin. Again this may prove difficult. If so, remove the bolt at the top of the shock absorber, and then it will come off easily. Keep the shock absorber upright to keep its air content at the top. Lower the jack, taking care as the hub unit goes down that the flexible brake pipe is not pulled. The weight of the hub itself should be left on the jack. Now the bolts should drive out the springs.
16 Having removed the two bolts from the spring eyes the spring itself should be unbolted from the car
17 Remove the rear squab trim pad: six screws and cup washers. Remove the panel in the body over the central spring mounting: two screws.
18 Remove the four nuts clamping the spring to the top of the differential casing. Take off the clamp, and using its nuts as nut and locknut take out the four studs.
19 The spring is now clear. Wriggle it out of the vertical links, and pull it clear of the car. If you were able to get the bolts out of the vertical links and spring eyes with the shock absorbers still in place it may prove difficult to squeeze the spring out. If so remove the shock absorber on the side towards which you extract the spring.
20 When reassembling note the spring clips have 'Front' stamped on them. The ground edge of the main leaf should be to the rear.
21 Do not tighten the bolts through the vertical links and spring eyes till the car weight is on them, so that the bushes will be squeezed in the correct position.
22 Seal the panel in the floor against wet with non-setting sealant.

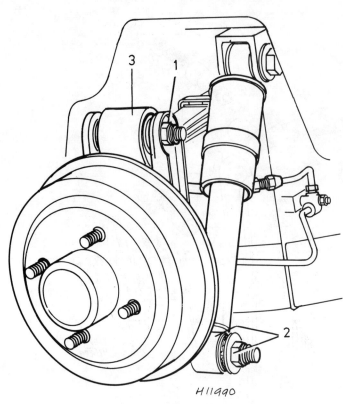

Fig. 11.14 Roadspring and shock absorber attachment to rear suspension vertical link (Sec 14)

1 Eye bolt
2 Shock absorber lower mounting
3 Spring safety leaf

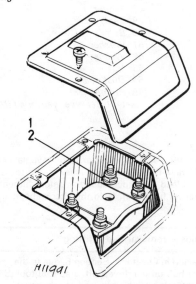

Fig. 11.15 Rear spring centre mounting (Sec 14)

1 Self-locking nut   2 Washer

removing the clevis pin from the lever on the backplate. Remove the small pull-off spring from the backplate.
3 Remove the rear shock absorbers as described in Section 12.
4 Mark a mating line across the drive shaft universal joint flanges and undo the four nuts and bolts which secure the universal joint to the differential drive shaft.
5 Undo the nut and remove the bolt from each road spring eye.
6 From behind the front seats inside the car remove the two screws

15 Rear roadspring – overhaul

1 With the spring removed as previously described, take off the spring plate.
2 Compress the spring in a vice and remove the spring bracket, rubber pad, nut and spacer.
3 Release the spring from the vice and remove the nuts and bolts from the spring leaf clips.

Chapter 11 Suspension and steering

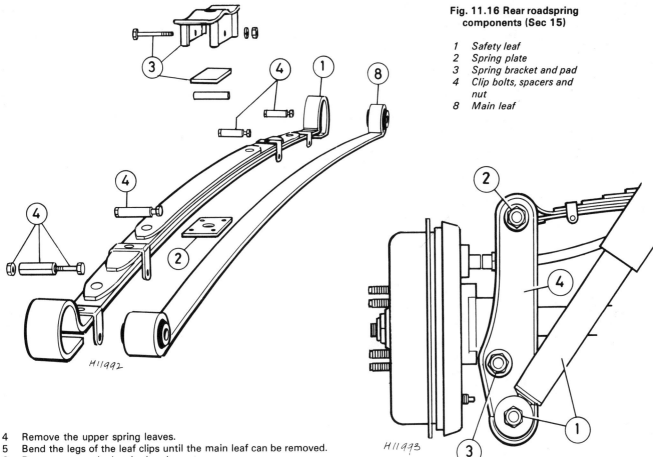

Fig. 11.16 Rear roadspring components (Sec 15)

1. Safety leaf
2. Spring plate
3. Spring bracket and pad
4. Clip bolts, spacers and nut
8. Main leaf

4 Remove the upper spring leaves.
5 Bend the legs of the leaf clips until the main leaf can be removed.
6 Renew any cracked or broken leaves.
7 Clean the spring leaves and smear them with graphite grease.
8 Commence reassembly by inserting the main leaf into the safety leaf so that the ground edge at the centre of the main leaf faces the plain unstamped legs of the spring clips (towards the rear of the car when spring installed).
9 The remainder of the reassembly procedure is a reversal of dismantling, but make sure that the rubber insulating bellows are correctly located in the recesses at the ends of the spring leaves.

16 Rear suspension vertical link – removal and refitting

1 Raise the rear of the car and support securely under the chassis members.
2 Remove the rear roadwheel.
3 Unbolt and separate the radius rod from its bracket on the vertical link.
4 Unbolt the shock absorber lower mounting.
5 Unbolt the upper end of the vertical link from the end of the roadspring.
6 Unbolt the vertical link from the trunnion housing and remove the link.
7 Refitting is a reversal of removal. Tighten all nuts and bolts to the specified torque except the spring eye bolts which should be tightened only after the weight of the car has been lowered onto its roadwheel.

17 Radius rod – removal and refitting

1 Raise the rear of the car and support securely. Remove the roadwheel.
2 Unbolt both ends of the radius rod, but note the position of the rear wheel alignment adjusting shims (see Section 28).
3 The flexible bushes may be renewed using a press or by employing a bolt with nut, washers and distance piece to draw out the old bushes and to pull the new ones into position.

Fig. 11.17 Rear suspension vertical link attachment (Sec 16)

1. Shock absorber and lower mounting bolt
2. Spring eyebolt
3. Trunnion bolt
4. Vertical link

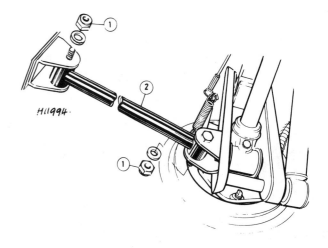

Fig. 11.18 Radius rod attachment (Sec 17)

1. Self-locking nuts
2. Rod

18.1 Using a balljoint splitter tool

20.1 Removing steering wheel centre cap (later models)

20.2 Steering wheel securing nut (later models)

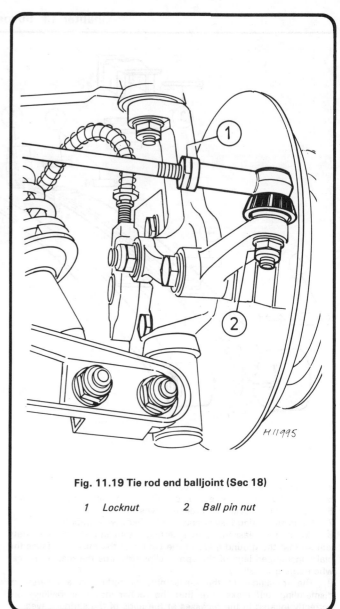
Fig. 11.19 Tie rod end balljoint (Sec 18)

1 Locknut   2 Ball pin nut

20.5 Steering wheel removed

Chapter 11 Suspension and steering

18 Tie-rod end balljoints – renewal

1 If the tie-rod outer balljoints are worn it will be necessary to renew the whole balljoint assembly as they cannot be dismantled and repaired. To remove a balljoint, free the balljoint shank from the steering arm using a suitable balljoint extractor (photo).
2 Measure or mark the position of the locknut on the tie-rod accurately to ensure near accurate 'toe-in' on reassembly.
3 Slacken off the balljoint locknut, and holding the tie-rod by its flat with a spanner, to prevent it from turning, unscrew the complete ball assembly from the rod. Replacement is a straightforward reversal of this process.
4 Check the front wheel alignment (see Section 27).

19 Steering rack bellows – renewal

1 If as the result of inspection the rack bellows are found to be split or leaking then they must be renewed immediately.
2 Mark the position of the tie-rod balljoint locknut. This can be done by counting the number of exposed threads on the tie-rod.
3 Release the locknut and then disconnect the balljoint from the steering arm using a suitable balljoint extractor.
4 Remove the balljoint and the locknut from the tie-rod.
5 Remove the bellows fixing clips and pull the bellows from the rack housing.
6 Refitting is a reversal of removal, but make sure that the bellows is secured by its clips when partially-extended to accommodate steering at full lock conditions.
7 However accurately the tie-rod end balljoints may have been refitted, still check the front wheel alignment (see Section 27).

20 Steering wheel – removal and refitting

Early models
1 Carefully prise off the horn push button (photo).
2 Lift the horn button away to expose the steering wheel securing nut (photo).
3 Next pull out the horn contact brush.
4 Undo and remove the steering wheel securing nut and the special clip which lies underneath it.
5 The wheel is then simply pulled off the splines (photo).

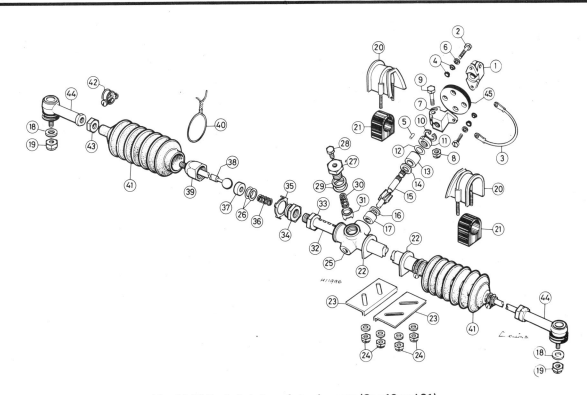

Fig. 11.20 Exploded view of steering gear (Sec 19 and 21)

1 Flexible coupling upper clamp	11 Retaining ring	23 Locking plates	35 Lockplate
2 Bolt	12 Shims	24 Self-locking nuts	36 Spring
3 Earth cable	13 Pinion upper bush	25 Pinion housing	37 Ball socket cap
4 Flexible bushes	14 Thrust washer	26 Shims	38 Tie-rod
5 Dowel	15 Pinion shaft	27 Damper plug	39 Cup nut
6 Washer	16 Thrust washer	28 Grease plug	40 Locking wire
7 Flexible coupling lower clamp	17 Pinion lower bush	29 Shims	41 Bellows
8 Self-locking nut	18 Washer	30 Damper spring	42 Clip
9 Pinch-bolt	19 Self-locking nut	31 Plunger	43 Locknut
10 Circlip	20 Mounting U bolts	32 Rack	44 Tie-rod end balljoint
	21 Insulators	33 Locknut	45 Flexible coupling
	22 Abutment plates	34 Sleeve nut	

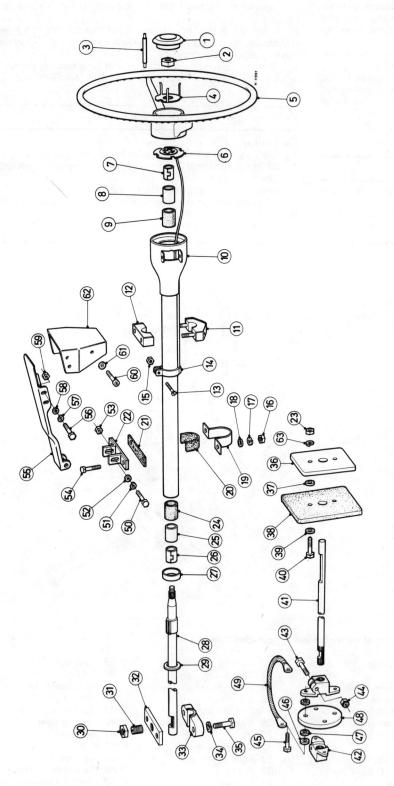

Fig. 11.21 Exploded view of typical steering column (Sec 20)

1 Horn button
2 Nut
3 Horn contact brush
4 Clip
5 Steering wheel
6 Horn contact ring
7 Nylon brush
8 Steel bush
9 Flexible bush
10 Column tube
11 Column clamp
12 Column clamp
13 Screw
14 Cable clip
15 Nut
16 Nut
17 Spring washer
18 Washer
19 Column clamp
20 Insulator
21 Felt pad
22 Bracket
23 Nut
24 Flexible bush
25 Steel bush
26 Nylon bush
27 End cap
28 Upper steering shaft
29 Nylon washer
30 Nut
31 Socket screw
32 Clamp
33 Safety clamp
34 Spring washer
35 Bolt
36 Retaining plate
37 Washer
38 Rubber seal
39 Washer
40 Bolt
41 Steering lower column
42 Coupling clamp
43 Pinch-bolt
44 Coupling clamp
45 Bolt
46 Washer
47 Washer
48 Flexible coupling
49 Earth cable
50 Bolt
51 Spring washer
52 Washer
53 Nut
54 Bolt
55 Support bracket
56 Bolt
57 Spring washer
58 Washer
59 Nut
60 Screw
61 Washer
62 Bracket
63 Spring washer

Chapter 11 Suspension and steering

6 Replacement is a simple reversal of this process. To ensure the wheel is correctly aligned on refitment it is helpful to mark the boss and a corresponding spline before pulling the wheel off.
7 Tighten the securing nut down firmly and peen the metal of the nut to the inner column to prevent the nut working loose.

Later models.
8 The horn switch is incorporated in the combination stalk switch, otherwise the operations are as described for earlier models.

21 Rack-and-pinion backlash – adjustment

1 Backlash between the pinion and the rack can be taken up by means of an adjustment between the rack damper cap, and the rack housing. If backlash is present adjust the rack damper in the following manner.
2 Disconnect the outer ends of the steering tie-rods by knocking out the tie-rod ball joint shanks from the holes in the steering arms.
3 Unscrew the damper cap and remove the spring and shims.
4 Refit the damper cap together with the plunger, but without the spring and the shims.
5 Tighten the damper cap until it requires about a 2 lb pull at the steering wheel rim to turn the wheel.
6 Measure the gap between the underside of the damper gap and the rack housing with a feeler gauge, and add to this figure a clearance figure of 0.002 to 0.005 in (0.05 to 0.127 mm). The total figure represents the thickness of the shims that must be fitted under the cap. Shims are available in thicknesses of 0.003 in and 0.010 in (0.76 and 0.254 mm).
7 Remove the cap, replace the spring, fit the necessary shims and tighten the cap down firmly. Reconnect the tie-rod balljoints to the steering arms.

22.2 Steering column flexible coupling

22 Steering column flexible coupling – removal and refitting

1 With the steering in the straight-ahead position, slacken the two bolts on the steering column safety clamp.
2 Working within the engine compartment, unscrew the two pinchbolts from the flexible coupling (photo).
3 Pull the universally-jointed steering shaft out of the flexible coupling.
4 Pull the flexible coupling from the pinion shaft.
5 Refitting is a reversal of removal, but set the steering in the straight-ahead position before connecting the coupling and position it so that the pinch bolts pass easily through the cut-outs in the shafts.
6 When fitting the column safety clamp, push the clamp upwards until it abuts against the steering upper shaft, then tighten the bolts.

23 Rack-and-pinion steering gear – removal and refitting

1 Unscrew the nut and remove the bolt from the bottom of the clamp on the lower end of the steering column. Undo the bolt from the chassis which holds the earth lead running to the grease plug on the rack-and-pinion steering unit.
2 Remove the nut from each of the two tie-rod to steering arm balljoints and unscrew the nuts two turns. Hit the head of the nut with a soft faced hammer to drive the shank of the balljoint out of the steering arm. Remove each nut and pull the tie-rods complete with balljoints away.
3 Undo the nuts and washers from the inverted U-bolts, remove the locating plates and the rubber bushes.
4 Pull the steering gear sufficiently forward to disengage the pinion shaft from the clamp and remove the steering gear from the side of the car.
5 Replacement commences by centralising the pinion shaft. Turn the shaft from lock to lock counting the number of turns and then rotate it half this number. Ensure the steering wheel is in the straight ahead position.
6 Offer the unit up to the chassis, connect the clamp to the splined pinion shaft, fit the rubber bushes, U-bolts, plates and nyloc nuts loosely and ensure the steering gear is positioned as shown. Then slide

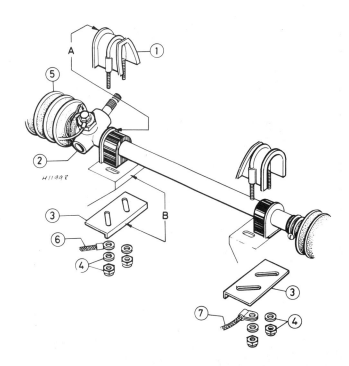

Fig. 11.22 Rack-and-pinion housing installation diagram (Sec 23)

1 U-bolt
2 Pinion housing
3 Locating plates
4 Self-locking nuts
5 Bellows
6 Earth bonding cables
7 Engine earth cable
A Flange clearance $\frac{1}{8}$ in (3.18 mm)
B Plate flange in contact with chassis

the U-bolt assemblies outwards until a clearance of $\frac{1}{8}$ in (3.18 mm) is present between the flanges on the U-bolt retainers and the rack-and-pinion flanges.
7 The U-bolts must then be held in this position while an assistant slides the plates inwards until their flanged faces abut the chassis. Then tighten the U-bolt securing nuts.
8 Reconnect the earth strap, fit the clamp nut and bolt, refit the tie-rods to the steering arms.
9 Check the front wheel alignment as described in Section 27.

24 Steering gear – overhaul

1 Undo the clips on the rubber gaiters and pull the gaiters back to expose the inner balljoint assemblies.
2 Loosen the locknuts and completely unscrew both the tie-rod assemblies from the rack.
3 Pull out the coil springs, free the tab washer, unscrew the nut and take off the tab washer, shims and cup.
4 Mark the positions of the locknuts on the tie-rods so the 'toe-in' is approximately correct on reassembly, undo the locknuts and screw off the balljoints.
5 Pull off the rubber gaiter and undo the cup nut from the tie-rods, and then remove the locknuts.
6 Unscrew and remove the damper cap, spring, plunger and shims.
7 The pinion can be removed after taking out the circlip. Take great care not to lose the small locating peg.
8 The rack can now be pulled from the housing tube and the thrust washer and bush taken from the pinion housing.
9 Clean all the parts thoroughly and examine the teeth of the rack-and-pinion for wear or damage. Also examine the pinion thrust washers, and bushes, thrust pad, and inner and outer balljoints and replace as necessary.
10 Reassembly commences by sliding the rack into place. Then fit the bush and thrust washer and adjust the pinion endfloat after assembling the thrust washer, bush, and retaining ring to the pinion, and fitting and securing the pinion to the housing with the circlip.
11 By trial and error fit a different number of shims (available in thicknesses of 0.004 in and 0.016 in) until the minimum amount of endfloat exists when the pinion is pushed in and out commensurate with free rotation of the pinion.
12 When the correct shim thickness has been ascertained fit a new rubber O-ring to the retaining ring and after the pinion assembly has been fitted, fit the dowel and circlip.
13 Then adjust the rack-and-pinion backlash as described in Section 21. Reassembly is now a straightforward reversal of the dismantling sequence. Check that the rack balljoints fit tightly but are free to move. If they are excessively loose or tight, adjustment can be made by varying the number and thickness of the shims between the balljoint cup and the ball housing.
14 After fitting the rack-and-pinion in place fill it with 2 oz (56 g) multi-purpose grease.

25 Steering column – removal, overhaul and refitting

1 Extract the securing screws and remove the two halves of the steering column shrouds. Undo and remove the bolt at the bottom of the flexible coupling clamp on the column immediately adjacent to the rack housing.
2 Separate the cables on the steering column under the facia at their snap connectors.
3 Remove the outer column support clamp lower down the column after undoing the securing bolts (photo).
4 Then undo the two bolts which hold the upper clamp in place and remove the clamp from the column.
5 Remove the special clip which holds the cables to the column.
6 To avoid damage to the switches on the sides of the column detach them after removing the switch covers by simply undoing the two small screws which hold each switch in place.
7 Carefully lift the switches away from the column.
8 Undo and remove the steeing wheel as described in Section 20.
9 At the base of the column inside the car, remove the two bolts which hold the two halves of the impact clamp together.
10 If there is any play in the top of the steering column it will be necessary to replace the rubber and nylon bush at the top and possibly also the similar bush at the bottom of the outer column. Pull the inner from the outer column.

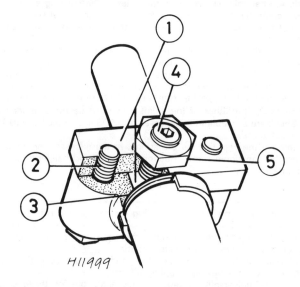

Fig. 11.23 Safety clamp at joint of steering shaft upper and lower column (Sec 25)

1 Clamp halves
2 Bolts
3 Plastic washer
4 Socket headed grubscrew
5 Locknut

25.1 Removing steering column shrouds

25.3 Steering column lower support

Chapter 11 Suspension and steering

11 At the lower end of the column take off the end cap, depress the buttons on the rubber bushes where they protrude through the hole in the bottom and the top of the column and with a steel bar carefully drift the bushes out.
12 As soon as the bushes are removed from the top and bottom of the column, the metal and nylon inserts can be taken from them, unless completely new units (strongly advised) are fitted on reassembly.
13 Refitting is a straightforward reversal of the removal sequence but note the following points.
14 When fitting the steering wheel the two lips on the direction indicator cancelling lug on the column must be parallel and in line with the horizontal spokes of the wheel. With the wheel at the desired height tighten the bolts on the impact clamp. Note that the column will not be able to telescope if adjusted to its lowest position. Tighten the grub screw with an Allen key to the specified torque. Then tighten the locknut.

26 Steering lock/ignition switch – removal and refitting

1 Remove the steering column as described in Section 25.
2 Remove the two nuts and withdraw the lock shroud and steering column tie bar.
3 The two shear-headed bolts, which hold the steering column lock clamp bracket, must now be withdrawn by centre punching them and drilling a hole in them so that an extractor can be used (photo).
4 Remove the lock.
5 Fit the new lock by offering it up to the column so that the locating dowel in the lock engages in the hole in the column.
6 Do not shear the bolt heads until lock operation has been checked by inserting the ignition key.

27 Steering angles and front wheel alignment

1 Accurate front wheel alignment is essential for good steering and even tyre wear. Before considering the steering angles, check that the tyres are correctly inflated, that the front wheels are not buckled, the hub bearings are not worn or incorrectly adjusted and that the steering linkage is in good order, without slackness or wear at the joints.
2 Wheel alignment consists of four factors:

Camber is the angle at which the road wheels are set from the vertical when viewed from the front or rear of the vehicle. Positive camber is the angle (in degrees) that the wheels are tilted outwards at the top from the vertical.

Castor is the angle between the steering axis and a vertical line when viewed from each side of the vehicle. Positive castor is indicated when the steering axis is inclined towards the rear of the wheels at its upper end.

Steering axis inclination, is the angle, when viewed from the front or rear of the vehicle, between the vertical and an imaginary line drawn between the upper and lower suspension swivel balljoints.

Toe is the amount by which the distance between the front inside edges of the roadwheel rims differs from that between the rear inside edges. If the distance between the front edges is less than that at the rear, the wheels are said to toe-in. If the distance between the front inside edges is greater than that at the rear, the wheels toe-out.
3 Due to the need for precision gauges to measure the small angles of the steering and suspension settings, it is preferable that adjustment of camber is left to a service station having the necessary equipment. Camber is adjusted by adding or removing shims at the lower wishbone brackets. One shim will make an alteration of 1° to the camber angle. Castor is set in production and cannot be adjusted.
4 To check the front wheel alignment, first make sure that the lengths of both tie-rods are equal when the steering is in the straight ahead position.
5 Obtain a tracking gauge. These are available in various forms from

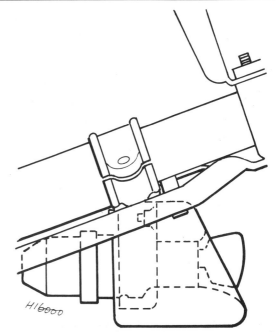

Fig. 11.24 Diagrammatic view of steering column lock/switch location (Sec 26)

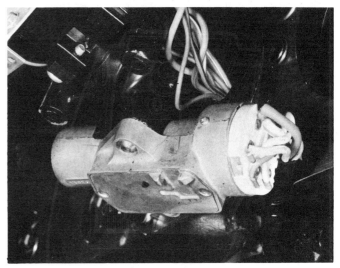

26.3 Steering column lock/ignition switch

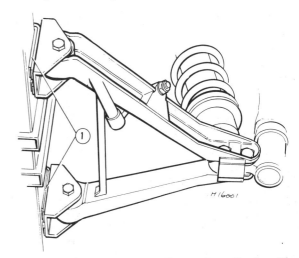

Fig. 11.25 Location of front suspension camber adjusting shims (1) (Sec 27)

Chapter 11 Suspension and steering

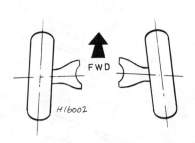

Fig. 11.26 Front wheel alignment diagram (exaggerated toe-in shown) (Sec 27)

27.6 Typical wheel alignment gauge

29.1 Fitting a roadwheel

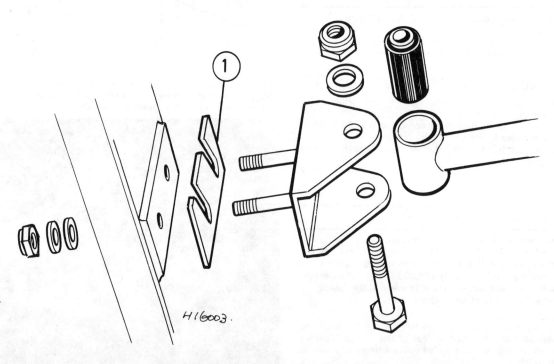

Fig. 11.27 Location of rear wheel alignment shims (Sec 28)

accessory stores. A gauge can be fabricated from a length of steel tubing suitably cranked to clear the sump and bellhousing and having a setscrew and locknut at one end.

6   With the gauge, measure the distance between the two wheel inner rims (at hub height) at the rear of the wheel. Push the vehicle forward to rotate the wheel through 180° (half a turn). Measure the distance between the wheel inner rims, again at hub height, at the front of the wheel. This last measurement should differ from the first by the appropriate toe-in (photo), according to specification (see Specifications Section).

7   Where the toe-in is found to be incorrect, release the tie-rod balljoint locknuts and turn the tie-rods equally. Only turn them a quarter of a turn at a time before re-checking the alignment. Do not grip the threaded part of the tie-rod during adjustment and make sure that the bellows outboard clip is released otherwise the bellows will twist as the tie-rod is rotated. Always turn both rods in the same direction when viewed from the centre line of the vehicle otherwise the rods will become unequal in length. This would cause the steering wheel spoke position to alter and cause problems on turns with tyre scrubbing.

8   On completion, tighten the tie-rod locknuts without disturbing their setting, check that the balljoint is at the centre of its arc of travel and then retighten the bellows clip.

28 Rear wheel alignment

1   Rear wheel alignment is adjusted by the insertion of shims behind the radius arm pivot brackets.
2   Any adjustment will apply to one particular wheel only and will not influence the total toe as is the case with the front roadwheels.
3   Unless a suitable gauge is available it is recommended that this work is left to your dealer.

29 Roadwheels and tyres

1   The standard roadwheels are of pressed-steel type retained by nuts (photo).
2   During earlier production runs on certain marks, wire wheels with knock-on hubs were available as an option.
3   Tyres on earlier models were of crossply construction while those on later models are of radial ply construction.
4   Check the tyre pressures weekly, including the spare.
5   The wheel nuts should be tightened to the appropriate torque as shown in the Specifications, and it is an advantage if a smear of grease is applied to the wheel stud threads.

Chapter 11 Suspension and steering

6 Every 6000 miles (10 000 km) the roadwheels should be moved round the vehicle (this does not apply where the wheels have been balanced on the vehicle) in order to even out the tyre tread wear. With radial types it is recommended that the wheels are moved front-to-rear and rear-to-front, not from side-to-side of the car. To do this, remove each wheel in turn, clean it thoroughly (both sides) and remove any flints which may be embedded in the tyre tread). Check the tread wear pattern which will indicate any mechanical or adjustment faults in the suspension or steering components. Examine the wheel bolt holes for elongation or wear. If such conditions are found, renew the wheel.
7 Renewal of the tyres should be carried out when the thickness of the tread pattern is worn to a minimum of 1 mm or the wear indicators (if incorporated) are visible.
8 Always adjust the front and rear tyre pressures after moving the wheels round as previously described.
8 All wheels are balanced initially, but have them done again halfway through the useful life of the tyres.
10 Never mix crossply and radial tyres on the same axle. It is permissible, however, to have crossplies on the front and radials on the rear, but neither this combination nor a mixture of textile and steel ply is recommended.

30 Fault diagnosis – suspension and steering

Symptom	Reason(s)
Steering feels vague, car wanders and floats at speed	Tyre pressures uneven Shock absorbers worn Spring broken Steering gear balljoints badly worn Suspension geometry incorrect Steering mechanism free play excessive Front suspension and rear axle pick-up points out of alignment
Stiff and heavy steering	Tyre pressures too low Steering gear unlubricated Front wheel toe-in incorrect Suspension geometry incorrect Steering gear incorrectly adjusted too tightly Steering column badly misaligned
Wheel wobble and vibration	Seized balljoints or swivels Wheel nuts loose Front wheels and tyres out of balance Steering balljoints badly worn Hub bearings badly worn or incorrectly adjusted Steering gear free play excessive Front springs weak or broken

Chapter 12 Bodywork and chassis

Contents

Bonnet (excluding 1500 models) – removal, refitting and adjustment	6
Bonnet (Mk IV and 1500 models) – removal, refitting and adjustment	7
Carpets – removal and refitting	33
Door – removal and refitting	28
Door exterior handle – removal and refitting	27
Door glass – removal and refitting	22
Door lock remote control – removal and refitting	26
Door lock striker – adjustment	25
Door locks – removal and refitting	24
Door trim panel – removal and refitting	21
Fascia crash roll – removal and refitting	38
Fascia panel – removal and refitting	37
Front bumper (excluding N. American models) – removal and refitting	10
Front bumper (N. American models) – removal and refitting	13
Front overrider (excluding N. American models) – removal and refitting	9
Front underrider (excluding N. American models) – removal and refitting	8
General description	1
Hardtop – removal and refitting	35
Hood – removal and refitting	36
Interior mirror – removal and refitting	40
Luggage boot lid – removal and refitting	29
Luggage boot lid lock and striker – removal, refitting and adjustment	30
Maintenance – bodywork and chassis	2
Maintenance – upholstery and carpets	3
Major body or chassis damage – repair	5
Minor body damage – repair	4
Parcels shelf – removal and refitting	34
Radiator grille (with overrider) – removal and refitting	17
Radiator grille (with underrider) – removal and refitting ...	16
Rear bumper (excluding N. American models) – removal and refitting	11
Rear bumper (earlier N. American models) – removal and refitting	14
Rear bumper (later N. American models) – removal and refitting	15
Rear overrider (excluding N. American models) – removal and refitting	12
Safety belts – maintenance, removal and refitting	39
Seats – removal and refitting	32
Spoiler – removal and refitting	18
Spoiler/air scoop (later N. American models) – removal and refitting	19
Transmission tunnel cover – removal and refitting	20
Windscreen glass – removal and refitting	31
Window regulator – removal and refitting	23

Specifications

Details of body dimensions, weights etc are given in the Introductory Chapter of this Manual.

Torque wrench settings	lbf ft	Nm
Bonnet hinge pivot bolt	30	41
Bumper end bolts	14	19
Bumper to body or brackets	30	41
Door hinge bolts	15	20
Seat belt anchorage bolts	30	41

1 General description

1 The all-steel welded body is bolted to a separate box section chassis frame, which consists of two longitudinal members joined together at three points. The chassis frame members at the centre of the car lie close together forming a strong 'backbone'. At both the front and rear ends the members fork so cradling the engine/gearbox unit and the differential respectively.

2 Short outriggers provide support for the body which is attached to the chassis by twelve nuts and bolts at six points. Two bolts are located in the engine compartment close to the dash panel; one bolt is positioned at the base of the front toe board under the accelerator pedal, with a further bolt in the same position on the passenger's side

Chapter 12 Bodywork and chassis

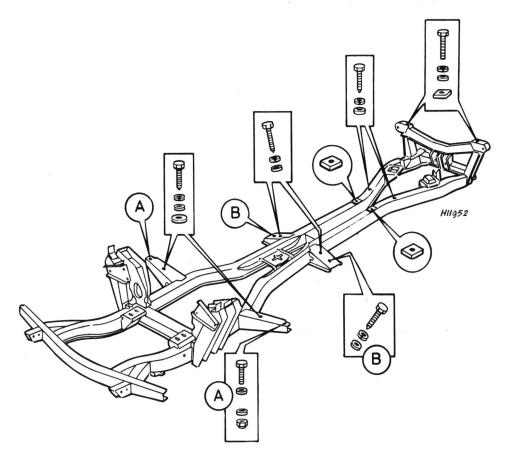

Fig. 12.1 Location of body mounting bolts (Sec 1)

of the car; two bolts can be found on each side of the body in line with the rear end of the gearbox cover; one bolt is located on each side of the front end of the rear seat pan; and finally one bolt can be found on each side of the spring access cover on the raised portion of the boot floor nearest the fuel tank and adjacent to each wheel arch.

3 An optional detachable hard-top has been available for all except the earliest models.

4 Over the years of production, certain design changes have been made to the grille, lamps, bumpers and overriders.

2 Maintenance – bodywork and underframe

1 The general condition of a vehicle's bodywork is the one thing that significantly affects its value. Maintenance is easy but needs to be regular. Neglect, particularly after minor damage, can lead quickly to further deterioration and costly repair bills. It is important also to keep watch on those parts of the vehicle not immediately visible, for instance the underside, inside all the wheel arches and the lower part of the engine compartment.

2 The basic maintenance routine for the bodywork is washing – preferably with a lot of water, from a hose. This will remove all the loose solids which may have stuck to the vehicle. It is important to flush these off in such a way as to prevent grit from scratching the finish. The wheel arches and underframe need washing in the same way to remove any accumulated mud which will retain moisture and tend to encourage rust. Paradoxically enough, the best time to clean the underframe and wheel arches is in wet weather when the mud is thoroughly wet and soft. In very wet weather the underframe is usually cleaned of large accumulations automatically and this is a good time for inspection.

3 Periodically, it is a good idea to have the whole of the underframe of the vehicle steam cleaned, engine compartment included, so that a thorough inspection can be carried out to see what minor repairs and renovations are necessary. Steam cleaning is available at many garages and is necessary for removal of the accumulation of oily grime which sometimes is allowed to become thick in certain areas. If steam cleaning facilities are not available, there are one or two excellent grease solvents available which can be brush applied. The dirt can then be simply hosed off.

4 After washing paintwork, wipe off with a chamois leather to give an unspotted clear finish. A coat of clear protective wax polish will give added protection against chemical pollutants in the air. If the paintwork sheen has dulled or oxidised, use a cleaner/polisher combination to restore the brilliance of the shine. This requires a little effort, but such dulling is usually caused because regular washing has been neglected. Always check that the door and ventilator opening drain holes and pipes are completely clear so that water can be drained out. Bright work should be treated in the same way as paintwork. Windscreens and windows can be kept clear of the smeary film which often appears, by adding a little ammonia to the water. If they are scratched, a good rub with a proprietary metal polish will often clear them. Never use any form of wax or other body or chromium polish on glass.

3 Maintenance – upholstery and carpets

1 Mats and carpets should be brushed or vacuum cleaned regularly to keep them free of grit. If they are badly stained remove them from the vehicle for scrubbing or sponging and make quite sure they are dry before refitting. Seats and interior trim panels can be kept clean by wiping with a damp cloth. If they do become stained (which can be more apparent on light coloured upholstery) use a little liquid detergent and a soft nail brush to scour the grime out of the grain of the material. Do not forget to keep the headlining clean in the same way as the upholstery. When using liquid cleaners inside the vehicle do not over-wet the surfaces being cleaned. Excessive damp could get into the seams and padded interior causing stains, offensive odours or even rot. If the inside of the vehicle gets wet accidentally it is

Chapter 12 Bodywork and chassis

worthwhile taking some trouble to dry it out properly, particularly where carpets are involved. *Do not leave oil or electric heaters inside the vehicle for this purpose.*

4 Minor body damage – repair

The photographic sequences on pages 294 and 295 illustrate the operations detailed in the following sub-sections.

Repair of minor scratches in bodywork

If the scratch is very superficial, and does not penetrate to the metal of the bodywork, repair is very simple. Lightly rub the area of the scratch with a paintwork renovator, or a very fine cutting paste, to remove loose paint from the scratch and to clear the surrounding bodywork of wax polish. Rinse the area with clean water.

Apply touch-up paint to the scratch using a fine paint brush; continue to apply fine layers of paint until the surface of the paint in the scratch is level with the surrounding paintwork. Allow the new paint at least two weeks to harden: then blend it into the surrounding paintwork by rubbing the scratch area with a paintwork renovator or a very fine cutting paste. Finally, apply wax polish.

Where the scratch has penetrated right through to the metal of the bodywork, causing the metal to rust, a different repair technique is required. Remove any loose rust from the bottom of the scratch with a penknife, then apply rust inhibiting paint to prevent the formation of rust in the future. Using a rubber or nylon applicator fill the scratch with bodystopper paste. If required, this paste can be mixed with cellulose thinners to provide a very thin paste which is ideal for filling narrow scratches. Before the stopper-paste in the scratch hardens, wrap a piece of smooth cotton rag around the top of a finger. Dip the finger in cellulose thinners and then quickly sweep it across the surface of the stopper-paste in the scratch; this will ensure that the surface of the stopper-paste is slightly hollowed. The scratch can now be painted over as described earlier in this Section.

Repair of dents in bodywork

When deep denting of the vehicle's bodywork has taken place, the first task is to pull the dent out, until the affected bodywork almost attains its original shape. There is little point in trying to restore the original shape completely, as the metal in the damaged area will have stretched on impact and cannot be reshaped fully to its original contour. It is better to bring the level of the dent up to a point which is about $\frac{1}{8}$ in (3 mm) below the level of the surrounding bodywork. In cases where the dent is very shallow anyway, it is not worth trying to pull it out at all. If the underside of the dent is accessible, it can be hammered out gently from behind, using a mallet with a wooden or plastic head. Whilst doing this, hold a suitable block of wood firmly against the outside of the panel to absorb the impact from the hammer blows and thus prevent a large area of the bodywork from being 'belled-out'.

Should the dent be in a section of the bodywork which has a double skin or some other factor making it inaccessible from behind, a different technique is called for. Drill several small holes through the metal inside the area – particularly in the deeper section. Then screw long self-tapping screws into the holes just sufficiently for them to gain a good purchase in the metal. Now the dent can be pulled out by pulling on the protruding heads of the screws with a pair of pliers.

The next stage of the repair is the removal of the paint from the damaged area, and from an inch or so of the surrounding 'sound' bodywork. This is accomplished most easily by using a wire brush or abrasive pad on a power drill, although it can be done just as effectively by hand using sheets of abrasive paper. To complete the preparation for filling, score the surface of the bare metal with a screwdriver or the tang of a file, or alternatively, drill small holes in the affected area. This will provide a really good 'key' for the filler paste.

To complete the repair see the Section on filling and re-spraying.

Repair of rust holes or gashes in bodywork

Remove all paint from the affected area and from an inch or so of the surrounding 'sound' bodywork, using an abrasive pad or a wire brush on a power drill. If these are not available a few sheets of abrasive paper will do the job just as effectively. With the paint removed you will be able to gauge the severity of the corrosion and therefore decide whether to renew the whole panel (if this is possible) or to repair the affected area. New body panels are not as expensive as most people think and it is often quicker and more satisfactory to fit a new panel than to attempt to repair large areas of corrosion.

Remove all fittings from the affected area except those which will act as a guide to the original shape of the damaged bodywork (eg headlamp shells etc). Then, using tin snips or a hacksaw blade, remove all loose metal and any other metal badly affected by corrosion. Hammer the edges of the hole inwards in order to create a slight depression for the filler paste.

Wire brush the affected area to remove the powdery rust from the surface of the remaining metal. Paint the affected area with rust inhibiting paint; if the back of the rusted area is accessible treat this also.

Before filling can take place it will be necessary to block the hole in some way. This can be achieved by the use of zinc gauze or aluminium tape.

Zinc gauze is probably the best material to use for a large hole. Cut a piece to the approximate size and shape of the hole to be filled, then position it in the hole so that its edges are below the level of the surrounding bodywork. It can be retained in position by several blobs of filler paste around its periphery.

Aluminium tape should be used for small or very narrow holes. Pull a piece off the roll and trim it to the approximate size and shape required, then pull off the backing paper (if used) and stick the tape over the hole; it can be overlapped if the thickness of one piece is insufficient. Burnish down the edges of the tape with the handle of a screwdriver or similar, to ensure that the tape is securely attached to the metal underneath.

Bodywork repairs – filling and re-spraying

Before using this Section, see the Sections on dent, deep scratch, rust holes and gash repairs.

Many types of bodyfiller are available, but generally speaking those proprietary kits which contain a tin of filler paste and a tube of resin hardener are best for this type of repair. A wide, flexible plastic or nylon applicator will be found invaluable for imparting a smooth and well contoured finish to the surface of the filler.

Mix up a little filler on a clean piece of card or board – measure the hardener carefully (follow the maker's instructions on the pack) otherwise the filler will set too rapidly or too slowly.

Using the applicator apply the filler paste to the prepared area; draw the applicator across the surface of the filler to achieve the correct contour and to level the filler surface. As soon as a contour that approximates to the correct one is achieved, stop working the paste – if you carry on too long the paste will become sticky and begin to 'pick up' on the applicator. Continue to add thin layers of filler paste at twenty-minute intervals until the level of the filler is just proud of the surrounding bodywork.

Once the filler has hardened, excess can be removed using a metal plane or file. From then on, progressively finer grades of abrasive paper should be used, starting with a 40 grade production paper and finishing with 400 grade wet-and-dry paper. Always wrap the abrasive paper around a flat rubber, cork, or wooden block – otherwise the surface of the filler will not be completely flat. During the smoothing of the filler surface the wet-and-dry paper should be periodically rinsed in water. This will ensure that a very smooth finish is imparted to the filler at the final stage.

At this stage the 'dent' should be surrounded by a ring of bare metal, which in turn should be encircled by the finely 'feathered' edge of the good paintwork. Rinse the repair area with clean water, until all of the dust produced by the rubbing-down operation has gone.

Spray the whole repair area with a light coat of primer – this will show up any imperfections in the surface of the filler. Repair these imperfections with fresh filler paste or bodystopper, and once more smooth the surface with abrasive paper. If bodystopper is used, it can be mixed with cellulose thinners to form a really thin paste which is ideal for filling small holes. Repeat this spray and repair procedure until you are satisfied that the surface of the filler, and the feathered edge of the paintwork are perfect. Clean the repair area with clean water and allow to dry fully.

The repair area is now ready for final spraying. Paint spraying must be carried out in a warm, dry, windless and dust free atmosphere. This condition can be created artificially if you have access to a large indoor working area, but if you are forced to work in the open, you will have to pick your day very carefully. If you are working indoors, dousing the floor in the work area with water will help to settle the dust which would otherwise be in the atmosphere. If the repair area is confined to one body panel, mask off the surrounding panels; this will help to minimise the effects of a slight mis-match in paint colours. Bodywork fittings (eg chrome strips, door handles etc) will also need to be masked off. Use genuine masking tape and several thicknesses of newspaper for the masking operations.

Before commencing to spray, agitate the aerosol can thoroughly, then spray a test area (an old tin, or similar) until the technique is mastered. Cover the repair area with a thick coat of primer; the thickness should be built up using several thin layers of paint rather than one thick one. Using 400 grade wet-and-dry paper, rub down the surface of the primer until it is really smooth. While doing this, the work area should be thoroughly doused with water, and the wet-and-dry paper periodically rinsed in water. Allow to dry before spraying on more paint.

Spray on the top coat, again building up the thickness by using several thin layers of paint. Start spraying in the centre of the repair area and then, using a circular motion, work outwards until the whole repair area and about 2 inches of the surrounding original paintwork is covered. Remove all masking material 10 to 15 minutes after spraying on the final coat of paint.

Allow the new paint at least two weeks to harden, then, using a paintwork renovator or a very fine cutting paste, blend the edges of the paint into the existing paintwork. Finally, apply wax polish.

5 Major body or chassis damage – repair

1 Rectification of major body or chassis damage should be entrusted to a specialist repairer having the necessary alignment jigs.
2 Failure to correct chassis distortion will affect roadholding, steering and tyre wear.

6 Bonnet (excluding 1500 models) – removal, refitting and adjustment

1 Remove the earth lead from the battery and pull apart the wires for the front lights and horn at the snap connections located inside the front of the bonnet above the top centre of the radiator grille.
2 One bolt holds each of the two front overriders in place. Undo the bolts and remove the overriders.
3 Undo the nut and bolt which hold the bonnet stay to the top of the coil spring/damper unit mounting bracket.
4 With the bonnet closed undo the bolt on Mk 1 and Mk 2 models from each of the two hinges.
5 On Mk 3 models undo and remove bolt from each of the hinges.
6 Replacement is a straightforward reversal of the removal sequence, but it will probably be necessary to adjust the bonnet so it closes correctly. A gap of $\frac{3}{16}$ in (5 mm) should exist between the bonnet, scuttle and doors. To achieve this loosen the bolts on Mk 1 and Mk 2 models and also on Mk 3 models and move the bonnet as required. Height adjustment so that a parallel gap exists between the rear vertical edge of the bonnet and the doors is made by slackening the bolts on both models and also by raising or lowering the rubber cones on the trailing edge of the bonnet. When all is correct tighten the bolts, recheck the gap, and reset the bonnet side catches on the side of the scuttle so the bonnet does not rattle and is held tightly shut.

7 Bonnet (Mk IV and 1500 models – removal, refitting and adjustment

1 Disconnect the battery.
2 Disconnect the lamp and horn leads at their connectors inside the front grille (photo).
3 Unbolt the bonnet stay from the wheel arch (photo).
4 Close the bonnet and except on N. American models, remove the overriders (see Section 8).
5 On N. American models, remove the front bumper (see Section 13).
6 With the help of an assistant to support the bonnet, remove the bonnet hinge pivot bolts and lift the bonnet from the car.
7 Refitting is a reversal of removal, but the following adjustment operations may be required.
8 Fore and aft movement can be carried out if the two hinge pivot bolts are first released (photo).

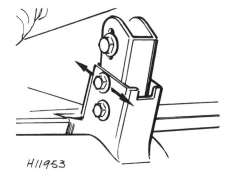

Fig. 12.2 Bonnet hinge bolts (Mk I and II) (Sec 6)

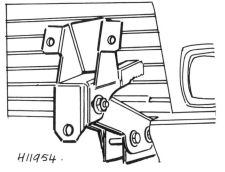

Fig. 12.3 Bonnet hinge bolts (Mk III) (Sec 6)

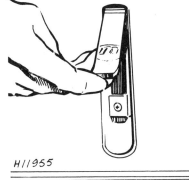

Fig. 12.4 Bonnet release lever (Sec 7)

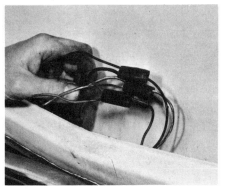

7.2 Wiring connectors inside front grille

7.3 Bonnet stay at wheel arch

7.8 Bonnet hinge pivot bolt

7.9 Bonnet buffer stop plate

7.10 Bonnet vertical adjustment bolts

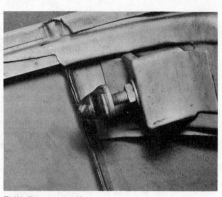
7.11 Bonnet buffer

9 Adjust the position of the stop plates to accept the buffers centrally (photo).
10 The bonnet can be raised or lowered if the two bolts on each hinge bracket are first released (photo).
11 The buffers may also be adjusted for height once the locknut is released (photo).
12 The bonnet catches may need slight adjustment of position on completion of bonnet alignment to provide positive rattle-free closures.

8 Front underrider (excluding N. American models) – removal and refitting

1 Remove the three screws from each underrider and lift them from the car.
2 Refit by reversing the removal operations.

9 Front overrider (excluding N. American models) – removal and refitting

1 From underneath and to the rear of the overrider, remove the bolt, spring washer and plain washer.
2 Open the bonnet and working within the engine compartment unclip the overrider from the bumper support bracket.

10 Front bumper (excluding N. American models) – removal and refitting

1 Remove the front underrider or overrider as previously described.
2 Open the bonnet and remove the bumper securing bolts.
3 Close the bonnet again and lift off the bumper.
4 Refitting is a reversal of removal.

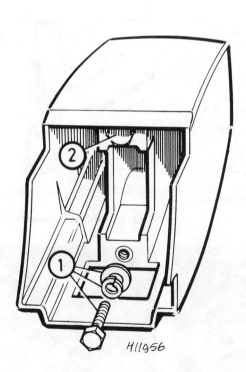

Fig. 12.5 Front overrider (Sec 9)

1 Lower fixing bolt 2 Bumper bracket bolt

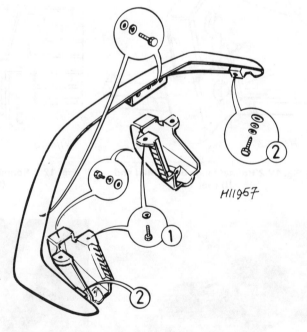

Fig. 12.6 Front bumper fixings (typical) (Sec 10)

1 Underrider bolt 2 End bolts

Chapter 12 Bodywork and chassis

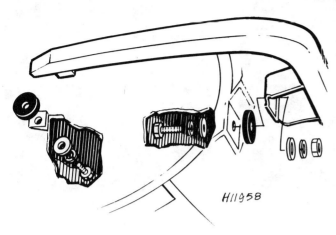

Fig. 12.7 Rear bumper fixings (typical) (Sec 11)

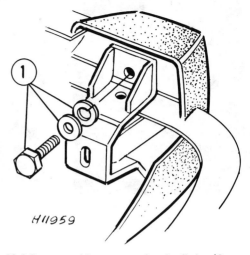

Fig. 12.8 Rear overrider – removal and refitting (Sec 12)

11 Rear bumper (excluding N. American models) – removal and refitting

1 Disconnect the leads to the rear number plate lamps at their connectors.
2 Working within the luggage boot, remove the nuts, bolts and spring washers. An assistant to support the bumper during removal will prove helpful.

12 Rear overrider (excluding N. American models) – removal and refitting

1 Remove the rear bumper as described in the preceding Section.
2 Unbolt the overrider from the bumper support bracket.
3 Refitting is a reversal of removal.

13 Front bumper (N. American models) – removal and refitting

1 Unscrew the two end retaining bolts.
2 Open the bonnet and unscrew the five upper retaining bolts.
3 Slacken, but do not remove, the two bolts from the centre of the lower edge of the bumper.
4 Slacken, but do not remove, the bolt from the lower edge of each overrider.
5 Remove the parking lamp lenses.
6 Remove the parking lamp body screws and push the lamp bodies through their bumper cut-outs.
7 Pull out the bumper from the car.
8 The energy absorbing buffer can be removed if the overrider and centre blocks are unbolted followed by removal of the bolts from the number plate reinforcement.
9 Refitting is a reversal of removal.

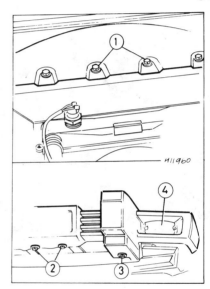

Fig. 12.9 Front bumper (later N. American) – attachment (Sec 13)

1 Upper bolts
2 Centre lower bolts
3 Overrider lower bolts
4 Lamp lens

14 Rear bumper (earlier N. American models) – removal and refitting

1 Prise out the two plastic clips which retain the overriders, then remove the bumper securing bolts and nuts.
2 Lift the complete bumper/overrider assembly from the car.
3 Refitting is a reversal of removal, but make sure that the rubber washers are correctly located between the bumper and overrider and the body panel.

15 Rear bumper (later N. American models) – removal and refitting

1 Working under the rear of the car, unbolt the bumper extension brackets.

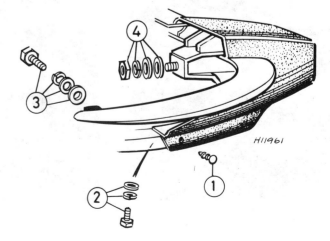

Fig. 12.10 Rear bumper (earlier N. American) attachment (Sec 14)

1 Plastic retainer
2 Overrider bolt
3 Bumper end bolt
4 Bumper/overrider mounting bolt

This photographic sequence shows the steps taken to repair the dent and paintwork damage shown above. In general, the procedure for repairing a hole will be similar; where there are substantial differences, the procedure is clearly described and shown in a separate photograph.

First remove any trim around the dent, then hammer out the dent where access is possible. This will minimise filling. Here, after the large dent has been hammered out, the damaged area is being made slightly concave.

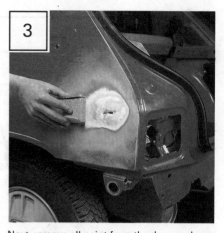
Next, remove all paint from the damaged area, by rubbing with coarse abrasive paper or using a power drill fitted with a wire brush or abrasive pad. 'Feather' the edge of the boundary with good paintwork using a finer grade of abrasive paper.

Where there are holes or other damage, the sheet metal should be cut away before proceeding further. The damaged area and any signs of rust should be treated with Turtle Wax Hi-Tech Rust Eater, which will also inhibit further rust formation.

For a large dent or hole mix Holts Body Plus Resin and Hardener according to the manufacturer's instructions and apply around the edge of the repair. Press Glass Fibre Matting over the repair area and leave for 20-30 minutes to harden. Then ...

... brush more Holts Body Plus Resin and Hardener onto the matting and leave to harden. Repeat the sequence with two or three layers of matting, checking that the final layer is lower than the surrounding area. Apply Holts Body Plus Filler Paste as shown in Step 5B.

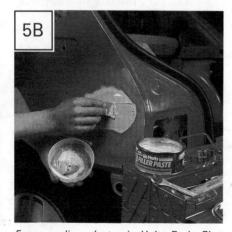
For a medium dent, mix Holts Body Plus Filler Paste and Hardener according to the manufacturer's instructions and apply it with a flexible applicator. Apply thin layers of filler at 20-minute intervals, until the filler surface is slightly proud of the surrounding bodywork.

For small dents and scratches use Holts No Mix Filler Paste straight from the tube. Apply it according to the instructions in thin layers, using the spatula provided. It will harden in minutes if applied outdoors and may then be used as its own knifing putty.

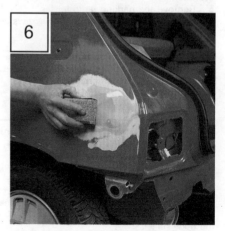
Use a plane or file for initial shaping. Then, using progressively finer grades of wet-and-dry paper, wrapped round a sanding block, and copious amounts of clean water, rub down the filler until glass smooth. 'Feather' the edges of adjoining paintwork.

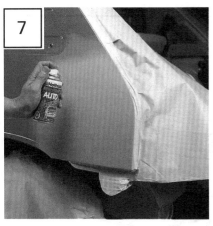

Protect adjoining areas before spraying the whole repair area and at least one inch of the surrounding sound paintwork with Holts Dupli-Color primer.

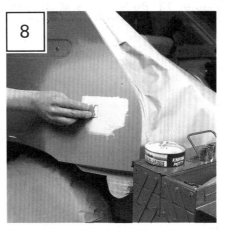

Fill any imperfections in the filler surface with a small amount of Holts Body Plus Knifing Putty. Using plenty of clean water, rub down the surface with a fine grade wet-and-dry paper – 400 grade is recommended – until it is really smooth.

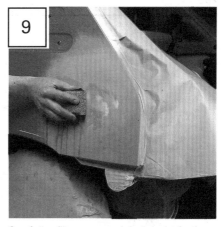

Carefully fill any remaining imperfections with knifing putty before applying the last coat of primer. Then rub down the surface with Holts Body Plus Rubbing Compound to ensure a really smooth surface.

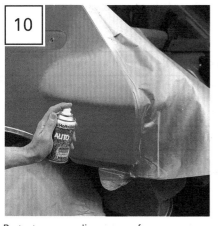

Protect surrounding areas from overspray before applying the topcoat in several thin layers. Agitate Holts Dupli-Color aerosol thoroughly. Start at the repair centre, spraying outwards with a side-to-side motion.

If the exact colour is not available off the shelf, local Holts Professional Spraymatch Centres will custom fill an aerosol to match perfectly.

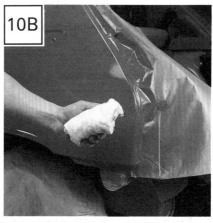

To identify whether a lacquer finish is required, rub a painted unrepaired part of the body with wax and a clean cloth.

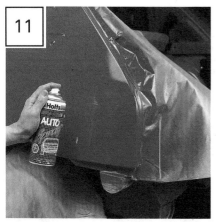

If *no* traces of paint appear on the cloth, spray Holts Dupli-Color clear lacquer over the repaired area to achieve the correct gloss level.

The paint will take about two weeks to harden fully. After this time it can be 'cut' with a mild cutting compound such as Turtle Wax Minute Cut prior to polishing with a final coating of Turtle Wax Extra.

When carrying out bodywork repairs, remember that the quality of the finished job is proportional to the time and effort expended.

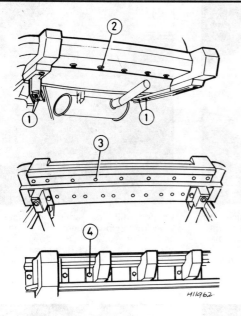

Fig. 12.11 Rear bumper (later N. American) attachment (Sec 15)

1 Bumper extension bracket bolts
2 Lower socket headed screws
3 Upper socket headed screws
4 Buffer fixing screws

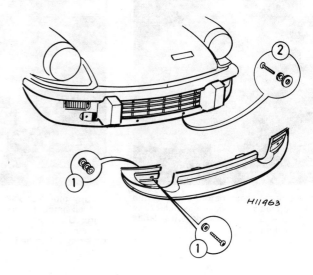

Fig. 12.12 Spoiler fixing bolts (1 and 2) (Sec 18)

2 Unscrew the two bumper end bolts.
3 Working within the luggage boot remove the two bumper securing bolts.
4 Lift the bumper assembly from the car.
5 If it is necessary to remove the energy absorbing buffers carry out the following operations.
6 Remove the five socket-headed screws from beneath the bumper.
7 Remove the nine socket-headed screws from the top retaining strip at the rear of the bumper.
8 Remove the four bumper side bracket screws.
9 Pull off the bumper cover and unscrew the two fixing screws from each buffer.

16 Radiator grille (with underrider) – removal and refitting

1 Remove the screw and plain washer from the centre of the grille (photo).
2 Remove the underrider bolts from one underrider only and remove the underrider with grille which can be eased out from under the second underrider (photos).

17 Radiator grille (with overrider) – removal and refitting

1 The operations are similar to those described in the preceding Section except that overriders not underriders are fitted.

18 Spoiler – removal and refitting

1 Remove the screws and spacers from both ends of the spoiler.
2 Support the spoiler and remove the inboard screws which hold it to the front crossmember.
3 Refitting is a reversal of removal, but fully tighten the inboard screws before the outboard screws.

19 Spoiler/air scoop (later N. American models) – removal and refitting

1 Lever the four clips, which are located under the air scoop, away from the body.
2 Remove the two end retaining screws.
3 Lower the air scoop and reversal of removal.

20 Transmission tunnel cover – removal and refitting

1 These operations are covered in Chapter 6, Section 3.

16.1 Radiator grille centre screw

16.2a Underrider bolts

16.2b Grille fixing at underrider

Chapter 12 Bodywork and chassis

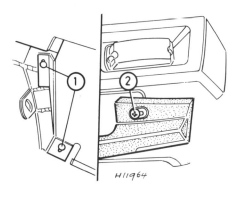

Fig. 12.13 Spoiler/air scoop fixing bolts (Sec 19)

1 Clips   2 End screws

21 Door trim panel – removal and refitting

1 Open the door wide, push the window regulator handle bezel against the trim panel and push out the handle retaining pin (photo).
2 Remove the window regulator handle.
3 Extract the screw from the door lock remote control handle escutcheon plate. Slide the plate off after having rotated it through 90° (photo).
4 Insert the fingers or a broad flat blade between the trim panel and the door and release the securing clips using a sharp, jerking motion (photos).
5 Refitting is a reversal of removal.

22 Door glass – removal and refitting

1 Remove the trim panel as described in Section 21.
2 Take off the conical spring from the window regulator and then temporarily push the regulator handle onto its shaft. The glass height will require adjustment to bring the regulator arms into view.
3 Remove the guide packing pieces (two bolts).
4 Disconnect the window regulator arms from the glass frame by

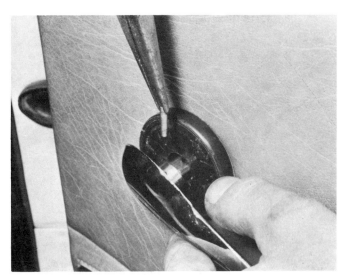

21.1 Window regulator handle retaining pin

21.3 Remote control handle plate screw

21.4a Removing door trim panel

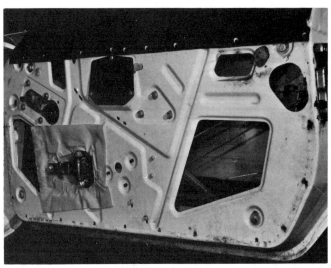

21.4b View of door inner panel

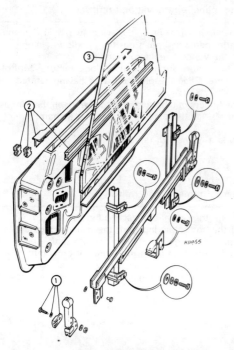

Fig. 12.14 Door glass details (Sec 22)

1 Guide packing pieces
2 Door waist weatherseals
3 Glass

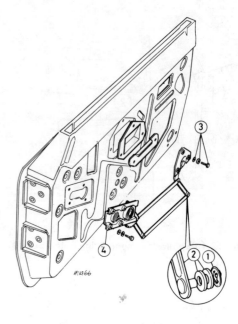

Fig. 12.15 Glass regulator components (Sec 23)

1 Retainer
2 Link and leather washers
3 Pivot plate bolts
4 Regulator

prising off the retainers and washers. Support the glass as necessary.
5 Carefully lower the glass.
6 Prise off the door waist inner and outer weatherseals from their clips.
7 Pull the glass from the door cavity.
8 Refitting is a reversal of removal.

23 Window regulator – removal and refitting

1 Remove the trim panel (Section 21).
2 Remove the conical spring from the regulator and temporarily refit the regulator handle.
3 Wind the door glass fully up and either wedge it or have an assistant retain it in this position.
4 Disconnect the regulator arms from the glass frame by prising off the retainers and washers.
5 Remove the interconnecting link with four leather washers.
6 Remove the three bolts which hold the triangular pivot plate and then the four bolts which secure the regulator unit (photos).
7 Withdraw the regulator assembly from the door cavity.
8 Refitting is a reversal of removal.

24 Door lock – removal and refitting

1 Remove the trim panel (Section 21).
2 Working through the small aperture in the door panel, disconnect the link rods (photo).
3 Working at the edge of the door, remove the three screws which hold the lock assembly.
4 Remove the lock and linkage from the door cavity (photo).
5 Refitting is a reversal of removal.

25 Door lock striker – adjustment

1 The door lock striker may require adjustment to compensate for door sag or after removal or refitting of a door.
2 To adjust, release but do not remove the striker securing screws

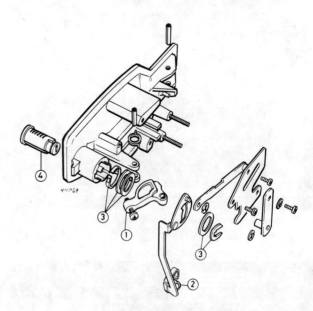

Fig. 12.16 Door exterior handle components (Sec 27)

1 Pivot bracket
2 Spring clip
3 Pivot bracket washers and clips
4 Lock cylinder

and move the striker as necessary so that its spigot passes centrally through the opening in the lock and gives positive closure without any tendency to rattle.

26 Door lock remote control – removal and refitting

1 Remove the trim panel as described in Section 21.
2 Take off the foam insulating pad from around the remote control lever assembly.
3 Detach the clip from the control rod at the remote control.

Chapter 12 Bodywork and chassis

299

23.6a Regulator pivot plate bolts

23.6b Regulator unit bolts

24.2 Door lock link rods

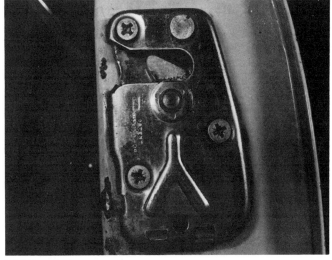

24.4 Door lock

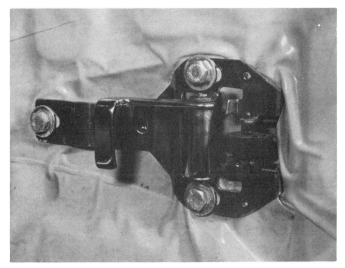

26.4 Door lock remote control

4 Unscrew the three remote control fixing bolts (photo).
5 Disconnect the control rod from the locking lever.
6 Withdraw the remote control assembly.
7 Refitting is a reversal of removal.

27 Door exterior handle – removal and refitting

1 Raise the glass fully and remove the trim panel (Section 21).
2 Remove the conical spring from the regulator, and the foam insulating pad from the remote control.
3 Working from the inner door panel using a long box spanner or socket with extension, unscrew the two handle securing nuts.
4 Disconnect the door lock control rod and then withdraw the exterior handle assembly with gasket.
5 The lock cylinder may be removed from the handle after removing the spring clip and the pivot bracket.
5 Reassembly and refitting are reversals of removal and dismantling.

28 Door – removal and refitting

1 Open the door fully and support on blocks suitably protected with pads of rag or other insulating material.

28.3 Door hinges

29.1 Luggage boot lid stay showing hinge

29.2 Luggage boot lid hinge

2 Drill out the rivet that secures the door check strap. Mark the outline of the hinges on the body.
3 Remove the six bolts which hold the door hinges to the body and lift the door away (photo).
4 If the hinges are to be renewed, unbolt them from the door edge.
5 Refitting is a reversal of removal, but before fully tightening the hinge bolts, check the alignment of the door within the body opening and for the door panel being flush with adjacent body panels when it is closed. Move the door as necessary within the tolerance provided at the door pillar bolt holes.
6 Finally adjust the door lock striker (Section 25).

29 Luggage boot lid – removal and refitting

1 Open the lid fully and with an assistant supporting it, unbolt the upper end of the stay (photo).
2 Mark the position of the hinges on the underside of the boot lid and then unscrew and remove the hinge bolts (photo).
3 Lift the lid from the car.
4 Refitting is a reversal of removal, but do not fully tighten the hinge bolts until the alignment of the lid has been checked in the closed position. The lid can be moved as necessary within the tolerance provided at the hinge bolt holes.
5 Reconnect the stay.
6 If closure is not positive, adjust the lock and striker as described in the next Section.

30 Luggage boot lid lock and striker – removal, refitting and adjustment

1 Open the lid fully and remove the striker sleeve (two screws 1) (photos).
2 Remove the screw (2).
3 Remove the four screws (3).
4 Remove the lock.
5 Pull out the lock barrel and shaft.
6 Push out the lock barrel escutcheon and sealing washer.
7 The striker is retained by two screws.
8 Refitting is a reversal of removal, but do not tighten the striker sleeve or striker screws until their mating action has been checked. Both components may be adjusted within the tolerance provided at their screw holes.

31 Windscreen glass – removal and refitting

1 If you are unfortunate enough to have a windscreen crack, fitting a replacement windscreen is one of the few jobs which the average owner is advised to leave to a professional mechanic. The owner who wishes to do the job himself will need the help of an assistant.
2 Take off the windscreen wiper arms and blades, the interior rear view mirror and remove the bright metal windscreen surround trims. The ends of the trims will be exposed by sliding the escutcheon at the middle of the bottom portion of the moulding to one side. The mouldings and escutcheon should then be carefully eased out of the rubber channel.
3 Work all round and under the outside edge of the windscreen sealing rubber with a screwdriver to break the mastic seal between the rubber and the windscreen frame flange.
4 Spread a blanket over the bonnet and with an assistant steadying the glass from the outside sit in the passenger's seat and press out the glass and rubber surround with one foot. Place rag between your foot

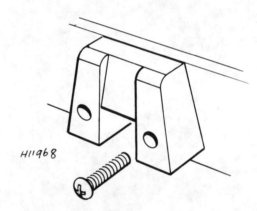
Fig. 12.17 Luggage boot lid lock striker (Sec 30)

30.1 Luggage boot lid lock
1 Striker sleeve screws   3 Lock fixing screws
2 Lock barrel shaft screw

Chapter 12 Bodywork and chassis

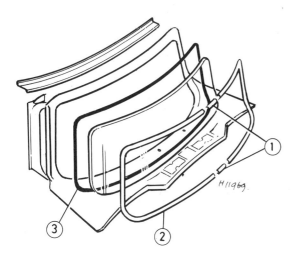

Fig. 12.18 Windscreen components (Sec 31)

1 Trim finisher covers
2 Bright trim
3 Rubber weathersealing surround

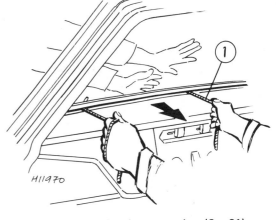

Fig. 12.19 Fitting the screen glass (Sec 31)

1 Cord

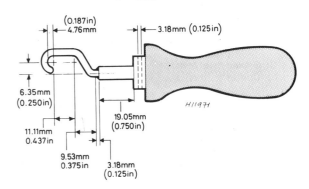

Fig. 12.20 Windscreen bright trim fitting tool (Sec 31)

and the glass. The glass is started most easily at one of the corners.
5 Clean off all the old mastic from the windscreen flange.
6 The windscreen is replaced after having smeared mastic sealing compound on the outside edge of the glass where it is covered by the rubber surround.
7 With the rubber surround fitted to the windscreen so that the joint is at the bottom, insert a length of cord all round the channel in the rubber which will sit over the windscreen aperture flange. Allow the two free ends of the cord to overlap slightly at the bottom centre.
8 Fit the windscreen from outside the car and with an assistant pressing the rubber surround hard against the body flange, slowly pull one end of the cord out moving round the windscreen and so drawing the lip of the rubber over the windscreen flange on the body.
9 Refit the bright trim. This is probably the most difficult job and a tool made up in accordance with the diagram will make fitting much easier by sliding the rubber lips over the trim.
10 Fit the trim escutcheon covers, the wiper arms and the mirror. Where applicable retrieve the vehicle licence disc or other labels from the old screen.

32 Seats – removal and refitting

1 Push the seat fully rearward and unscrew the front seat slide bolt.
2 Move the seat to the tilt position and unscrew the seat slide rear bolt (photos).
3 If a safety belt warning system is fitted, disconnect the wiring at the floor connector (photo).
4 Remove the seat from the car.
5 Refitting is a reversal of removal.

33 Carpets – removal and refitting

1 The front carpet can be removed only after first having withdrawn the transmission tunnel cover as described in Chapter 6, Section 3.
2 The rear carpet is removed after withdrawing the seats (Section 32), the safety belt anchorages, the centre arm rest and trim pads.
3 Refit by reversing the removal operations.

34 Parcels shelf – removal and refitting

1 This is simply a matter of removing the fixing nuts and bolts and lifting the shelf away.
2 Refitting is a reversal of removal.

32.2a Seat tilted

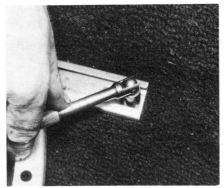

32.2b Removing seat slide bolt

32.3 Seat belt warning switch wiring connector

Chapter 12 Bodywork and chassis

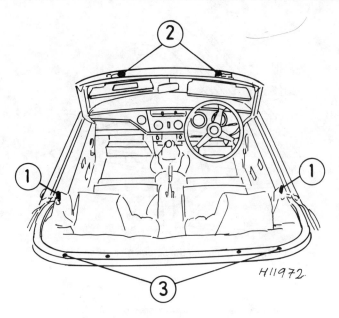

Fig. 12.21 Hard top attachment points (Sec 35)

1 Domed-head bolts
2 Domed-head bolts
3 Bolts and washers

35 Hardtop – removal and refitting

1 Disconnect the three fasteners which hold the hood stowage cover to the rear deck.
2 Remove the two domed-headed bolts (2) with their washers towards the rear.
3 Remove the remaining dome-headed bolts (3) on the windscreen rail.
4 Remove the two bolts from the rear edge.
5 With the help of an assistant, lift the hardtop from the car.
6 Refitting is a reversal of removal, but check that the spacer sleeves are located at the windscreen rail.

36 Hood – removal and refitting

1 Disconnect the eight fasteners around the bottom edge of the hood.
2 Remove the bolts which secure the hood hinges and stowage brackets (if fitted).
3 Remove the four dome-headed bolts which secure the hood retainer and the hood.
4 Remove the hood complete.
5 Refitting is a reversal of removal.

37 Fascia panel – removal and refitting

1 Disconnect the battery.

Centre section

2 Slacken the two grub screws and pull the knobs from the heater control levers.
3 Extract the four screws and their cup washers from the corners of the centre panel.
4 Carefully lower the panel and then pull out the instrument panel lamp bulb holders and disconnect the electrical spade terminals (photo).
5 Remove the centre panel from the car.

Passenger side section

6 Remove the crash padding from the lower edge of the fascia panel. This is held by four screws.

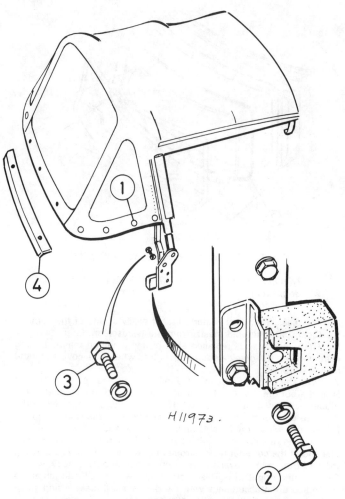

Fig. 12.22 Hood attachment points (Sec 36)

1 Fastener
2 Stowage bracket bolts
3 Hinge bolt
4 Hood retainer

37.4 Centre fascia panel lowered

Chapter 12 Bodywork and chassis

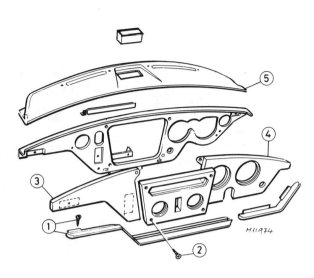

Fig. 12.23 Fascia panel (Sec 37)

1. Lower crash padding
2. Centre panel screw
3. Passenger side section
4. Driver's side section
5. Crash roll

7 Remove the screw from the top corner of the centre panel at the side nearest the passenger.
8 Reach behind the passenger side fascia and unscrew the two retaining nuts. Withdraw the fascia section.

Driver's side section

9 Remove the longer section of the crash padding from the lower edge of the fascia. This is held by four screws.
10 Release and lower the fascia centre section as previously described.
11 Refer to Chapter 10 and remove the speedometer.
12 Remove the hazard warning switch. Do this by unscrewing the switch knob and then the switch locking ring.
13 Refer to Chapter 10 and remove the tachometer.
14 Disconnect the choke control cable from the carburettor. Pull the choke control knob to withdraw the inner cable and then unscrew the outer cable bezel to release the outer cable from the fascia panel.
15 Remove the wiper switch (early models). To do this refer to Chapter 10.
16 Pull out the direction indicator warning lamp bulb holder.
17 Reach behind the fascia panel and unscrew the two panel securing nuts and remove the panel.
18 Refitting is a reversal of removal. Reconnect the battery.

38 Fascia crash roll – removal and refitting

1 Remove the three fascia panel sections as described in the preceding Section.
2 Remove the two bolts which hold the air flow control to the fascia bracket.
3 Unbolt and remove the demister vents.
4 Unscrew and remove the six nuts which hold the crash roll in position.
5 Pull out the ashtray.
6 Straighten the retaining tags and withdraw the ashtray retainer.
7 Prise the crash roll upwards and remove it.
8 Refitting is a reversal of removal.

39 Safety belts – maintenance, removal and refitting

1 Seat belts should be checked regularly for cuts, fraying or other damage and if evident, renewed immediately.

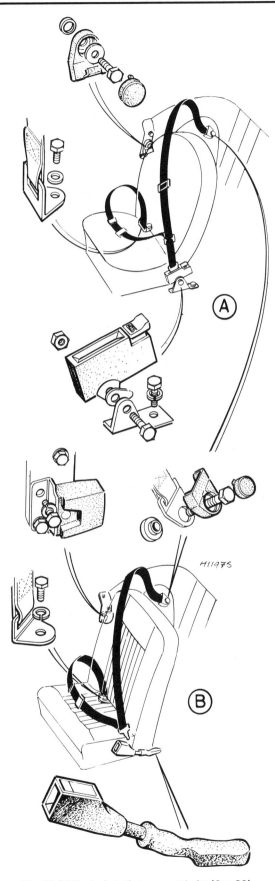

Fig. 12.24 Typical static type seat belts (Sec 39)

A Earlier type   B Later type

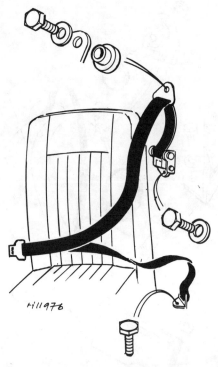

Fig. 12.25 Typical inertia reel seat belt (Sec 39)

2 To clean a seat belt use warm water and a little detergent, nothing else.
3 It is most important to retain the original sequence of fitting of the spacers, washers and anchor plates when refitting after removal.
4 Never alter the maker's points of attachment.
5 Earlier cars were fitted with static seat belts while later models have the inertia reel type.

40 Interior mirror – removal and refitting

1 The interior rear view mirror is designed with a breakable base in order to avoid injury to occupants in the event of a front-end collision.
2 The mirror may be slid out of its retaining base if one of the end blocks is unscrewed and removed (photos).

40.2a Mirror end retaining block

40.2b Removing mirror base

Conversion factors

Length (distance)
Inches (in)	X	25.4	= Millimetres (mm)	X 0.0394	= Inches (in)
Feet (ft)	X	0.305	= Metres (m)	X 3.281	= Feet (ft)
Miles	X	1.609	= Kilometres (km)	X 0.621	= Miles

Volume (capacity)
Cubic inches (cu in; in³)	X	16.387	= Cubic centimetres (cc; cm³)	X 0.061	= Cubic inches (cu in; in³)
Imperial pints (Imp pt)	X	0.568	= Litres (l)	X 1.76	= Imperial pints (Imp pt)
Imperial quarts (Imp qt)	X	1.137	= Litres (l)	X 0.88	= Imperial quarts (Imp qt)
Imperial quarts (Imp qt)	X	1.201	= US quarts (US qt)	X 0.833	= Imperial quarts (Imp qt)
US quarts (US qt)	X	0.946	= Litres (l)	X 1.057	= US quarts (US qt)
Imperial gallons (Imp gal)	X	4.546	= Litres (l)	X 0.22	= Imperial gallons (Imp gal)
Imperial gallons (Imp gal)	X	1.201	= US gallons (US gal)	X 0.833	= Imperial gallons (Imp gal)
US gallons (US gal)	X	3.785	= Litres (l)	X 0.264	= US gallons (US gal)

Mass (weight)
Ounces (oz)	X	28.35	= Grams (g)	X 0.035	= Ounces (oz)
Pounds (lb)	X	0.454	= Kilograms (kg)	X 2.205	= Pounds (lb)

Force
Ounces-force (ozf; oz)	X	0.278	= Newtons (N)	X 3.6	= Ounces-force (ozf; oz)
Pounds-force (lbf; lb)	X	4.448	= Newtons (N)	X 0.225	= Pounds-force (lbf; lb)
Newtons (N)	X	0.1	= Kilograms-force (kgf; kg)	X 9.81	= Newtons (N)

Pressure
Pounds-force per square inch (psi; lbf/in²; lb/in²)	X	0.070	= Kilograms-force per square centimetre (kgf/cm²; kg/cm²)	X 14.223	= Pounds-force per square inch (psi; lbf/in²; lb/in²)
Pounds-force per square inch (psi; lbf/in²; lb/in²)	X	0.068	= Atmospheres (atm)	X 14.696	= Pounds-force per square inch (psi; lbf/in²; lb/in²)
Pounds-force per square inch (psi; lbf/in²; lb/in²)	X	0.069	= Bars	X 14.5	= Pounds-force per square inch (psi; lbf/in²; lb/in²)
Pounds-force per square inch (psi; lbf/in²; lb/in²)	X	6.895	= Kilopascals (kPa)	X 0.145	= Pounds-force per square inch (psi; lbf/in²; lb/in²)
Kilopascals (kPa)	X	0.01	= Kilograms-force per square centimetre (kgf/cm²; kg/cm²)	X 98.1	= Kilopascals (kPa)

Torque (moment of force)
Pounds-force inches (lbf in; lb in)	X	1.152	= Kilograms-force centimetre (kgf cm; kg cm)	X 0.868	= Pounds-force inches (lbf in; lb in)
Pounds-force inches (lbf in; lb in)	X	0.113	= Newton metres (Nm)	X 8.85	= Pounds-force inches (lbf in; lb in)
Pounds-force inches (lbf in; lb in)	X	0.083	= Pounds-force feet (lbf ft; lb ft)	X 12	= Pounds-force inches (lbf in; lb in)
Pounds-force feet (lbf ft; lb ft)	X	0.138	= Kilograms-force metres (kgf m; kg m)	X 7.233	= Pounds-force feet (lbf ft; lb ft)
Pounds-force feet (lbf ft; lb ft)	X	1.356	= Newton metres (Nm)	X 0.738	= Pounds-force feet (lbf ft; lb ft)
Newton metres (Nm)	X	0.102	= Kilograms-force metres (kgf m; kg m)	X 9.804	= Newton metres (Nm)

Power
Horsepower (hp)	X	745.7	= Watts (W)	X 0.0013	= Horsepower (hp)

Velocity (speed)
Miles per hour (miles/hr; mph)	X	1.609	= Kilometres per hour (km/hr; kph)	X 0.621	= Miles per hour (miles/hr; mph)

*Fuel consumption**
Miles per gallon, Imperial (mpg)	X	0.354	= Kilometres per litre (km/l)	X 2.825	= Miles per gallon, Imperial (mpg)
Miles per gallon, US (mpg)	X	0.425	= Kilometres per litre (km/l)	X 2.352	= Miles per gallon, US (mpg)

Temperature
Degrees Fahrenheit = (°C x 1.8) + 32   Degrees Celsius (Degrees Centigrade; °C) = (°F - 32) x 0.56

*It is common practice to convert from miles per gallon (mpg) to litres/100 kilometres (l/100km), where mpg (Imperial) x l/100 km = 282 and mpg (US) x l/100 km = 235

Index

A

About this manual – 2
Acknowledgements – 2
Air cleaner
   general – 87
   servicing – 87, 89
Antifreeze – 77
Anti-roll bar (front)
   removal and refitting – 272
Alternator
   description – 216
   drivebelt adjustment – 218
   removal and refitting – 218

B

Battery
   maintenance – 212
   removal and refitting – 213
Bearings (front hub)
   adjustment – 270
   renewal – 270
Big-end and main bearings
   examination – 50
Bleeding the brakes – 207
Bleeding the clutch – 136
Bodywork and chassis – 288 *et seq*
Bodywork and chassis
   body damage
      major – 291
      minor – 290
   description – 288
   front bumpers – 292
   front underrider – 292
   maintenance
      bodywork and underframe – 289
      upholstery and carpets – 289
   rear bumper – 293
   rear overrider – 293
   specifications – 288

Bodywork repair sequence (colour) – 294 and 295
Bonnet
   removal, refitting and adjustment – 291
Brake drum
   inspection and renewal – 201
Brake pedal – 209
Brake pipes
   flexible – 205
   rigid – 205
Braking system – 196 *et seq*
   bleeding – 207
   description – 197
   fault diagnosis – 209
   maintenance – 197
   specifications – 196
Bulbs – 211
Buying spare parts – 10

C

Caliper (brakes)
   removal, overhaul and refitting – 200
Camshaft (1147 cc & 1296 cc)
   removal and refitting – 45
   examination – 51
Camshaft (1493 cc)
   removal and refitting – 65
Capacities – 8
Carburettor (CD4)
   choke removal – 106
   deceleration and by-pass valve
      checking and adjusting – 106
   overhaul – 107
   removal and refitting – 107
   tuning – 105
Carburettor (CD4T)
   automatic choke – 106
   deceleration and by-pass valve
      checking and adjusting – 106
   overhaul – 109

Index

removal and refitting – 109
tuning – 105
Carburettors (general)
description – 92
specifications – 85
throttle and choke controls – 113
Carburettor (Stromberg 150 CDSE)
overhaul – 104
removal and refitting – 102
tuning – 100
Carburettor (SU HS2)
overhaul – 96
removal and refitting – 95
tuning – 94
Carburettor (SU HS4)
overhaul – 100
removal and refitting – 100
tuning – 99
Carburettor (SU HS2E)
overhaul – 100
removal and refitting – 100
tuning – 98
Carpets – 301
Clutch – 133 *et seq*
Clutch
bleeding – 136
description – 133
fault diagnosis – 138
maintenance – 135
pedal removal – 136
refitting – 137
release bearing – 138
removal – 136
specifications – 133
Coil spring/shock absorber (front)
removal and refitting – 272
Contact breaker gap
Delco Remy distributor – 121
Lucas distributor – 121
Conversion factors – 305
Cooling and heating systems – 75
Cooling system
control valve – 83
description – 75
draining and refilling – 76
fault diagnosis – 83
maintenance – 76
sender unit – 80
specifications – 75
Courtesy lamp switch – 224
Crankcase ventilation system
1147 cc & 1296 cc – 54
1493 cc – 65
Cylinder bores (1147 cc & 1296 cc)
examination – 50
Cylinder head (1147 cc & 1296 cc)
decarbonising – 52
dismantling – 49
removal and refitting – 39
Cylinder head (1493 cc)
decarbonising – 65
dismantling – 65
refitting – 62
removal – 61

D

Differential and final drive – 188 *et seq*
Differential and final drive
carrier
removal and refitting – 190
description – 188
fault diagnosis – 195
maintenance – 188

mounting plate cushions
renewal – 194
output shaft oil seal
renewal – 192
pinion oil seal
renewal – 190
removal and refitting – 189
Dimensions – 8
Disc (brakes)
inspection and renewal – 201
Disc pads
inspection and renewal – 197
Distributor
dwell angle – 121
overhaul
mechanical type – 125
breakerless type – 127
removal and refitting – 124
Doors
exterior handle
removal and refitting – 299
glass
removal and refitting – 297
lock
remote control – 298
removal and refitting – 298
striker – 298
removal and refitting – 299
trim panel
removal and refitting – 297
Drivebelt
adjustment – 80
renewal – 81
Driveshaft
removal and refitting – 185
Dynamo
overhaul – 213
removal and refitting – 213
testing in vehicle – 213

E

Electric fan – 77
Electric system – 210 *et seq*
Electrical system
description – 212
fault diagnosis – 238
front parking/flasher lamp – 225
headlamp bulb – 225
number plate lamp – 227
rear lamps – 225
service interval indicator – 230
side marker lamp – 227
specifications – 210
steering column switch
removal and refitting – 222
Electronic ignition
air gap adjustment – 124
description – 122
dwell angle – 123
Emission control systems
air injection system (AIS) – 110, 112
air temperature control air cleaner – 109
anti run-on valve – 109, 111, 113
catalytic converter – 110, 113
description – 109
exhaust gas recirculation (EGR) system – 110, 112
fuel evaporative control system – 110, 112
Engine – 24 *et seq*
Engine (all models)
description – 30
dismantling (general) – 36
major operations
engine in car – 33
engine removed – 33

Index

removal – 33
Engine (1147 cc & 1296 cc)
   ancillaries removal – 37
   dismantling – 36, 48
   examination and renovation – 50
   fault diagnosis – 59
   lubrication system – 53
   reassembly – 55
   refitting – 35, 36
   removal – 33, 36
   start-up after overhaul – 59
   ventilation system – 54
Engine (1493 cc)
   ancillaries removal – 38
   dismantling – 65
   examination – 65
   fault diagnosis – 74
   reassembly – 68
   refitting – 35, 36
   removal – 34
   start-up after overhaul – 74
Exhaust system – 117

F

Fan blades
   removal and refitting – 79
Fan viscous coupling
   removal and refitting – 80
Fascia
   crash roll – 303
   panel – 302
Fault diagnosis – 20 *et seq*
Fault diagnosis
   clutch – 138
   cooling system – 83
   differential/final drive – 195
   electrical system – 238
   emission control systems – 120
   fuel and exhaust systems – 120
   ignition system – 132
   propshaft, driveshaft, UJ – 187
   suspension and steering – 287
Flywheel starter ring (1147 cc & 1296 cc)
   examination – 52
Front wheel alignment – 285
Fuel, carburation and emission control systems – 85 *et seq*
Fuel, carburation and emission control systems
   general description – 86
   maintenance – 86
Fuel gauge – 231
Fuel line filter
   renewal – 91
Fuel pump
   cleaning – 89
   description – 89
   overhaul – 89
   removal and refitting – 89
   testing – 89
   type – 85
Fuel tank
   capacities – 85
   general description – 86
   maintenance – 86
   removal and refitting – 92
   transmitter
      removal and refitting – 92
Fuses – 211

G

Gearbox and overdrive – 139 *et seq*
Gearbox and overdrive
   description – 141
   dismantling (general) – 144
   fault diagnosis – 181
   oil capacity – 140
   removal and refitting – 141
   specifications – 139
Gearbox (up to 1970)
   dismantling – 144
   examination and renovation – 147
   gear selectors overhaul – 151
   input shaft overhaul – 150
   mainshaft overhaul – 151
   reassembly – 154
   remote control assembly overhaul – 152
Gearbox (1970 to 1974)
   description – 159
   dismantling – 160
   examination – 160
   input shaft overhaul – 160
   mainshaft overhaul – 160
   reassembly – 160
   selector and remote control mechanism
      overhaul – 160
Gearbox (Dec 1974 on)
   description – 160
   dismantling – 163
   gearchange and mechanism
      overhaul – 166
   input shaft overhaul – 166
   inspection – 168
   mainshaft overhaul – 164
   reassembly – 168
   removal and refitting – 160
General dimensions, weights and capacities – 8
Glossary – 9

H

Handbrake
   adjustment – 207
   cables renewal – 209
   lever – 208
   pawl and ratchet – 209
Hardtop – 302
Headlamp
   bulb renewal – 224
Heater
   booster motor
      removal and refitting – 83
      removal and refitting – 81
Hood – 302
Horn – 235
Hydraulic hoses (flexible)
   inspection and renewal – 205

I

Ignition system – 121 *et seq*
Ignition
   coil – 121, 130
   contact breaker – 121
   dwell angle – 121
   electronic
      air gap adjustment – 124
      description – 122
      drive resistor – 131
      dwell angle – 123
   fault diagnosis – 132
   mechanical breaker
      capacitor – 130
      description – 122
      points adjustment – 122
      renewal – 123
   spark plugs – 131

Index

timing
    mechanical type distributor – 128
    breakerless type distributor – 128
Interior lamp bulb renewal – 227
Interior mirror – 304
Introduction to the Triumph Spitfire – 2

J

Jacking – 13

L

Lamps
    front parking/flasher – 225
    headlamp – 225
    number plate – 227
    rear – 225
    side marker – 227
Light switch – 222
Lubricants and fluids – 15
Lubrication system
    1147 cc & 1296 cc – 53
    1493 cc – 65
Luggage boot lid
    lock and striker
        removal, refitting, adjustment – 300
    removal and refitting – 300

M

Manifolds – 117
Master cylinder (brake – single type)
    removal, overhaul and refitting – 202
Master cylinder (brake – tandem type)
    removal, overhaul and refitting – 203
Master cylinder (clutch)
    removal, overhaul and refitting – 135

O

Oil (engine)
    capacities – 8
    topping-up – 33
Oil pump
    1147 cc & 1296 cc – 48, 52
    1493 cc – 65
Overdrive (type D)
    description – 171
    operating lever adjustment – 174
    overhaul – 172
    removal and refitting – 172
    valves
        removal, inspection and refitting – 175
Overdrive (type J)
    description – 175
    filter cleaning – 176
    overhaul – 179
    pump valve
        removal and refitting – 176
    relief valve
        removal and refitting – 176
    relief valve
        removal and refitting – 176
    removal and refitting – 176

P

Parcels shelf – 301
Pistons and piston rings examination – 51

Pistons, conrods and big-end bearings (1147 cc & 1296 cc)
    removal and refitting – 41
    dismantling – 49
Piston, conrods and big-end bearings (1493 cc)
    removal and refitting – 62
    dismantling – 65
Points gap – 121
Pressure differential warning actuator (PDWA) – 205
Propeller shaft
    balancing – 187
    fault diagnosis – 187
    removal and refitting – 183
Propeller shaft, driveshaft and universal joints – 182
    et seq

R

Radiator
    capacity – 75
    pressure cap rating – 75
    removal, repair and refitting – 78
Radiator grille
    removal and refitting – 296
Radio
    installation – 235
    suppression of interference – 236
Rear brakes
    adjustment – 197
    shoes
        inspection and renewal – 198
Relays – 221
Remote control assembly
    overhaul – 152
Roadwheels – 286
Rockers and rocker shaft (1147 cc and 1296 cc)
    examination – 51
Rocker gear (1147 cc & 1296 cc)
    dismantling – 48
Rocker gear (1493 cc)
    dismantling and reassembly – 62, 65
Routine maintenance – 17

S

Safety belts – 303
Safety first! – 16
Seat belt warning system – 235
Seats – 301
Selectors (gear) – 152
Slave cylinder (clutch)
    removal, overhaul and refitting – 136
Spark plugs
    conditions (colour) – 129
    gap – 121, 131
    type – 121, 131
Speedometer
    cable renewal – 228, 229
    removal and refitting – 228
Spoiler – 296
Starter motor
    description and testing – 218
    overhaul – 219
    removal and refitting – 291
Steering
    angles and front wheel alignment – 285
    column
        flexible coupling – 283
        removal, overhaul and refitting – 284
    fault diagnosis – 287
    lock/ignition switch
        removal and refitting – 285
    rack-and-pinion
        backlash adjustment – 283

overhaul – 284
    removal and refitting – 283
  rack bellows – 281
  rear wheel alignment – 286
  tie-rod end balljoints
    renewal – 281
  wheel
    removal and refitting – 281
Sump (1147 cc & 1296 cc)
  examination – 52
  removal and refitting – 40
Sump (1493 cc)
  removal and refitting – 62
Suspension and steering – 268 et seq
Suspension and steering
  description – 269
  maintenance – 269
  specifications – 268
Suspension (front)
  lower wishbone
    removal and refitting – 276
  stub axle carrier
    removal and refitting – 276
  upper balljoint renewal – 276
  upper wishbone
    removal and refitting – 276
Suspension (rear)
  hubs
    removal and refitting – 277
  radius rod – 279
  roadspring
    overhaul – 278
    removal and refitting – 277
  shock absorbers
    removal and refitting – 277
  vertical link – 279

T

Tachometer – 228, 230
Tappets (1147 cc & 1296 cc)
  examination – 52
Temperature gauge – 231
Thermostat – 75, 77
Throttle and choke controls – 113
Timing chain, gears and oil seal (1147 cc & 1296 cc)
  examination – 51
  removal and refitting – 43
Timing chain tensioner (1147 cc & 1296 cc)
  examination – 51
Timing cover, sprockets and chain (1493 cc)
  removal and refitting – 62

Timing (ignition)
  electronic type distributor – 128
  mechanical type distributor – 128
Tools and working facilities – 11
Towing – 13
Transmission cover – 296
Trunnion (front)
  removal, overhaul and refitting – 274
Tyres – 286

U

Universal joints (driveshaft)
  overhaul – 186
Universal joints (propshaft)
  overhaul – 183
Use of English – 9

V

Valves and valve seats (1147 cc & 1296 cc)
  examination – 51
Valve clearances adjustment
  1147 cc & 1296 cc – 59
  1493 cc – 74
Valve guides (1147 cc & 1296 cc)
  examination – 52
Vehicle identification numbers – 10
Voltage regulator unit – 216

W

Water pump
  description – 79
  belt adjustment – 80
Weights (kerb) – 8
Wheel cylinder (rear)
  overhaul – 201
  removal and refitting – 202
Windscreen washer system – 234
Windscreen washer/wiper switch
  removal and refitting – 222
Windscreen glass
  removal and refitting – 300
Window regulator
  removal and refitting – 298
Wiper motor
  rack tubing – 234
  removal, overhaul and refitting – 233
  wheelboxes – 234
Wiring diagrams – 240 to 267

Printed by
J H Haynes & Co Ltd
Sparkford Nr Yeovil
Somerset BA22 7JJ England